Chapman & Hall/CRC
Interdisciplinary Statistics Series

Age-Period-Cohort Analysis

New Models, Methods, and Empirical Applications

Yang Yang and Kenneth C. Land

CRC Press
Taylor & Francis Group
Boca Raton   London   New York

CRC Press is an imprint of the
Taylor & Francis Group, an **informa** business

A CHAPMAN & HALL BOOK

CHAPMAN & HALL/CRC
Interdisciplinary Statistics Series

Series editors: N. Keiding, B.J.T. Morgan, C.K. Wikle, P. van der Heijden

Published titles

AGE-PERIOD-COHORT ANALYSIS: NEW MODELS, METHODS, AND EMPIRICAL APPLICATIONS	Y. Yang and K. C. Land
AN INVARIANT APPROACH TO STATISTICAL ANALYSIS OF SHAPES	S. Lele and J. Richtsmeier
ASTROSTATISTICS	G. Babu and E. Feigelson
BAYESIAN ANALYSIS FOR POPULATION ECOLOGY	Ruth King, Byron J.T. Morgan, Olivier Gimenez, and Stephen P. Brooks
BAYESIAN DISEASE MAPPING: HIERARCHICAL MODELING IN SPATIAL EPIDEMIOLOGY, SECOND EDITION	Andrew B. Lawson
BIOEQUIVALENCE AND STATISTICS IN CLINICAL PHARMACOLOGY	S. Patterson and B. Jones
CLINICAL TRIALS IN ONCOLOGY, THIRD EDITION	S. Green, J. Benedetti, A. Smith, and J. Crowley
CLUSTER RANDOMISED TRIALS	R.J. Hayes and L.H. Moulton
CORRESPONDENCE ANALYSIS IN PRACTICE, SECOND EDITION	M. Greenacre
DESIGN AND ANALYSIS OF QUALITY OF LIFE STUDIES IN CLINICAL TRIALS, SECOND EDITION	D.L. Fairclough
DYNAMICAL SEARCH	L. Pronzato, H. Wynn, and A. Zhigljavsky
FLEXIBLE IMPUTATION OF MISSING DATA	S. van Buuren
GENERALIZED LATENT VARIABLE MODELING: MULTILEVEL, LONGITUDINAL, AND STRUCTURAL EQUATION MODELS	A. Skrondal and S. Rabe-Hesketh

Published titles

CRC Press
Taylor & Francis Group
6000 Broken Sound Parkway NW, Suite 300
Boca Raton, FL 33487-2742

© 2013 by Taylor & Francis Group, LLC
CRC Press is an imprint of Taylor & Francis Group, an Informa business

No claim to original U.S. Government works

Printed on Acid-free paper
Version Date: 20130114

International Standard Book Number-13: 978-1-4665-0752-4 (Hardback)

Visit the Taylor & Francis Web site at
http://www.taylorandfrancis.com

and the CRC Press Web site at
http://www.crcpress.com

Preface

This book is based on a decade of our collaborative work on new models, methods, and empirical applications of age-period-cohort (APC) analysis. The identification and statistical estimation of classical APC multiple classification/accounting models—often termed the APC "conundrum"—have been a challenging analytic problem in demography, epidemiology, sociology, and the other social sciences for about four decades. The last great synthesis of APC methodology for the social sciences and demography was based on the work of William M. Mason and Stephen E. Fienberg in the 1970s and 1980s and presented in their 1985 book *Cohort Analysis in Social Research: Beyond the Identification Problem* (New York: Springer-Verlag).

The Mason–Fienberg synthesis so dominated these disciplines in the 1980s and 1990s that relatively few new contributions to APC methodology were published in these disciplines during these years. Some APC methodological work continued in epidemiology, however, and around the year 2000, new interest emerged in demography and the social sciences. One of us entered the doctoral program at Duke University in that year, and the other became aware of Wenjiang Fu's initial work in the early 2000s on the intrinsic estimator as a new approach to the identification and estimation of the APC accounting model. We then teamed up with Fu in a 2004 article on statistical properties and empirical applications of the intrinsic estimator.

This initial work on the intrinsic estimator led us to think more generally about APC analysis. The classical accounting model was formulated for a research design typically consisting of an age-by-time period table of population rates or proportions with a single observation per cell. However, new research designs that permit new classes of statistical models had emerged and produced new datasets for APC analysis by the year 2000. One of these is the repeated cross-sectional sample survey design in which data are obtained from individual members of a representative sample of a population repeatedly in a sequence over a number of years. When we initially studied some published APC analyses of data from repeated sample surveys, we found that they applied the classical APC accounting model. But this model does not take full advantage of the statistical power of the numerous individual observations within a specific cohort and time period in a repeated survey design.

To do so, we were driven toward a hierarchical APC (HAPC) specification in the form of cross-classified models in which the individual observations in repeated cross-sectional surveys are nested within time periods and cohorts. These models can be specified in mixed (fixed and random) effects or purely fixed effects forms. However, the mixed effects forms of HAPC models have both statistical and substantive advantages. Importantly, HAPC models avoid the underidentification problems of the classical APC accounting model and can be specified as linear mixed models (LMMs) for continuous, relatively bell-shaped (Gaussian) outcome variables or as generalized linear mixed models (GLMMs) for discrete, nonnormally distributed (non-Gaussian) outcomes. These specifications permit us to take advantage of the many developments in the statistical theory and methodology of mixed models and associated computer software in the past three decades, developments that were not available to APC analysts in the 1970s and the 1980s. Our initial articles on the statistical methodology and empirical applications of HAPC models of the LMM and GLMM classes were published in 2006 and 2008. Most recently, we extended the reach of the HAPC approach to many other areas of research using APC analysis, such as the joint application of the mixed effects models and heteroscedastic regression in a study of trends in self-reported health with Hui Zheng and the use of HAPC models for the aggregate population rates data design in the case of cancer incidence and mortality that we illustrate in the book. These extensions to different directions and datasets are opening up a new genre of APC analyses with great potential.

On recognizing the nested nature of the individual-level observations in repeated cross-section survey designs and the HAPC modeling framework to which it led us, we turned our sights to a third research design from which a number of datasets began to emerge in the 1990s and 2000s: the accelerated longitudinal panel study design in which an initial wave of study participants is repeatedly surveyed across a number of subsequent time periods. What makes this design "accelerated" is the presence of study participants from a number of cohorts in the initial and subsequent waves. This permits the analysis of age-by-cohort and other cross-level interactions within the HAPC-GLMM framework that we developed for the repeated cross-sectional study design and also avoids the classical identification conundrum.

In sum, the approaches that we have developed synthesize APC models and methods for these three research designs—age-by-time period tables of population rates or proportions, repeated cross-section sample surveys, and accelerated longitudinal panel studies—within a single, consistent HAPC-GLMM statistical modeling framework. Many approaches to APC analysis, including pure fixed effects approaches such as that of the APC accounting model, are special cases of this general system. And, by recognizing this, analyses of datasets can be conducted by application of alternative specifications within this frame with the resulting empirical estimates compared for consistency across models, a form of sensitivity analysis. We emphasize that we do not claim to have "solved" the APC analysis problem in any of the work we have done. On the other hand, approaches to APC analysis can be arrayed according to their statistical properties, with some models and methods having better properties than others. By this criterion, the models we have developed and describe in this book are relatively good. We believe that their empirical application to many different substantive problems will lead to many fascinating new findings about how various outcome variables develop along the age, period, and cohort dimensions. And additional developments in APC statistical models and methods will be forthcoming, including variations in the HAPC-GLMM family of models, as the analytic problems posed by APC analysis continue to stimulate new approaches and as new models, methods, and computational algorithms are developed in statistics.

The general objective of this book is to bring our work together in one place. We build on our prior articles and include new technical discussions of statistical issues and many new empirical applications. Additional details on many of the published articles and empirical analyses cited in the book as well as computer software and sample programs to estimate the models can be found on the web page http://www.unc.edu/~yangy819/apc/index.html.

Finally, we thank our collaborators on issues of APC analysis, including those who contributed to prior publications, especially Wenjiang J. Fu, Sam Schulhofer-Wohl, and Hui Zheng, and those who have assisted with data analyses featured in this book that are part of ongoing research projects, including Ting Li, a mathematical demographer and specialist in the bio-demography of aging; and Steven Frenk, a medical sociologist with diverse interests. Both of them joined the Lineberger Comprehensive Cancer Center and Carolina Population Center at the University of North Carolina in 2011 as postdoctoral fellows working with the lead author (Y.Y.) and have contributed with the highest levels of rigor and dedication to the synergy of the research team and various projects associated with the APC analysis, cancer, and aging. We thank Igor Akushevich, senior research scientist in the Center for Population Health and Aging of the Duke Population Research Institute, who provided assistance with cancer incidence and mortality data preparation. We also thank the students who have taken courses on cohort analysis and demographic methods that we taught over the years, asked interesting questions that prompt us to do a better job at explicating various methods with examples and additional materials, and provided their new perspectives both conceptually and analytically on this old problem. It has truly been intellectually stimulating and a pleasure to work with them.

Yang Yang
University of North Carolina at Chapel Hill

Kenneth C. Land
Duke University

Contents

1

Introduction

Demographers, epidemiologists, and social scientists often deal with temporally ordered datasets, that is, population or sample survey data in the form of observations or measurements on individuals or groups/populations of individuals that are repeated or ordered along a time dimension. In this context, a long-standing analytic problem is the conceptualization, estimation, and interpretation of the differential contributions of three time-related changes to the phenomena of interest, namely, the effects of differences in the ages of the individuals at the time of observation on an outcome of interest, termed *age (A) effects*; the effects of differences in the time periods of observation or measurement of the outcome, termed *period (P) effects*; and the effects of differences in the year of birth or some other shared life events for a set of individuals, termed *cohort (C) effects*. To address this problem, researchers need to compare age-specific data recorded at different points in time and from different cohorts. A systematic study of such data is termed *age-period-cohort (APC) analysis*. APC analysis has the unique ability to depict parsimoniously the entire complex of social, historical, and environmental factors that simultaneously affect individuals and populations of individuals. It has thus been widely used to address questions of enduring importance to the studies of social change, etiology of diseases, aging, and population processes and dynamics.

The distinct meanings of A, P, and C effects will be elaborated and become more concrete in specific contexts. As a first specification, consider the definition of these terms in the context of aging and human development across the life course, health, and chronic disease epidemiology (Yang 2007, 2009, 2010). In this context, the following applies:

Age effects are variations associated with chronological age groups. They can arise from physiological changes, accumulation of social experience, social role or status changes, or a combination of these. Age effects therefore reflect biological and social processes of aging internal to individuals and represent developmental changes across the life course. This can clearly be seen in the considerable regularities of age variations across time and space in many outcomes, such as fertility, schooling, employment, marriage and family structure, disease prevalence and incidence, and mortality.

Period effects are variations over time periods or calendar years that influence all age groups simultaneously. Period effects subsume a complex set of historical events and environmental factors, such as world wars, economic expansions and contractions, famine and pandemics of infectious diseases, public health interventions, and technology breakthroughs. Shifts in social, cultural, economic, or physical environments may in turn induce similar changes in the lives of all individuals at a point in time. Thus, period effects are evident from a correspondence in timing of changes in events and social and epidemiologic conditions that influence these events. For example, the decrease in lung cancer mortality in the United States after 1990 followed reductions in tar and nicotine yield per cigarette and increases in smoking cessation in earlier years (Jemal, Chu, and Tarone 2001). In addition to these direct effects, there may also be changes in disease classification or diagnostic techniques that affect the incidence of, or mortality from, certain diseases. For example, the increase in the slope of the period trend of U.S. female breast cancer mortality in the 1980s coincided with the marked increase in breast cancer incidence due to expanded use of diagnosis via mammography (Tarone, Chu, and Gaudette 1997).

Cohort effects are changes across groups of individuals who experience an initial event such as birth or marriage in the same year or years. Birth cohorts are the most commonly examined unit of analysis in APC analysis. A birth cohort moves through life together and encounters the same historical and social events at the same ages. Birth cohorts that experience different historical and social conditions at various stages of their life course therefore have diverse exposures to socioeconomic, behavioral, and environmental risk factors. Cohort effects are evident in many cancer sites, chronic diseases, and human mortality. An in-depth discussion of the concept of cohort effects is given in the next chapter.

The challenges posed by APC analysis are well known. Whether observed time-related changes can be distilled out and separated into aging, time period, and cohort components is a question usually deemed conceptually important but empirically intractable. It has been termed the "conundrum" of APC analysis (Glenn 2005: 20) for two reasons. The first is data limitations. Using cross-sectional data at one point in time, for example, aging and cohort effects are intermingled and confounded. Using longitudinal panel data for a single cohort, on the other hand, aging and period effects are intermingled and confounded. The second reason is the use of conventional linear regression models that suffer from either specification errors or an identification problem and consequently are incapable of distinguishing A, P, and C effects.

The identification problem has been a topic of intense discussion and research since the 1970s. This led to a synthesis of APC methodology for the social sciences and demography based on the work of William M. Mason and Stephen E. Fienberg in the 1970s and 1980s (Fienberg and Mason 1979; Mason and Fienberg 1985). The Mason-Fienberg synthesis so dominated these disciplines in the 1980s and 1990s that relatively few new contributions to APC methodology were published in these decades. By comparison, APC methodology continued to be of interest in epidemiology, within which several new graphical and analytic methods were published during this period.

Although a variety of approaches has been proposed to solve the APC conundrum, each has limitations. Yet another challenge is a criticism often lodged against general-purpose methods of APC analysis, namely, they provide no avenue for testing specific, substantive, and mechanism-based hypotheses and thus are mere accounting devices of algebraic convenience that may be misleading. This leads to the question: What should an analyst do to model APC data in empirical research to further an understanding of the social and biological mechanisms generating the data? Since the year 2000, new interest in APC models and methods has emerged in the social sciences to address this question. This includes a series of studies by us as well as works by others exemplified in a special issue of the *Sociological Methods & Research* (36(3) February 2008).

The major objective of this book is to present new APC models, methods, and empirical applications. Statistics has continued to develop as a discipline since the Mason-Fienberg synthesis of 1985. New statistical models and new computationally intensive estimation methods have been developed (e.g., mixed [fixed and random] effects models, Markov chain Monte Carlo methods). For another, datasets with new research designs that invite or even require the analysis of separate age, period, and cohort components of change are available. Accordingly, we seek to show some ways in which these statistical models and methods and research designs can be applied to open new possibilities for APC analysis. We aim to articulate and compare new and extant models and methods that can be widely used by analysts. We also aim to provide some useful guidelines on how to conduct APC analysis. In doing so, this book intends to make two essential contributions to quantitative studies of time-related change. First, through the introduction of the generalized linear mixed model (GLMM) framework, we show how innovative estimation methods and new model specifications resolve the "model identification problem" that has hampered the development of APC analysis for the past decades. Second, we address the major criticism against the utility of APC analysis by explaining and demonstrating the use of new models within the GLMM framework to uncover the mechanisms underlying age patterns and temporal trends in phenomena of interest to researchers. We achieve these goals through both methodological expositions and empirical studies. For empirical illustrations, we draw examples on a wide variety of disciplines, such as sociology, demography, and epidemiology but focus on aging, longevity, and health disparities. We do not, however, claim that the new models and methods presented here are "solutions" to the APC analysis problem in any absolute sense. As articulated in Chapter 4, the classical APC identification problem in tabular arrays of population rates or proportions is a member of a class of structural underidentification problems for which there can never be a "complete" resolution.

The contents of the volume are as follows: Chapter 2 discusses the conceptualization of cohort effects and theoretical rationale for the importance of cohort analysis. Chapter 3 introduces prototypical datasets to be analyzed in further detail in subsequent chapters that characterize the application of APC analysis in three common research designs. Chapter 4 lays out the formal algebra of the APC analysis conundrum, reviews some conventional approaches to this problem, and sketches a GLMM framework that we use to organize the new families of models and methods.

Chapter 5 focuses on an innovation within the conventional linear regression models: the Intrinsic Estimator (IE) as a new method of coefficient estimation. Chapter 6 introduces a three-step procedure for APC analysis through empirical studies of U.S. cancer incidence and mortality trends by sex and race. It also illustrates the utility of APC models in demographic projections and forecasts through an empirical APC analysis and construction of the associated implied projections of cancer mortality in the period 2010–2029. As part of the methodological exposition of the nature and utilities of the IE method, we include in this chapter algebraic details of its statistical properties with proofs (Section 5.3; Appendices 5.1–5.3) and model validation through Monte Carlo simulation analysis (Section 5.5). We also include computational algorithms for obtaining the prediction intervals for forecasting (Appendix 6.1). Readers not adept with or interested in advanced statistical methods can skip these sections.

Chapters 7 and 8 introduce the mixed effects models for APC analysis using the hierarchical APC (HAPC) models. We emphasize two breakthroughs of this type of models compared to the linear fixed effects models classically used in APC analysis: *contextualization* of individual lives within cohorts and periods, which avoids the model identification problem, and incorporation of additional covariates, which allows for *mechanism-based hypothesis testing*. We illustrate in Chapter 7 the application of these models in studies of verbal ability trends in the United States and changing sex and race disparities in obesity. In Chapter 8 we analyze the social inequalities of happiness in relation to macroeconomic conditions and cohort characteristics and cancer mortality rates in relation to known risk factors and diagnostic and treatment factors. We also discuss in Chapter 8 extensions to HAPC models such as the full Bayesian estimation for small sample size problems and conjunction with the heteroscedastic regression for ascertainment of between-group and within-group variations. Readers who are not statistically sophisticated can skip these extensions in Sections 8.4 and 8.5.

Chapter 9 develops a similar GLMM approach to the analysis of prospective panel data using accelerated longitudinal cohort designs. Through empirical examples in studies of social stratification of aging and health, we show how to model age trajectories and cohort variations using HAPC-growth curve models. Chapter 10 concludes the volume with recaps of new avenues for APC analysis presented in previous chapters and suggestions for future directions of methodological research and data collection.

To facilitate the application of the methods described in the volume (in Chapters 5–9), we have developed a companion World Wide Web page on APC analysis (http://www.unc.edu/~yangy819/apc/index.html). This page provides links to PDF files of major methodological and substantive articles on APC analysis we reference in the book. It also provides sample codes using existing general-purpose statistical software packages, including R, SAS, and Stata. These are connected to the empirical analyses reported in the book.

References

Fienberg, S. E., and W. M. Mason. 1979. Identification and estimation of age-period-cohort models in the analysis of discrete archival data. *Sociological Methodology* 10:1–67.

Glenn, N. D. 2005. *Cohort analysis*. 2nd ed. Thousand Oaks, CA: Sage.

Jemal, A., K. C. Chu, and R. E. Tarone. 2001. Recent trends in lung cancer mortality in the United States. *Journal of the National Cancer Institute* 93:277–283.

Mason, W. M., and S. E. Fienberg, Eds. 1985. *Cohort analysis in social research: Beyond the identification problem*. New York: Springer-Verlag.

Tarone, R. E., K. C. Chu, and L. A. Gaudette. 1997. Birth cohort and calendar period trends in breast cancer mortality in the United States and Canada. *Journal of the National Cancer Institute* 89:251–256.

Yang, Y. 2007. Age/period/cohort distinctions. In *Encyclopedia of health and aging*, ed. K. S. Markides, 20–22. Los Angeles: Sage.

Yang, Y. 2009. Age, period, cohort effects. In *Encyclopedia of the life course and human development*, ed. D. Carr, R. Crosnoe, M. E. Hughes, and A. M. Pienta, 6–10. New York: Gale.

Yang, Y. 2010. Aging, cohorts, and methods. In *Handbook of aging and the social sciences*, ed. R. H. Binstock and L. K. George, 17–30. Burlington, VT: Academic Press.

2

Why Cohort Analysis?

2.1 Introduction

Although studies of time-related change have long existed in the history of science, those that consider cohort change as distinct from age and time period variations appear in scholarly literature only relatively recently. Why is cohort analysis useful? Examples of the utility of cohort analysis can be found in demographic studies of human mortality. Descriptive analyses of nineteenth century English death rates clearly indicated stronger regularities in birth cohort changes relative to period variations (Derrick 1927; Kermack, McKendrick, and McKinlay 1934). The relevance of this approach was then recognized in subsequent epidemiologic investigations, the earliest of which is the well-known study of tuberculosis mortality conducted by Frost (1940) that emphasized the influence of early life conditions, rather than current conditions, on development of a disease that has long latency. The usefulness of cohort analysis demonstrated by these early studies and the convenience of using simple indicators that are widely available in many kinds of data facilitated the quick spread of cohort analysis in demography and epidemiology. Although age-period-cohort (APC) analysis took root in these two fields relatively independently of one another, their common interests in health, mortality, and longevity and similarities in the development of analytic techniques unite them as one cottage industry.

2.2 The Conceptualization of Cohort Effects

The applicability of APC analysis relies on the substantive importance of cohort influences (Hobcraft, Menken, and Preston 1982). APC analysis is, in this sense, synonymous with cohort analysis (Smith 2008). Among various cohorts defined by different initial events (such as marriage and college entrance), birth cohorts are the most commonly examined unit of analysis.

We introduce here the conceptualization of birth cohort effects and the importance of identifying such effects in the contexts of different fields of studies that are concerned with time change.

Norman Ryder, in a seminal article published in 1965, articulated the conceptual relevance of birth cohort to the study of social historical change. First, a birth cohort moves through life together and encounters the same historical and social events at the same ages. Cohort effects then reflect formative experiences resulting from the intersection of individual biographies and macrosocial influences. Second, the succession of birth cohorts with different life experiences, termed *demographic metabolism* by Ryder, constantly changes the composition of the population and transforms the society. Therefore, cohorts can be conceived as the essence of social change. Third, cohort membership could be considered as a social structural category that has an analytic utility similar to that of social class. They both have explanatory power because they are surrogate indices of common characteristics of individuals in each category. Comparisons of historical cohorts can thus be useful in addressing an extraordinary range of substantive issues in social research. For instance, a recent analysis of Census 2000 data by Hughes and O'Rand (2004) compared the baby boomers (born 1946–1964) with their predecessors born earlier in the twentieth century who were defined by events and experiences unique to their times: young Progressives (1906–1915), Jazz Age babies (1916–1925), Depression kids (1926–1935), and war babies (1936–1945). Differences in socioeconomic attainments, marriage, fertility, and family structure across these cohorts are substantial and reflect the post-World War II transformation of American society.

Cohort effects may also arise from differentials in early life conditions. This conceptualization is particularly relevant in the examination of health and illness wherein long-term exposure to risk factors is the major cause of the disease. Chronic disease epidemiology has long noted the importance of early life exposures in explaining the susceptibility to disease and mortality later in adulthood (Ben-Shlomo and Kuh 2002). The *fetal origin hypothesis*, for instance, argues that malnutrition *in utero* and during infancy adversely affects intrauterine growth and postnatal development and may increase the risks of cardiovascular and respiratory diseases, cancer, non-insulin-dependent diabetes, metabolic syndromes, and mortality (Barker 1998). The *cohort morbidity phenotype hypothesis* further links reductions in early life exposures to inflammatory infection to cohort declines in cardiovascular and overall mortalities (Finch and Crimmins 2004). The theory of *technophysio evolution* (Fogel and Costa 1997) also implies that throughout the twentieth century, individuals' health capital has changed with the year of birth. More recent cohorts fared substantially better in the initial endowment of health capital at birth and have lower depreciation rates in that stock of health capital. These led to improved physiological capacities in later cohorts that also bode well for effectiveness of medical treatments. This is consistent with mounting evidence in the recent demographic literature that shows successive birth cohorts experience later onset of chronic diseases and disabilities (Crimmins, Reynolds, and Saito 1999; Freedman and Martin 1998).

Studies of aging use cohort analysis to assess changes in individual outcomes in relation to aging within or across birth cohorts. Sociological theories of aging and the life course emphasize that the way individual lives unfold with age is largely shaped by social historical context (Elder 1974). Examining cohort membership as a contextual characteristic has important implications for a better understanding of the heterogeneous experiences of aging. Riley (1987) articulated the cohort-specific aging process and advanced the *principle of cohort differences in aging*. A birth cohort shares the same birth year and ages together. Because members in different birth cohorts age in unique ways shaped by the disparate sociohistorical and epidemiologic conditions, each cohort experiences a distinct life course. For example, recent studies on old-age depression found evidence of cohort heterogeneity in mental health and an apparent age-by-cohort interaction effect whereby depression declined with age more rapidly for earlier cohorts (Kasen et al. 2003; Yang 2007). Just as the cohort perspective enhances the understanding of the aging process, the aging perspective also sheds light on the mechanism underlying cohort effects. That is, cohort differences in social propensities and biological capacities are affected not only by early life but also lifelong accumulation of exposures. This conceptualization of cohort effects has been advanced by Ryder (1965) and referred to as the *continuously accumulating cohort effects* by Hobcraft, Menken, and Preston (1982). Hobcraft and colleagues further suggested the use of cohort analysis to capture the process by which the imprint of past events is differentiated by age and becomes embodied in cohorts differentially. We discuss this cohort theory in further detail and its mathematical realization in the final chapter of this book.

2.3 Distinguishing Age, Period, and Cohort

Because the objective of cohort analysis is to identify the source of variations attributable to birth cohort, it is imperative first to distinguish cohort from potential confounding factors, including age and period. To understand problems associated with inadequate data and study designs, we give an example in Table 2.1 that illustrates the relationship of birth cohort with chronological age and calendar year in any given dataset. In this example, data are arrayed by four ages (a60, a70, a80, and a90) for each of the four periods p1980, p1990, p2000, and p2010. Birth cohort membership (c) is shown by diagonals within the age-by-period array and ranges between 1890 and 1950.

TABLE 2.1

Hypothetical Data Arrayed by Age,
Period and Cohort

	p1980	p1990	p2000	p2010
a60	c1920	c1930	c1940	c1950
a70	c1910	c1920	c1930	c1940
a80	c1900	c1910	c1920	c1930
a90	c1890	c1900	c1910	c1920

The problem with using cross-sectional data gathered at one point in time is the confounding of age and cohort changes. This can be seen by moving down the rows within columns of Table 2.1. For example, in year 1980, changes across ages 60 to 90 are indistinguishable from changes across cohorts 1920 to 1890. If these changes are solely age related, then the same age pattern should hold in subsequent years 1990 to 2010, which subsume different sets of cohorts. If these changes are also a function of cohort differences, then the age pattern of 1980 should not resemble those of the other years containing different cohorts. Therefore, the detection of age and cohort effects can only be achieved when comparing data collected at different points in time. Similarly, comparing data from multiple time periods for the same age group precludes distinguishing between period and cohort effects. This kind of analysis is less common in practice, however.

The consequence of using a cross-sectional design to infer age, period, or cohort effects can lead to erroneous inferences. For example, early epidemiologic studies of age patterns of breast cancer incidence in the 1940s suggested a "break" in the age curve produced by a fall in the incidence between ages 45 and 50 known as the "Clemmesen's hook" (Clemmesen 1948). While some later authors interpreted this as an age-related change due to hormonal or other physiological factors, MacMahon (1957) suggested that this break was an artifact produced by cross-sectional studies of age changes that actually were due to cohort changes. Comparing age-specific incidence data from female populations in Connecticut and England arrayed along the period and cohort lines, MacMahon found the break in the increasing age pattern of incidence rates arrayed by periods (MacMahon 1957: Figure 3) but no break during the menopausal age span for any cohort (MacMahon 1957: Figure 5). This, suggested by the study, is because changes in incidence between successive cohorts occurred in just those cohorts necessary to explain the observation of the break at certain ages in particular time periods.

The problem with using a single-cohort longitudinal panel is the confounding of age and period effects. This can be seen by moving across cells within the cohort diagonals of the array shown. Suppose the cohort of interest is that born in 1920 first assessed in year 1980 at the age of 60. The cohort aged to 70, 80, and 90 at each subsequent 10-year follow-up in years 1990,

2000, and 2010, respectively. Therefore, within each cohort it is difficult to disentangle whether changes across the cells are age related, time period related, or both. It also is not possible to ascertain the extent to which the age trajectories identified for a single cohort have patterns shared with other cohorts as compared to being unique to the cohort studied.

This confounding of age, period, and cohort effects in single-cohort longitudinal studies is problematic whether the data studied are in the form of tables of population occurrence/exposure rates often studied by demographers and epidemiologists or in the form of individual-level observations of the cohort members measured repeatedly over time as the cohort ages. As an illustration of the latter, consider the Hamil-Luker, Land, and Blau (2004) study of latent life course trajectories of cocaine use among a national representative sample of youths aged 14–16 initially surveyed in 1979 that was followed as a cohort with surveys in 1984, 1988, 1992, 1994, and 1998. Figure 2.1 depicts three different trajectories of cocaine usage among the cohort of adolescents grouped into the "delinquents" latent cluster. One of these shows a clear peak of reported cocaine use in the 1980s when sample members were in their 20s. Members of this trajectory grouping then "age out" to lower rates of cocaine use in their 30s. A second cluster exhibits a steady decline of reported cocaine use from the late teens through the 20s into the 30s. The third trajectory consists of youths with comparatively low rates of reported cocaine use throughout the ages/time periods studied.

While this study was the first to conduct such a thorough latent class and latent trajectory analysis of cocaine use across the adolescent and young

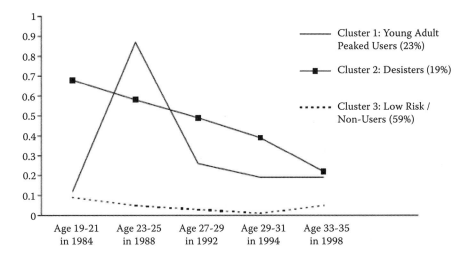

FIGURE 2.1
Delinquents' predicted probability of cocaine usage by latent cluster membership, $N = 244$.

adult years, it could not distinguish age, period, and cohort components in the trajectories it identified and estimated. The reason is that the age and time dimensions on the horizontal axis of Figure 2.1 are identical to each other, and thus age/life course developmental aspects of the trajectories cannot be separated from those of the time periods of the follow-up surveys. In addition, because the study was based on a single cohort, the possible effects of different cohort socialization and life course experiences cannot be deciphered. Thus, while it may be the case that all recent cohorts of youths in the United States have similar latent clusters and latent trajectories of cocaine use, the extent to which the trajectories were affected by the societal milieu of cocaine experimentation and use in the 1980s and early 1990s cannot be ascertained.

2.4 Summary

In empirical investigations confined to changes in age and/or period trends, the cohort effect frequently is ignored or assumed nonexistent. The assumption of no cohort effects can greatly simplify analysis, but its tenability can be called into question when such effects are present.

Cohort analysis aims at distinguishing age, period, and cohort effects and is theoretically important in three ways. First, it is crucial for attributions of etiology or social causation. Conceptually, age effects represent aging-related developmental changes within individuals, whereas temporal trends across time periods reflect exogenous changes in social and epidemiologic conditions. And, cohort changes reflect the intersection of both as a result of differential accumulation of lifetime exposures to environmental conditions. Second, cohort analysis also relates to the generalizability of research findings. In the absence of period and cohort effects, age changes are broadly applicable across individuals in different time periods and/or cohorts. The presence of either or both of these effects, however, indicates the existence of exogenous forces or exposures that are period and cohort specific. Third, to the extent that these effects serve as aggregates and proxies for different sets of structural correlates, analyses that allow for their distinction are especially valuable for better understanding and identifying the underlying social and environmental factors that are amenable to modifications.

In sum, because cohort analysis has the capacity to depict the entire complex of social, historical, and environmental factors that shape individual life courses parsimoniously, its importance for constructing and refining theory, measurement, and analysis can hardly be overstated. In spite of these theoretical merits and conceptual relevance, empirical cohort analysis has been hampered by data limitations and methodological challenges. We now turn to these issues in the chapters that follow.

References

Barker, D. J. P. 1998. *In utero* programming of chronic disease. *Clinical Science* 95:115–128.

Ben-Shlomo, Y., and D. Kuh. 2002. A life course approach to chronic disease epidemiology: Conceptual models, empirical challenges and interdisciplinary perspectives. *International Journal of Epidemiology* 31:285–293.

Clemmesen, J. 1948. I. Results from statistical research. *British Journal of Radiology* 21:583–590.

Crimmins, E. M., S. L. Reynolds, and Y. Saito. 1999. Trends in health and ability to work among the older working-age population. *Journal of Gerontology: Social Sciences* 54B:S31–S40.

Derrick, V. P. A. 1927. Observations on (1) errors on age on the population statistics of England and Wales and (2) the changes in mortality indicated by the national records. *Journal of the Institute of Actuaries* 58:117–159.

Elder, G. H., Jr. 1974. *Children of the Great Depression: Social change in life experience.* Chicago: University of Chicago Press.

Finch, C. E., and E. M. Crimmins. 2004. Inflammatory exposure and historical change in human life-spans. *Science* 305:1736–1739.

Fogel, R. W., and D. L. Costa. 1997. A theory of technophysio evolution, with some implications for forecasting population, health care costs, and pension costs. *Demography* 34:49–66.

Freedman, V. A., and L. G. Martin 1998. Understanding trends in functional limitations among older Americans. *American Journal of Public Health* 88:1457–1462.

Frost, W. H. 1940. The age selection of mortality from tuberculosis in successive decades. *The Milbank Memorial Fund Quarterly,* 18(1, Jan., 1940):61-66.

Hamil-Luker, J., K. C. Land, and J. Blau. 2004. Diverse trajectories of cocaine use through early adulthood among rebellious and socially conforming youth. *Social Science Research* 33:300–321.

Hobcraft, J., J. Menken, and S. Preston. 1982. Age, period, and cohort effects in demography: A review. *Population Index* 48:4–43.

Hughes, M.E., and A.M. O'Rand. 2004. The lives and times of the baby boomers. In R. Farley and J. Haaga (eds.) *The American People Census 2000*, pp. 224–258.

Kasen, S., P. Cohen, H. Chen, and D. Castille. 2003. Depression in adult women: Age changes and cohort effects. *American Journal of Public Health* 93:2061–2066.

Kermack, W. O., A. G. McKendrick, and P. L. McKinlay. 1934. Death-rates in Great Britain and Sweden: Expression of specific mortality rates as products of two factors, and some consequences thereof. *Journal of Hygiene* 34:433–457.

MacMahon, B. 1957. Breast cancer at menopausal ages: An explanation of observed incidence changes. *Cancer* 10:1037–1044.

Riley, M. W. 1987. On the significance of age in sociology. *American Sociological Review* 52:1–14.

Ryder, N. B. 1965. The cohort as a concept in the study of social change. *American Sociological Review* 30:843–861.

Smith, H. L. 2008. Advances in age-period-cohort analysis. *Sociological Methods & Research* 36:287–296.

Yang, Y. 2007. Is old age depressing? Growth trajectories and cohort variations in late-life depression. *Journal of Health and Social Behavior* 48:16–32.

3

APC Analysis of Data from Three Common Research Designs

3.1 Introduction

As previously indicated, the goal of age-period-cohort (APC) analysis is to distinguish and statistically estimate the unique effects associated with age, period, and cohort. The extent to which this goal can be realized depends on research designs and modeling strategies. In this chapter, we focus on three research designs that allow analysts to distinguish age, period, and cohort effects and thus are more suitable for APC analysis than cross-sectional or single-cohort panel designs. We identify two to three prototypical datasets that characterize the application of APC analysis in each of three designs. In each case, we commence with statements of substantive problems that arise from the scientific literature and then describe data that will be analyzed in subsequent chapters to address these problems. We draw examples on a wide variety of topics throughout the book but focus on aging, longevity, and health disparities, as these are problems that have long histories of temporal analysis and for which APC analysis is highly salient.

3.2 Repeated Cross-Sectional Data Designs

We illustrated in Chapter 2 how age and cohort changes are confounded in cross-sectional data collected at one point in time and therefore do not permit cohort analysis. Cross-sectional data collected repeatedly across time, however, are well suited for APC analysis. Pooling data of all years, one can formulate a rectangular age-by-period array of observations, where columns correspond to age-specific observations in each year, and rows are observations from each age across years. Linking the diagonal cells of the array yields the observations belonging to people born during the same calendar years who age together.

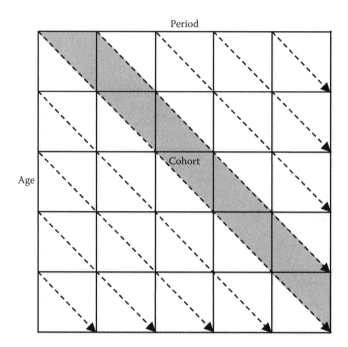

FIGURE 3.1
Age-by-time period data structure.

   A generic form of the age-by-time period array is shown in Figure 3.1. In this rectangular array, age-specific observations such as mortality rates or proportions of a population with a certain attribute (e.g., having a "very good" or "excellent" self-reported health status) are summarized in an age-by-period table for age groups and time periods that are of equal interval lengths (e.g., 5 or 10 years). The diagonal elements of the matrix correspond to observations of birth cohorts. One cohort is indicated by the shaded area bounded by the dashed arrows. A numerical example of the relationship between A, P, and C in this kind of data structure, namely, C = P − A, has been shown previously in Chapter 2. Although only a longitudinal panel study design provides data from true birth cohorts that follow identical individuals over time, the design illustrated in Table 3.1, if based on complete population data or repeated representative sample surveys thereof, allows for a classic demographic analysis using the synthetic cohort approach (Mason and Fienberg 1985; Preston, Heuveline, and Guillot 2001) that traces essentially the same groups of people from the same birth cohorts over a large segment of the life span. The composition of cohorts in this setting may be affected somewhat by international migration.

TABLE 3.1

Age-Specific Lung Cancer Incidence Rates (per 100,000 Population) for White Males and White Females, United States 1973–2008

	White Males							
	Period							
Age	1973–74	1975–79	1980–84	1985–89	1990–94	1995–99	2000–04	2005–08
20–24	0.39	0.22	0.22	0.27	0.20	0.14	0.27	0.33
25–29	0.67	0.52	0.43	0.23	0.38	0.35	0.58	0.51
30–34	2.15	2.04	1.58	1.68	1.35	1.08	1.07	1.01
35–39	7.61	7.43	6.64	5.28	4.37	3.46	3.42	2.63
40–44	23.33	20.83	19.58	16.93	11.89	10.27	10.80	8.71
45–49	56.24	60.38	51.09	43.76	33.59	26.09	26.39	22.10
50–54	104.38	115.14	111.54	103.73	78.85	62.54	57.13	50.63
55–59	182.69	187.61	201.56	194.98	154.78	130.15	118.75	99.07
60–64	272.44	298.26	301.73	311.32	280.51	235.78	227.68	187.87
65–69	383.62	407.04	435.12	418.41	419.10	371.68	358.45	317.09
70–74	470.01	501.01	538.41	540.25	503.63	483.21	487.61	434.58
75–79	485.08	542.78	586.05	588.58	572.95	537.63	569.73	541.55
80–84	403.33	477.56	551.72	580.48	589.73	549.94	556.07	561.49
85+	324.35	375.34	424.38	457.22	473.46	455.13	480.50	472.68

	White Females							
	Period							
Age	1973–74	1975–79	1980–84	1985–89	1990–94	1995–99	2000–04	2005–08
20–24	0.23	0.15	0.22	0.27	0.35	0.31	0.31	0.34
25–29	0.68	0.61	0.53	0.62	0.50	0.43	0.59	0.65
30–34	1.75	1.77	1.20	1.23	1.35	1.14	1.18	1.37
35–39	5.66	5.89	5.67	4.23	3.60	4.22	4.07	3.07
40–44	13.86	16.56	16.48	12.75	10.81	9.03	11.65	9.08
45–49	26.63	34.08	33.33	33.82	27.67	23.09	23.89	25.01
50–54	40.64	52.43	65.78	66.66	62.08	51.71	46.67	44.81
55–59	63.75	81.27	98.54	115.71	113.78	102.12	96.61	78.07
60–64	76.88	105.93	135.84	164.30	172.65	166.89	170.44	149.49
65–69	84.30	115.08	165.17	200.91	234.38	248.80	254.54	249.04
70–74	72.64	110.21	167.20	230.46	273.19	307.13	326.30	327.83
75–79	71.32	96.72	142.09	210.70	282.92	315.38	353.70	375.94
80–84	64.87	76.25	120.73	168.55	236.59	292.49	322.44	350.92
85+	59.66	74.33	91.22	118.60	146.34	188.93	227.06	249.75

Repeated cross-sectional data designs have largely been analyzed as age-by-time period arrays in which age, time period, and birth cohort are considered same-level factors affecting the outcome of interest. The subsequent modeling approach suffers from a major problem, called the "model identification problem" or APC conundrum, induced by the algebraic relationship between A, P, and C noted. Different temporal groupings for the A, P, and C dimensions can be used to break the linear dependency. For example, one can use single years of age, time periods corresponding to years in which vital statistics are recorded or surveys are conducted (which may be several years apart) and cohorts defined by either 5- or 10-year intervals that are conventional in demography or application of substantive classifications (e.g., war babies, baby boomers, baby busters, etc.). We defer the discussion of limitations of using this approach in conventional linear regression models to Chapter 4. Here, we simply note that, when the data used to construct the age-by-time period data structure of Figure 3.1 are in such a form that different temporal intervals can be used for the age, time period, and cohort groupings, the same design can be considered alternatively as a multilevel data design whose structure is illustrated in Figure 3.2; each row is a cohort, and each column is a time period. Note that data balanced in age-by-time period arrays (as shown in Figure 3.1) are necessarily unbalanced in cohort-by-time period arrays. In this design, individual-level age-specific observations are nested in and cross-classified simultaneously by two higher-level *social contexts* defined by time period and birth cohort. This multilevel data structure then motivates the employment of a different modeling approach (the

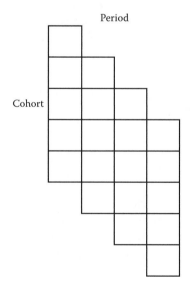

FIGURE 3.2
Cohort by period cross-classification data structure.

generalized linear mixed model [GLMM] approach described in Chapter 4) that avoids the model identification problem of classical linear APC models.

3.3 Research Design I: Age-by-Time Period Tabular Array of Rates/Proportions

In demographic and epidemiologic investigations, researchers typically are interested in aggregate population-level or tabular data such as rates of morbidity, disability, and mortality. As shown in Figure 3.1, the data structure is one in which rates or proportions are arranged in rectangular arrays with age intervals defining the rows and time periods defining the columns. The use of this research design is most widely seen in chronic disease epidemiology and demography. Our empirical analyses focus on the case of malignant neoplasm. Cancer has taken the place of heart disease to become the number 1 killer of Americans under age 65 (Jemal et al. 2005). Since President Richard Nixon signed the National Cancer Act in December 1971, broadening the scope and responsibilities of the National Cancer Institute (NCI), the nation has engaged itself in 40 years of the "War on Cancer" (Marshall 2011). Where do we stand in new cases of malignancies today compared to the earlier decades? And, how do we fare in survival and longevity in the face of the most invincible maladies known to humankind? As both direct and indirect indicators of the influences of myriad genetic, social behavioral, economic, and environmental factors on population health, the age patterns, temporal trends, and birth cohort variation of cancer incidence and mortality rates become crucial for understanding the longevity prospect of human beings (Manton, Akushevich, and Kravchenko 2009).

3.3.1 Understanding Cancer Incidence and Mortality Using APC Analysis: Biodemography, Social Disparities, and Forecasting

Although there is a vast literature on cancer morbidity and mortality, several questions remain that beg the utilization of APC analysis. First, what is the relationship between cancer and aging? Previous studies showed declines in tumor progression rates and cancer mortality with advanced age (Frank 2007; Manton, Akushevich, and Kravchenko 2009). While this may be contributed in part by selective survival that decreases population heterogeneity in older ages (Vaupel, Manton, and Stallard 1979), there also are biological explanations suggesting that the processes of cancer development and senescence might interact and be related (Pompei and Wilson 2001; Arbeev et al. 2005). It is not clear, however, whether or to what extent the observed declines with age are due to cohort dynamics that reflect differences in lifelong exposures to carcinogens and other risk factors unrelated to developmental changes. It also remains unknown whether the age declines persist or change across historical time. Therefore, APC analysis should be employed to better tease out the possible causes of this decline and inform the underlying biological processes of aging.

The second question builds on the first one and concerns social disparities in the age patterns and temporal dynamics. For instance, cancer incidence rates are higher for males than females in older ages (Arbeev et al. 2005), and the male excess in cancer mortality rates increases with age (Yang and Kozloski 2012). In addition to biological differences between men and women, the age changes in the sex gaps may be due to social behavioral factors such as cigarette smoking that have shown apparent birth cohort variations (Pampel 2005). Similarly, racial differences in cancer incidence and mortality trends can also be attributed to changes related to age, birth cohort, or period components. Tobacco use has been one of the best-established behavioral risk factors for cancer and many other degenerative diseases (Manton, Akushevich, and Kravchenko 2009). There is evidence of large cohort increases in cigarette smoking that initiated, peaked, and decelerated in a gender-specific fashion, with women lagging behind men by about 20 years (Wang and Preston 2009). There is also evidence that obesity, a major biobehavioral risk factor for cancer and other chronic diseases, shows large increases over time and in more recent cohorts born after 1955, and such period and cohort changes are most pronounced in black females (Reither, Hauser, and Yang 2009). How differential exposures and vulnerabilities to these risk factors lead to sex and racial disparities in trends and patterns of cancer morbidity and mortality merits a systematic APC analysis.

Third, what do current trends in cancer mortality tell us about the future? Conventional demographic forecasting techniques make use of only age and period variations to project mortality into the future and ignore birth cohort variation. Under the condition of no substantial cohort changes in mortality, this omission has no serious consequences. However, the accuracy of the age-period-based forecast may be significantly decreased when children and young adults alive today experience different epidemiologic conditions as they age and hence different mortality rates from their predecessors. The obesity epidemic is one example of how health conditions across cohorts can change quite rapidly and dramatically, and sometimes for the worse. While the idea of using APC models for mortality forecasting has been suggested by Osmond (1985) and hence is not new, the ever-changing cohort patterns of myriad health indicators today call for revisiting the forecasting method and its implementation.

We address these questions by examining data on U.S. cancer incidence and mortality rates for multiple sites to reveal their distinct clinical and public health characteristics and different implications for socioeconomic changes during the years 1973 to 2008.

3.3.2 Cancer Incidence Rates from Surveillance, Epidemiology, and End Results (SEER): 1973–2008

The NCI's Surveillance, Epidemiology, and End Results (SEER) Program collects information through cancer registries in various states on newly diagnosed invasive cancers in the United States. The population coverage of the SEER Program was 9.5% of the U.S. population from 1973 to 1991, increased to 13.8% from 1992 to 1999, and to 26.2% since 2000. We obtained incidence cancer and population count data from 1973 to 2008 by age, year of diagnosis, sex, and race from the SEER research data using the SEER*Stat Software. We analyzed the trends of the top 20 cancer sites for males and females (Jemal et al. 2008): lung and bronchus, prostate, breast, colon and rectum, pancreas, ovary, leukemia, non-Hodgkin lymphoma, uterine corpus, esophagus, urinary bladder, liver and intrahepatic bile duct, kidney and renal pelvis, stomach, brain and other nervous system, myeloma, uterine cervix, thyroid, oral cavity and pharynx, and melanoma of the skin. Site and histology were coded according to the *International Classification of Diseases for Oncology (ICD-O)* edition in use at the time of diagnosis and converted to the third edition coding. The SEER site recode is provided at http://seer.cancer.gov/siterecode/icdo3_d01272003/. Cases of cancer incidence were associated with the population estimates using racial and ethnic groups that changed over the period. These groups included white, black, and other (American Indian/Alaska Native and Asian/Pacific Islander combined) for the period before 1992 and include an additional Hispanic origin after 1992. For analyses of long-term trends, we used former data for the three major racial and ethnic populations. More details on population estimates used for the calculation of cancer incidence are provided at http://seer.canc er.gov/data/index.html.

Table 3.1 shows the age-specific lung cancer incidence rates for white males and females for ages 20 to 85+ and time periods from 1973 to 2008. Birth cohort-specific rates lie along the diagonals of the age-by-time period rectangular array. There are $a = 14$ five-year age groups, $p = 8$ five-year periods, and $c = 21$ ten-year birth cohorts that range from those born before 1885 to those born in 1985 (marked by the midinterval birth year). This rate table is representative of data widely used in population studies and all cancer incidence rate data used in subsequent analyses.

3.3.3 Cancer Mortality Rates from the National Center for Health Statistics (NCHS): 1969–2007

We obtained the U.S. cancer mortality data collected by the National Center for Health Statistics (NCHS) using the SEER*Stat software for the period 1969 to 2008 by sex and race. The *International Classification of Diseases (ICD)* versions changed over this time period (*ICD-8* to *ICD-10*). The underlying causes of death are based on the death certificate information reported to the

Centers for Disease Control and Prevention (CDC) National Vital Statistics and categorized according to SEER site groups to ensure comparability among *ICD* versions (Jemal et al. 2008). The SEER cause of death recode is listed at http://seer.cancer.g ov/codrecode/19 69+_d09172004/. Denominators in the death rate computation are from county-level population estimates that were summed to the state and national levels. More detailed descriptions of the methodologies used for the population estimates by the Census Bureau and the NCI are available elsewhere (Jemal et al. 2008). Similar to cancer incidence data, cancer mortality (or other cause-specific mortality) data from the NCHS were not consistently recorded for every racial and ethnic group for all periods. To facilitate comparison with the incidence data across time, we analyzed the trends for the three major racial groups: white, black, and other. All cancer mortality data can be accessed at the SEER website: http:// seer.cancer.gov/mortality/index.html.

Table 3.2 presents the age-specific lung cancer death rates for white males and females for ages 20 to 85+ and time periods from 1969 to 2007. Similar to the incidence rate data, there are $a = 14$ five-year age groups, $p = 8$ five-year periods, and $c = 21$ ten-year birth cohorts that range from those born before 1885 to those born in 1985 (marked by the midinterval birth year). Data on rates for other sites by sex and race take the same form and are presented graphically in subsequent analyses.

The examples given take the form of an age-by-period array using aggregate age and period groups in equal intervals (5 years). We can also utilize differential temporal groupings of the age, year, and cohort variables to create a multilevel data array of rates. Using the lung cancer mortality rate data as an example, we show the alternative data structure in Table 3.3; number of observations and mean are presented by cohort and period. Because the SEER website mentioned provides single-year age-specific data on cancer incidence rates but not mortality counts, we obtained the single-year age-specific data on mortality for a slightly shorter time period until 2002 from the National Bureau of Economic Research (NBER) website. We used data on national death certificate reports from the NCHS multiple causes of death file to obtain the number of cancer deaths. The cause of death was determined using *ICD-8* for 1969–1978, *ICD-9* for 1979–1998, and *ICD-10* for 1999 and later. The population estimates for 86 single-year age groups (0, 1, 2, ..., 84, 85+) are from the SEER U.S. population data as introduced previously. For each 5-year birth cohort in a 5-year time period, there are multiple single-year age-specific lung cancer death rates that can also be sex and race specific. We used data on black and white populations only for the multilevel analysis due to the poor quality of data on the other race group and lack of information on other covariates for this group. In this case, there can be up to 100 observations within each cohort-by-period cell. Note that birth cohorts defined in the previous age-by-period

TABLE 3.2

Age-Specific Lung Cancer Death Rates (per 100,000 Population) for White Males and White Females, United States 1969–2007

	White Males							
	Period							
Age	1969–74	1975–79	1980–84	1985–89	1990–94	1995–99	2000–04	2005–07
20–24	0.17	0.11	0.12	0.08	0.10	0.11	0.11	0.07
25–29	0.38	0.35	0.30	0.27	0.23	0.22	0.21	0.20
30–34	1.82	1.48	1.05	1.11	0.98	0.97	0.63	0.55
35–39	7.37	5.84	4.87	3.81	3.65	3.03	2.55	1.95
40–44	20.50	18.34	15.78	13.29	10.70	9.78	9.06	7.27
45–49	45.87	46.68	42.95	37.44	30.57	23.70	22.38	20.17
50–54	84.48	93.08	93.12	85.59	73.28	57.86	47.75	44.39
55–59	149.33	155.63	165.20	164.97	148.33	121.48	101.36	84.08
60–64	231.90	249.30	252.11	265.38	257.32	217.92	187.06	162.94
65–69	315.83	345.18	361.64	363.77	373.21	339.60	299.91	269.95
70–74	367.93	425.23	460.48	476.62	466.60	456.26	421.13	383.22
75–79	382.20	457.91	516.57	545.13	546.39	521.67	511.77	489.76
80–84	328.01	430.51	501.16	562.57	590.58	568.68	541.83	541.89
85+	240.67	325.83	408.14	466.82	522.39	535.45	493.98	464.79

	White Females							
	Period							
Age	1969–74	1975–79	1980–84	1985–89	1990–94	1995–99	2000–04	2005–07
20–24	0.09	0.08	0.05	0.06	0.06	0.03	0.04	0.08
25–29	0.23	0.57	0.18	0.21	0.21	0.21	0.16	0.18
30–34	1.01	2.61	0.79	0.77	0.83	0.83	0.61	0.44
35–39	3.39	11.77	3.34	2.66	2.71	2.91	2.60	1.87
40–44	8.45	41.49	9.59	8.72	7.20	7.35	8.04	7.34
45–49	17.19	97.72	23.63	23.00	20.66	16.46	16.77	17.74
50–54	27.27	180.81	43.64	47.17	44.99	38.82	32.42	31.30
55–59	39.01	270.18	68.31	78.57	82.34	75.24	67.29	57.06
60–64	45.45	343.71	93.42	116.05	129.06	127.23	121.26	110.17
65–69	49.61	400.83	117.06	148.40	175.60	183.74	183.07	174.83
70–74	51.37	431.75	123.19	172.18	211.65	235.27	244.42	241.36
75–79	55.14	385.16	113.02	165.96	223.94	254.68	280.02	289.41
80–84	56.36	377.58	102.21	143.53	202.69	250.54	279.78	289.47
85+	57.88	270.41	92.96	115.62	154.57	193.55	220.49	231.68

TABLE 3.3

Cohort by Period Cross-Classified Data Structure of Lung Cancer Incidence and Mortality Rates: Number of Observations and Mean (per 100,000 Population)

Incidence

	1973–74		1975–79		1980–84		1985–89		Period Year 1990–94		1995–99		2000–04		2005–07		Total	
Cohort	N	Mean	N	Mean	N	Mean	N	Mean	N	Mean	N	Mean	N	Mean	N	Mean	N	Mean
1884–1889	12	235.71	0	0	0	0	0	0	0	0	0	0	0	0	0	0	12	235.71
1890–1894	40	254.01	60	256.05	0	0	0	0	0	0	0	0	0	0	0	0	100	255.23
1895–1899	40	272.03	100	273.95	60	295.46	0	0	0	0	0	0	0	0	0	0	200	280.02
1900–1904	40	303.81	100	311.66	100	355.36	60	346.21	0	0	0	0	0	0	0	0	300	332.09
1905–1909	40	245.41	100	316.59	100	392.35	100	404.26	60	374.14	0	0	0	0	0	0	400	358.96
1910–1914	40	198.03	100	264.35	100	383.64	100	419.99	100	421.12	60	400.14	0	0	0	0	500	361.68
1915–1919	40	154.10	100	206.09	100	312.44	100	401.78	100	448.63	100	448.32	60	429.08	0	0	600	356.06
1920–1924	40	95.47	100	150.33	100	255.90	100	343.63	100	436.82	100	470.62	100	465.40	56	436.47	696	345.59
1925–1929	40	61.58	100	97.31	100	177.52	100	266.36	100	338.50	100	418.18	100	469.37	80	460.87	720	300.08
1930–1934	40	22.04	100	50.41	100	96.67	100	166.86	100	232.28	100	307.47	100	388.93	80	427.60	720	221.32
1935–1939	40	8.24	100	22.44	100	50.48	100	94.57	100	146.61	100	208.99	100	291.80	80	356.10	720	153.20
1940–1944	36	2.21	100	5.62	100	19.48	100	43.86	100	78.43	100	123.67	100	193.78	80	258.57	716	93.92
1945–1949	0	0	60	2.39	100	4.25	100	14.54	100	36.55	100	61.86	100	107.23	80	154.64	640	54.62
1950–1954	0	0	0	0	60	1.78	100	5.00	100	14.46	100	27.87	100	54.73	80	84.62	540	31.63
1955–1959	0	0	0	0	0	0	60	1.47	100	4.02	100	12.03	100	26.84	80	49.53	440	18.95
1960–1964	0	0	0	0	0	0	0	0	60	1.48	100	3.68	100	9.48	80	18.86	340	8.57
1965–1969	0	0	0	0	0	0	0	0	0	0	60	1.12	100	2.67	80	5.50	240	3.22
1970–1974	0	0	0	0	0	0	0	0	0	0	0	0	60	1.06	80	1.94	140	1.57
1975–1979	0	0	0	0	0	0	0	0	0	0	0	0	0	0	40	1.20	40	1.20
Total	448	150.66	1,120	165.52	1,120	198.79	1,120	211.56	1,120	212.75	1,120	207.45	1,120	202.53	896	189.67	8,064	195.92

Mortality

Period Year

Cohort	1969–74		1975–79		1980–84		1985–89		1990–94		1995–99		2000–02		2005–07		Total	
	N	Mean	N	Mean	N	Mean	N	Mean	N	Mean	N	Mean	N	Mean	N	Mean	N	Mean
1884–1889	84	151.69	0	0	0	0	0	0	0	0	0	0	0	0			84	151.69
1890–1894	120	186.34	60	207.68	0	0	0	0	0	0	0	0	0	0			180	193.46
1895–1899	120	201.26	100	237.94	60	255.50	0	0	0	0	0	0	0	0			280	225.98
1900–1904	120	200.72	100	257.89	100	309.25	60	317.09	0	0	0	0	0	0			380	262.70
1905–1909	120	166.67	100	240.97	100	308.15	100	356.25	60	371.80	0	0	0	0			480	276.76
1910–1914	120	135.44	100	211.63	100	295.19	100	371.26	100	419.91	60	410.37	0	0			580	294.27
1915–1919	120	96.42	100	165.16	100	245.71	100	328.94	100	395.90	100	422.61	48	410.80			668	280.12
1920–1924	120	59.26	100	117.17	100	191.01	100	276.27	100	359.46	100	408.95	60	419.88			680	246.45
1925–1929	120	33.30	100	73.55	100	127.44	100	202.79	100	281.14	100	353.82	60	388.05			680	192.87
1930–1934	120	13.90	100	36.47	100	72.89	100	125.45	100	194.70	100	252.07	60	305.57			680	129.65
1935–1939	120	4.41	100	14.44	100	34.51	100	68.64	100	117.47	100	167.69	60	213.26			680	78.82
1940–1944	60	1.45	100	4.07	100	12.64	100	30.83	100	61.42	100	97.40	60	136.11			620	46.60
1945–1949	0	0	60	1.37	100	3.63	100	10.81	100	26.52	100	49.01	60	73.03			520	25.89
1950–1954	0	0	0	0	60	0.83	100	3.25	100	9.43	100	21.36	60	36.39			420	13.42
1955–1959	0	0	0	0	0	0	60	0.96	100	3.09	100	8.62	60	17.38			320	7.10
1960–1964	0	0	0	0	0	0	0	0	60	1.07	100	2.43	60	5.88			220	3.00
1965–1969	0	0	0	0	0	0	0	0	0	0	60	0.63	60	1.28			120	0.95
1970–1974	0	0	0	0	0	0	0	0	0	0	0	0	24	0.40			24	0.40
Total	1,344	107.56	1,120	132.56	1,120	156.63	1,120	175.47	1,120	186.85	1,120	181.30	672	171.93			7,616	156.62

contingency tables are 10-year cohorts that are labeled by the midinterval birth year and overlap with adjacent neighbors. The cohorts in this table are actual 5-year groups defined by birth years and mutually exclusive. In sum, we can cross classify rates by birth cohort and time period and examine age-specific data that can be considered as nested in the cohort-by-time period contexts.

3.4 Research Design II: Repeated Cross-Sectional Sample Surveys

The previous section focused on the possibilities for APC analysis using repeated cross-sectional data at the population level. In this section, we show additional utilities of such design using microdata from repeated cross-sectional sample surveys that are increasingly available to social scientists. Compared to the longitudinal panel design, which usually spans a short time period, the synthetic cohort approach has the advantage of facilitating simultaneous tests of age and period effects because it is based on representative national surveys of all ages conducted regularly from one period to the next and often covering multiple decades. It suffers less from the difficulty in locating sample respondents across time in panel studies, although it is not exempt from attrition due to mortality. Pooling data of all ages and survey years yields the age-by-time period data structure as shown in Figure 3.1. And, arranging data by all birth cohorts and survey years yields the same cohort-by-period cross-classification data structure as shown in Figure 3.2. In this case, individual respondents (usually not the same individuals in more than one survey) of any birth cohort are interviewed in multiple replications of the survey, and individual respondents in any particular wave of the survey can be drawn from multiple birth cohorts. Although sample surveys share these characteristics with tabular data at the aggregate population level, they possess an additional and important feature. That is, repeated cross-sectional surveys provide individual-level data on both the responses and a wide range of covariates, which can be employed for much finer-grained analyses of explanatory hypotheses.

3.4.1 General Social Survey (GSS) 1972–2006: Verbal Test Score and Subjective Well-Being

We first used data from General Social Survey (GSS) conducted over three decades for detailed methodological expositions. The GSS are a typical example of repeated cross-sectional surveys that have monitored the attitudes and behaviors of adults in the United States since 1972 (Davis, Smith, and Marsden 2005). Each survey uses multistage stratified probability sampling and includes a nationally representative sample of noninstitutionalized adults age 18 and older in the United States. We focused on two substantive problems that have attracted attention in previous studies using APC analysis.

The first problem concerns controversy over trends in verbal ability. A series of articles published in the *American Sociological Review* in 1999 centered on the existence of an intercohort decline in verbal ability in the GSS 1974 to 1996. The debate was initiated by Alwin (1991) and Glenn's (1994) finding of a long-term intercohort decline in verbal ability beginning in the early part of the twentieth century. Wilson and Gove (1999) took issue with this finding and argued that the Alwin and Glenn analyses confused cohort effects with aging effects and ignored time period effects. In response, Glenn (1999) disagreed that the decline in GSS vocabulary scores resulted solely from period influences and argued against the Wilson and Gove claim that cohort differences actually reflected only age effects. The previous findings on trends in verbal scores are interesting and suggestive. But, until age, period, and cohort effects are simultaneously estimated, the question of whether the trends are due to age, period, or cohort components remains incompletely resolved. We used this specific example to motivate the statistical methodology we present in Chapter 6. The substantive results are therefore presented for illustration.

Table 3.4 summarizes the descriptive statistics of the verbal test score data we analyzed. This is an extension of the 1974–1996 data on which the controversy was based and includes 17 cross sections of the GSS: 1974, 1976, 1978, 1982, 1984, 1987, 1988, 1989, 1990, 1991, 1993, 1994, 1996, 1998, 2000, 2004, and 2006. In these surveys, a survey respondent's vocabulary knowledge

TABLE 3.4

Summary Statistics for GSS Vocabulary Test Data, 1974–2006

	Description and Coding	N	Mean	SD	Min	Max
Dependent Variable						
WORDSUM	A 10-item composite vocabulary scale score	22,042	6.05	2.13	0	10
Level 1 Variables						
AGE	Age at survey year	22,042	45.13	17.37	18	89
EDUCATION	Highest levels of education respondent completed	22,042	12.84	3.02	0	20
SEX	1 = female; 0 = male	22,042	0.57	0.5	0	1
RACE	1 = black; 0 = white	22,042	0.15	0.36	0	1
Level 2 Variables						
COHORT	Five-year birth cohorts	20	—	—	–1894	1985–89
PERIOD	Survey years	17	—	—	1974	2006

is measured by a composite scale score named WORDSUM, which is constructed by adding the correct answers to 10 verbal test questions and ranges from 0 to 10. WORDSUM has an empirical frequency distribution that is approximately bell shaped with a mean of about 6 and is reported in previous studies to have an internal reliability of .71 (Wilson and Gove 1999:258).* The data include 22,042 respondents who had measures on WORDSUM and other covariates across all survey years. Respondents' ages in the data pooled across all surveys varied from 18 to 89. The average years of education completed was around 12.8 years. Fifty-seven percent of respondents were female, and 15% were black. There were 20 five-year birth cohorts. The oldest cohort member was born before 1894 and the youngest born in 1985–1989. As has been introduced previously and will be clearer in chapters to follow, gender, race, and education are level 1 covariates, and survey years and birth cohorts are level 2 contextual variables in hierarchical model specifications. Table 3.5 further illustrates the two-way cross-classification structure of the verbal ability data in terms of mean verbal test scores of individual respondents in cohort by period cells.

A second problem of interest to APC analysts that can be studied with GSS data is trends in social inequalities of subjective well-being. As increases in life expectancies in the United States continue into the twenty-first century, there has been a growing need for research and policy to take into account both the quantity and the quality of life. A fundamentally important question for research communities, policy makers, and public authorities is, Are Americans living better as well as longer lives? The measure of subjective quality of life that has been examined most frequently is general happiness. Previous analyses of correlates of happiness have mostly been concerned with cross-sectional individual-level characteristics and attainments. We knew relatively little beyond the stratification of subjective quality of life at a static point in time. Specifically, there is little knowledge of the social heterogeneity of life course patterns, time trends, and birth cohort differences in general happiness. And, still less is known about individual- and macro-level mechanisms underlying these patterns and trends. We use this as an example to show how a systematic APC analysis can reveal the various ways in which social stratification operates over the life course and historical time, test the extent to which aging and life course theories apply to subjective well-being, and shed new light on the changes in the social distribution of quality of life in the United States.

Table 3.6 summarizes the descriptive statistics of the happiness data we analyzed. *Happiness* was assessed as a single-item scale reported from respondents. The data on happiness are available annually in 22 survey years

---

* In an item analysis of individual words in WORDSUM, Alwin (1991: 628) found that some of the words have become more difficult over time. The general conclusion in this series of articles (Alwin and McCammon 1999; Glenn 1994; Wilson and Gove 1999), however, was that word obsolescence does not account for observed changes in the test scores over time.

TABLE 3.5

Cohort by Period Cross-Classified Data Structure of GSS: Mean Verbal Test Score

Cohort	Year																	Mean
	1974	1976	1978	1982	1984	1987	1988	1989	1990	1991	1993	1994	1996	1998	2000	2004	2006	
–1899	5.24	5.26	5.50	4.64	4.89	5.00	5.67											5.23
1900–1904	5.61	5.81	5.65	5.74	6.03	4.86	4.33	5.07	6.29	5.67	4.00							5.59
1905–1909	5.81	5.61	4.96	5.05	5.45	5.82	4.61	4.82	5.42	6.17	4.92	5.11	5.94	5.50				5.42
1910–1914	6.17	6.56	6.21	5.27	6.17	5.52	4.89	5.70	6.61	5.38	5.16	5.71	5.45	5.88	5.88			5.85
1915–1919	6.06	6.46	6.08	5.94	6.41	5.46	5.57	6.27	6.62	6.02	6.60	5.99	5.22	6.11	5.93	6.19	6.00	6.05
1920–1924	6.10	6.02	6.24	5.93	5.67	5.21	5.63	6.39	6.57	5.84	5.59	5.94	5.62	5.73	4.96	7.00	6.08	5.89
1925–1929	6.33	5.94	6.34	5.88	5.92	5.47	6.31	5.63	6.19	6.40	6.31	6.68	6.06	5.88	6.63	6.57	5.89	6.12
1930–1934	6.18	6.44	6.12	5.97	5.83	5.95	5.58	5.63	6.30	6.00	6.00	6.37	5.99	5.98	5.88	6.03	6.44	6.06
1935–1939	6.19	6.33	6.03	5.96	6.36	5.81	5.88	5.72	6.65	5.45	6.22	6.15	6.22	6.36	6.22	6.96	6.19	6.16
1940–1944	6.22	6.26	6.28	6.27	6.19	5.94	5.81	6.26	6.48	6.83	6.01	6.18	6.48	6.79	6.75	6.62	6.14	6.30
1945–1949	6.50	6.05	6.33	6.26	6.67	6.33	6.37	6.60	6.68	6.37	6.57	6.72	6.73	6.77	6.60	6.30	6.87	6.50
1950–1954	5.47	5.88	5.77	6.01	6.13	5.88	6.40	6.16	6.42	6.38	6.32	6.53	6.41	6.55	6.18	6.49	6.44	6.16
1955–1959	5.10	5.33	5.37	5.28	5.72	5.70	6.03	5.93	5.74	6.29	6.30	6.25	6.06	6.34	5.95	6.00	6.44	5.90
1960–1964			5.14	4.93	5.50	5.68	5.62	5.98	5.92	6.00	6.02	6.08	6.03	6.09	6.23	6.23	6.04	5.89
1965–1969					4.95	4.94	4.69	5.22	5.06	5.95	5.65	5.81	5.98	5.59	5.66	6.16	6.22	5.64
1970–1974							6.50	5.06	5.52	4.69	5.21	5.56	5.45	5.72	5.98	6.06	5.82	5.65
1975–1979											4.50	5.17	5.11	5.75	5.51	6.01	5.86	5.68
1980–1984														6.00	5.00	5.94	5.83	5.73
1985–1989																4.67	5.38	5.09
Mean	6.02	6.04	5.96	5.74	5.99	5.69	5.76	5.94	6.14	6.09	6.03	6.16	6.04	6.13	6.02	6.21	6.15	6.00

TABLE 3.6

Summary Statistics for GSS Happiness Data, 1972–2004 ($N = 28,869$)

	Description and Coding	Mean	SD	Min	Max
Dependent Variable					
HAPPY	Level of happiness: 1 = very happy; 2 = pretty happy; 3 = not too happy	2.2	0.64	1	3
Level 1 Variables					
AGE	Respondent's age at survey year (centered around grand mean and divided by 10)	44.84	17.12	18	89
FEMALE	1 = female; 0 = male	0.55	0.50	0	1
BLACK	1 = black; 0 = white	0.14	0.34	0	1
EDUCATION	Respondent's years of schooling (reference = 12–15)	12.55	3.16	0	20
EDUC 1	1 = 0–11 years of schooling; 0 = otherwise	0.26	0.44	0	1
EDUC 2	1 = 16 or more years of schooling; 0 = otherwise	0.20	0.40	0	1
FAMILY INCOME	Family income in thousands adjusted for household size and converted to 1986 dollars (reference = 2nd and 3rd quartiles)	13.23	13.22	0.05	162.61
LOWER QUARTILE	1 = lower 25%; 0 = otherwise	0.25	0.43	0	1
UPPER QUARTILE	1 = upper 25%; 0 = otherwise	0.25	0.43	0	1
Marital status	Respondent's marital status (reference = married)				
DIVORCED	1 = divorced or separated; 0 = otherwise	0.15	0.36	0	1
WIDOWED	1 = widowed; 0 = otherwise	0.09	0.29	0	1
NEVERMAR	1 = never married; 0 = otherwise	0.18	0.39	0	1

Health	Respondent's self-rated health (reference = good)				
EXCELLENT	1 = excellent; 0 = otherwise	0.32	0.47	0	1
FAIR	1 = fair; 0 = otherwise	0.18	0.39	0	1
POOR	1 = poor; 0 = otherwise	0.05	0.23	0	1
Work status	Respondent's work status (reference = full time and other)				
PARTTIME	1 = part time; 0 = otherwise	0.10	0.30	0	1
UNEMPLOY	1 = unemployed; 0 = otherwise	0.03	0.17	0	1
RETIRED	1 = retired; 0 = otherwise	0.12	0.32	0	1
Children	Respondent's number of children				
CHILDS_0	1 = no children; 0 = 1 or more children	0.27	0.44	0	1
ATTEND	Frequency respondent attends religious activities	3.89	2.67	0	8
Level 2 Variables					
COHORT	Five-year birth cohort	—	—	−1899	1980–86
RCS	Relative cohort size: % age 15–19 by cohort	12.92	1.68	10.6	15.34
PERIOD	Survey year	—	—	1972	2004
GDP	National GDP per capita in thousands by year	26.27	5.05	19.56	36.59
UNEMP	Unemployment rate by year	6.44	1.42	4.00	9.70

from 1972 to 1994 (except for 1979, 1981, and 1992) and biannually from 1994 to 2004. In all years, the GSS item on overall happiness was the following: "Taken all together, how would you say things are these days—would you say that you are very happy, pretty happy, or not too happy?" The responses were coded as 1 (= very happy), 2 (= pretty happy), and 3 (= not too happy). Evidence of the psychometric adequacy of this measure has been provided elsewhere (Yang 2008). The data include 28,869 respondents who had measures on HAPPY and other covariates across all survey years. Respondents' ages ranged from 18 to 89. There were 18 five-year cohorts born between 1899 and 1986. The operational definitions and descriptive statistics of level 1 and level 2 covariates used in the analysis are reported in the table. In addition to period and cohort, three additional covariates at level 2 are included. The cohort-level covariate is relative cohort size (RCS) adopted from O'Brien (2000). Two period-level covariates are annual gross domestic product (GDP) per capita in thousands published by the U.S. Department of Commerce and the annual unemployment rate published by the U.S. Department of Labor, Bureau of Labor Statistics.

3.4.2 National Health and Nutrition Examination Surveys (NHANES) 1971–2008: The Obesity Epidemic

Studies of the U.S. population have consistently shown a disconcerting increase in obesity rates in recent decades (Flegal et al. 2002; Freedman et al. 2002; Mokdad et al. 2001; Ogden et al. 2002). According to estimates from the National Health and Nutrition Examination Survey (NHANES), the prevalence of obesity among U.S. adults more than doubled in the last 20 years of the twentieth century (Flegal et al. 2002). Although specific explanations vary, the vast majority share the common view that secular changes in U.S. society (i.e., period effects) lie at the root of the obesity epidemic. Since there has been insufficient research to disentangle period effects from age and birth cohort effects, the importance of secular change relative to cohort membership is yet to be examined analytically. In a study of 1.7 million participants in the 1976–2002 National Health Interview Survey (NHIS), Reither, Hauser, and Yang (2009) employed APC analysis to determine what role birth cohorts have played in the U.S. obesity epidemic. Results confirmed that period effects are principally responsible for the epidemic, but they also demonstrated that cohort membership is more influential than previously assumed. For example, relative to the 1955–1959 birth cohort, the odds of class 2 obesity (body mass index [BMI] $\geq$ 35) increased by 50% for the 1980–1984 birth cohort. They concluded that secular change and birth cohort membership have independently contributed to elevated odds of obesity among recent cohorts of Americans.

We built on this study to further examine the obesity epidemic and underlying mechanisms related to distinct age, period, and cohort influences. We addressed two remaining questions. First, because the BMI is based on self-reported weight and height in the NHIS, it potentially can deviate from actual BMI. Therefore, we use data on measured BMI from the NHANES to more directly compare results regarding period and cohort effects with estimates from the same data that first established the period trends in obesity (Flegal et al. 2002). Second, the Reither, Hauser, and Yang (2009) study estimated A, P, and C effects for each stratified sample by sex, race, and education. Although the results showed differences in these effects for men and women, black and white, and the college educated and less educated, the differences were not assessed quantitatively. We formally tested the hypothesis of social disparities in trends of obesity by estimating sex, race, and educational differences in A, P, and C effects. We also accounted for other individual-level risk factors for obesity, such as family income, in the analysis.

The NHANES, conducted by the NCHS, uses a multistage stratified sampling design and includes a representative sample of the noninstitutionalized U.S. population, with an oversample of older persons and minorities (CDC 2010). There are nine survey periods: NHANES I (1971–1975); NHANES II (1976–1980); NHANES III, phase 1 (1988–1991); NHANES III, phase 2 (1991–1994); 1999–2000; 2001–2002; 2003–2004; 2005–2006; and 2007–2008. The study includes respondents aged 25 to 74 who attended household interviews and clinical examinations between 1971 and 2008. We also examined all available ages that are as young as 2, but only report results on adults above the age of 25 who were more likely to have completed education and for whom meaningful measures of income are available. Birth years were calculated using survey year (midinterval value) minus respondents' age. This resulted in 17 five-year birth cohorts born between 1899 and 1982, with the earliest cohort spanning six years (1899–1904) and the most recent cohort spanning just three years (1980–1982). We excluded foreign-born respondents due to their lack of the same period and cohort experiences and exposures with the native respondents. We further excluded respondents who were pregnant when weighed. The final sample included 40,261 individuals. Obesity status was assessed using BMI calculated from examination data on measured height and weight. When examination data were missing, self-reported height and weight was used. For adult respondents in our sample, obesity was defined as BMI ≥ 30.0 kg/m^2.[*] Because data on Hispanic ethnicity were not available until 1988, we used the black-white race variable available for all survey years. For descriptive analysis, we used the sampling weight to adjust for the effects of survey design and nonresponse. Table 3.7 presents weighted summary statistics of the obesity outcome; five individual-level covariates (age, sex, race, education level, and income quartiles); and level 2 variables of time period and birth cohort.

---

[*] We thank Dr. Whitney Robinson for assistance with data preparation, including providing data on the measured weight and height for nine survey periods, construction of the obesity variable across ages, as well as substantive expertise on the measures and trends of obesity.

TABLE 3.7

Summary Statistics for NHANES Obesity Data, 1971–2008

	Description and Coding	N	Mean	SD	Min	Max
Dependent Variable						
OBESE	Measured BMI categorized into obesity status: 1 = obese (BMI ≥ 30); 0 = not obese	40,261	0.30	0.46	0	1
Level 1 Variables		40,261				
AGE	Age in years; centered around grand mean	40,261	46	15	25	74
SEX	Respondent's sex: 1 = male; 0 = female	40,261	0.48	0.50	0	1
RACE	Respondent's race: 1 = black; 0 = white	40,261	0.12	0.33	0	1
EDUCATION	Respondent's years of schooling (reference = 12 years or less)	40,261	0.50	0.50	0	1
SOME COLLEGE	1 = 13–15 years of schooling; 0 = otherwise	40,261	0.26	0.44	0	1
COLLEGE	1 = 16 or more years of schooling; 0 = otherwise	40,261	0.24	0.43	0	1
FAMILY INCOME	Total family income adjusted to 2000 dollars (reference = 2nd and 3rd income quartiles)	40,261	39,797.7	32,305.5	0	120,407.7
LOWER QUARTILE	1 = Lowest 25% in each wave; 0 = otherwise	40,261	0.21	0.41	0	1
UPPER QUARTILE	1 = Highest 25% in each wave; 0 = otherwise	40,261	0.28	0.45	0	1
Level 2 Variables						
COHORT	Five-year birth cohort	17	—	—	1899	1982
PERIOD	Survey year	9	—	—	1971	2008

3.4.3 National Health Interview Surveys (NHIS) 1984–2007: Health Disparities

Social disparities in health or simply health disparities are a long-standing research problem in medical sociology and social epidemiology. As data from more large surveys accumulate over time and become available to analysts, the question of trends and changes in health disparities also becomes more relevant. In addition to a large body of demographic and epidemiologic research on age variation and temporal trends in health, research on health disparities across the life course, cohorts, and time periods is also rapidly growing. We defer the review of research literature on changing health disparities to Chapter 6. In general, there is evidence for increases in gender, race, and socioeconomic inequalities across the life course, birth cohorts, and time periods in the United States. Two questions remain. First, to what extent do increasing disparities in specific domains of social stratification contribute to increasing overall health disparities (dispersion or variance of health outcomes)? Second, to what extent are disparities across ages, cohorts, and time periods independent of each other?

A major limitation of prior research is that it has treated these three time dimensions separately. However, they can be intertwined. For example, an increase in health disparities across time periods may result from cohort replacement, in which cohorts with larger within-cohort health disparities succeed cohorts with smaller within-cohort health disparities, or an aging society, wherein the elderly, who usually have larger within-age health disparities than younger people, increase their proportionate share in the population structure, or from some combination of the two. Similarly, a widening health disparity with age may be confounded with period or cohort patterns. Some studies have tried to disentangle age and cohort patterns in health disparities and have found distinct age effects and cohort variations in mean levels of health and changing health disparities by education, income, gender, and race over life course and across birth cohorts (Chen, Yang, and Liu 2010; Lauderdale 2001; Lynch 2003; Yang and Lee 2009). Lynch (2003) also found each pattern was suppressed when the other one was ignored. However, there is a lack of simultaneous assessment of the effects of age, period, and cohort in health disparities.

We addressed this gap in the literature using data on self-reported health. Self-rated health is a widely used measure of general health status that has been found to be highly predictive of mortality and strongly correlated with objective assessments of health, including physician diagnoses (Idler and Benyamini 1997). In fact, self-rated health is a good indicator of objective health and subclinical illness and has been found in some studies to be more predictive of mortality among the elderly than physician assessments (Schoenfeld et al. 1994). Close relationships between self-rated health and objective health indicators also hold across population subgroups (Bosworth et al. 2001). Based on these and related findings regarding its robustness as a single, summary index of an individual's health status, we studied self-rated health as a health outcome variable. But, we also caution that health is not a singular condition, and findings in this chapter may not always generalize to health disparity trends associated with specific health outcomes.

Our analysis was based on annual data from the NHIS for the 24-year period 1984 to 2007. The NHIS is a multistage probability sample survey of the civilian noninstitutionalized U.S. population conducted by the NCHS. NHIS collects health information for each member of a family or household sampled, as reported by one primary respondent. To reduce reporting/measurement errors, we limited our analysis to the primary respondent. The sample size for men was about 16,670 each year (in total 16,670*24 = 400,080),

TABLE 3.8

Summary Statistics for Self-Reported Health Data from NHIS, 1984–2007

	Description and Coding	N	Mean	SD	Min	Max
Dependent Variable						
HEALTH	Respondent's self-rated health: 1 = poor, 2 = fine, 3 = good, 4 = very good, 5 = excellent	701,888	3.76	1.13	1	5
Level 1 Variables						
SEX	1 = man, 0 = woman	701,888	0.57	0.49	0	1
RACE	1 = white, 0 = other races	701,888	0.82	0.38	0	1
AGE	Respondent's age at survey year	701,888	46.55	17.30	18	85
EDUCATION	Respondent's years of schooling	701,888	12.65	3.17	0	18
MARRIED	1 = married, 0 = others	701,888	0.55	0.50	0	1
EMPLOYED	1 = employed, 0 = others	701,888	0.67	0.47	0	1
INCOME	Household income in thousands	701,888	4.58	2.76	0.07	10.41
REDESIGN	Effect of survey redesign 1 = 1995–2007, 0 = 1984–1994	701,888			0	1
Level 2 Variables						
COHORT	Five-year birth cohorts	18	—	—	1899	1985
PERIOD	Survey year	24	—	—	1984	2007

and for women was about 12,575 each year (in total 12,575*24 = 301,800). The outcome variable, self-rated health, has remained largely unchanged across periodic revisions of the NHIS questionnaires, which facilitates the analysis of trends. It has five response categories: poor, fair, good, very good, and excellent. Table 3.8 shows the summary statistics for self-rated health and individual-level demographic and social variables that have been linked to health in previous research. The respondents were aged 18 to 85 and belonged to 18 five-year cohorts born between 1899 and 1985.

The sampling frame for the NHIS is redesigned every 10 years and was redesigned, during the period studied here, in 1995. Nonetheless, the fundamental design of the 1995–2007 NHIS was similar to that of the 1985–1994 NHIS. Three changes in the sampling design and weighting structure are notable. First, the number of primary sampling locations has increased from 198 to 358 since 1995. Second, both black and Hispanic populations were oversampled in the 1995–2007 NHIS, while only blacks were oversampled in the 1985–1994 NHIS. Third, the weighting structure changed after 1996. These three changes potentially affect the sample variances after 1995 and 1996. Therefore, we used the sample weights in all analyses to adjust for the multistage sampling design. We also created an indicator/dummy variable named "redesign" to adjust the regression model estimates for any effects of sampling design changes since 1995.

3.4.4 Birth Cohort and Time Period Covariates Related to Cancer Trends

As mentioned in Section 3.3.1, an extension of the cancer trend analysis is to examine the role of two prominent biobehavioral risk factors for cancer incidence and mortality, cigarette smoking and obesity, in explaining the temporal trends in cancer incidence and mortality. Diagnostic and screening techniques (such as colonoscopy, mammography, and prostate-specific antigen [PSA] screening) and treatment (such as hormonal replacement therapy [HRT]) changes have also been frequently associated with time period changes in cancer incidence. To facilitate more detailed statistical investigations of cancer trends in relation to hypothesized period and cohort-level covariates, we constructed period and cohort data on smoking and obesity, and period data on HRT and mammography from national sample surveys and other published sources.

Data on Americans' smoking behaviors came from the NHIS. Respondents in the surveys were first asked if they had smoked 100 cigarettes in their lifetime. If they responded in the affirmative, they were then asked if they currently smoked cigarettes. We adopted the cohort smoking variable used by Wang and Preston (2009) that was an estimate of the average number of years spent as a current smoker before the age of 40 based on the NHIS conducted between 1965 and 2005. Compared to single measures of smoking status, this measure indexes the past behavior and history of surviving cohort members and hence better captures the cumulative exposures to carcinogens that contribute to cohort differences. The estimates pertain to 5-year female and male birth cohorts from 1885 and 1984. We constructed period-specific rates of smoking based on three measures available in the NHIS: percentage of current smokers, percentage quit smoking (or smoking cessation), and percentage ever smoked. Based on the original data that are available for most years from 1970 to 2007, we created 5-year sex- and race-specific (black and white) smoking rates for the same eight periods of the cancer data and interpolated the data for the period 1980–1984 using the neighboring periods. We only present the prevalence of current smokers in Table 3.9 as it appeared to be the best period effect predictor among the three.

Data on obesity came from the NHANES as described. We created rates of obesity as defined in Table 3.7 based on measured BMI by both cohort and period. The cohort-specific obesity rates were calculated as percentage of obese members in 5-year birth cohorts by sex and race (black and white). Because the NHANES studies were not conducted every year, we used data from all survey years to construct 5-year obesity rates that applied to approximately the same 5-year periods for the cancer data. We interpolated the rates for 1980–1984 and 1995–1999 using estimates from adjacent periods.

Data on diagnostic and screening hardly exist for the entire period of the past 30 years on which the cancer trends analysis was based. We were able to

TABLE 3.9

Major Risk Factors for Cancer Incidence and Mortality by Birth Cohort
and Time Period

	Data Source	Description and Coding	Mean	SD	Min	Max
Cohort-Level Variables						
SMOKE	National Health Interview Surveys[a]	Average number of years spent as a cigarette smoker before age 40 among men and women in different birth cohorts	10.08	4.69	0.88	17.9
OBESITY	National Health and Nutrition Examination Survey, 1971–2008[b]	Percentage of respondents who are obese based on measurements obtained from a physical examination (1 = obese)	23.88	9.56	5.3	47.3
Period-Level Variables						
SMOKE	National Health Interview Survey Public Use Data File 1970–2007[c, d]	Percentage of respondents who currently smoke (1 = yes)	31.37	8.99	17.94	53.33
OBESITY	National Health and Nutrition Examination Survey, 1971–2008	Percentage of respondents who are obese based on measurements obtained from a physical examination (1 = yes)	26.36	10.27	12.2	52.8
HRT	Kaiser Permanente Northwest (KPNW) Outpatient Pharmacy System	Percentage of women aged 45 year and older who were dispensed at least one prescription of estrogen plus progestin in each period (1 = dispensed)	9.14	6.26	2.64	19.63
MAMMOGRAM	National Health Interview Survey Public Use Data File 2000, 2005	Percentage of women aged 40 years and older who reported a mammogram within the past 2 years in each period (1 = reported mammogram)	39.93	28.76	2.33	73.97

[a] *Source:* Data from Haidong Wang and Samuel H. Preston, *Proceedings of the National Academy of Sciences of the United States of America* 106, no. 2 (2009): 393–398; constructed by Burns et al., Cigarette smoking behavior in the United States," in *Changes in Cigarette-Related Disease Risk and Their Implications for Prevention and Control*, edited by D. M. Burns, L. Garfinkel, J. M. Samet, Bethesda, MD: National Institutes of Health, 1998: 113–304; updated estimates supplied by David M. Burns, June 29, 2005.

[b] Sample includes white and black respondents over the age of 19.

[c] Respondents first affirmed that they had smoked 100 cigarettes (i.e., 5 packs) in their lifetime.

[d] Sample includes white and black respondents over the age of 19.

find two covariates pertinent to breast cancer incidence based on studies previously published on most years during this period. Using the data from the Kaiser Permanente Northwest (KPNW) Outpatient Pharmacy System, Glass et al. (2007) reported that breast cancer incidence showed a sharp decline following reduced use of HRT in 2002–2003. And, it was documented that the rise and fall of HRT formulations containing estrogen were also followed by the increase and decrease in a hormonally related cancer in the 1970s (Jemal, Ward, and Thun 2007). The study by Glass et al. (2007) suggested similar overall period trends in the percentage of women dispensed at least one prescription per year from two different hormone therapies: unopposed estrogens and estrogen plus progestin. Because the latter has shown a stronger association with the risk of breast cancer according to the American Cancer Society's report (http://www.cancer.org/Cancer/BreastCancer/DetailedGuide/breast-cancer-risk-factors), we constructed a period variable of HRT usage based on this therapy only. The original data are for two separate age categories: 45–59 and 60 and older and only start from year 1988. We combined the two age groups' estimates using the total population size in each group in each period as the weights. We then imputed the usage data by a backward extrapolation to year 1970 based on the observation that the rates showed a strong log-linear increasing trend from 1988 to 1999. The final variable of period pattern of HRT usage indicates the percentage of women aged 45 and older who were dispensed at least one prescription of estrogen plus progestin in each of eight 5-year periods from 1970 to 2006. The second period-level variable for breast cancer analysis is the percentage of women aged 40 years and older who reported a mammogram within the past 2 years. The original estimates came from the study conducted by Smith et al. (2008) using the NHIS data 1987–2005. We conducted a similar backward extrapolation to year 1970 in log-linear scale for black and white women and grouped the period-specific percentages by 5 years.

3.5 Research Design III: Prospective Cohort Panels and the Accelerated Longitudinal Design

The former two designs using repeated cross-sectional data adopted a synthetic cohort approach. That is, there was no information about cross-time linkages within individuals, and each synthetic cohort contained individuals from different cohorts at each point in time. Inferences drawn from such designs therefore assume that synthetic cohorts mimic true cohorts, and changes over time across synthetic cohort members mimic the age trajectories of change within true cohorts. If the composition of cohorts does not change over time due to migration or other factors and sample sizes are large,

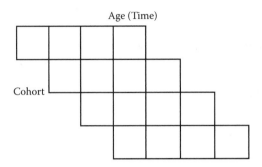

FIGURE 3.3
Accelerated longitudinal panel design.

these assumptions are generally met. However, longitudinal data obtained from the same persons followed over time are increasingly available. The primary advantage of longitudinal panel data designs is that they provide cross-time linkages within individuals and hence information pertaining to true birth cohorts.

The simplest longitudinal design follows one birth cohort for a period of time. As pointed out in Chapter 1, this is insufficient for cohort analysis due to lack of data on different cohorts. An *accelerated cohort design* follows multiple cohorts forward over multiple points in time (Tonry, Ohlin, and Farrington 1991) and is an important advance in aging and cohort research. This design allows a more rapid accumulation of information on age for multiple cohorts than does a single longitudinal cohort design and makes the cohort analysis possible. An example of this data structure is shown in Figure 3.3, where four birth cohorts were followed up at four time points. The columns represent their ages at each measurement, and the rows represent trajectories of change for each cohort. This design is especially useful for testing aging- and cohort-related hypotheses because it allows the distinction between intraindividual change with age within cohorts and intercohort differences.

We continue the examination of health disparities for empirical illustrations with a focus on cohort variations in social disparities in health and comorbidity over the life course. The process by which health deteriorates or is maintained as individuals age and its relationship to social status have attracted great attention in recent social and epidemiologic research. But, findings have been inconsistent regarding whether social disparities in health grow or diminish over the life course. One key issue that has contributed to such inconsistency is the confounding of aging and the cohort succession process. This is a long-standing problem that has hindered our ability to adequately test competing theories, such as the cumulative advantage, age-as-leveler, selective survival, and double-jeopardy hypotheses. Following Riley's lead (1987), we introduce an innovative component to this body of research—cohort variations in aging experiences—to resolve the inconsistency. An earlier study conducted by Yang (2007) using an accelerated longitudinal panel dataset on depression discovered independent age and cohort effects: Birth cohorts with different formative experiences had distinct trajectories of change in mental health with age. This highlights the relevance of social historical context represented by cohort membership to health outcomes above and beyond socioeconomic and more proximate behavioral correlates.

Building on this study, we further analyzed intercohort variations and intracohort disparities in health over the life course. Because health is quintessentially a multidimensional phenomenon, we expanded the health outcomes to include multiple indicators of physical and mental health. We also paid attention to its different manifestations by examining both individual and cumulative health indicators. We drew on studies published by Yang and Lee (2009, 2010) for empirical applications. And, we discuss cross-national health impacts of cohort change by comparing findings from these studies with a recent Chinese study (Chen, Yang, and Liu 2010). These studies jointly provide strong tests of the proposition that considering the process of cohort change is important for the theory, measurement, and analysis of social inequalities in health over the life course.

3.5.1 Americans' Changing Lives (ACL) Study 1986–2002: Depression, Physical Disability, and Self-Rated Health

The Americans' Changing Lives (ACL) study is a long-term nationally representative longitudinal survey of the adult noninstitutionalized U.S. population (House et al. 2005). It uses an accelerated longitudinal design wherein an initial sample of 3,617 individuals from a broad array of ages (25 and older), and thus of multiple birth cohorts (before 1905 to 1964), were interviewed in 1986 and monitored with three follow-up surveys in 1989, 1994, and 2001–2002. The analytic sample used by Yang and Lee (2009) consisted of all black and white respondents at the baseline ($N = 3,497$ in 1986) and at subsequent waves for which their data on all variables were available: $N = 2,780$ in 1989, $N = 2,331$ in 1994, and $N = 1,566$ in 2001. The fourth wave included fewer than 100 proxy interviews that were treated as self-reported interviews. The majority of attrition in subsequent waves is due to non-response and death. The number of nonrespondents was 557 at wave two, 476 at wave three, and 607 at wave four; the number of those who died was 160 by wave two, which increased to 530 by wave three and increased to 1153 by the final wave. A small number of observations in the last two waves were excluded due to missing values for one or more covariates. Together, these yielded 10,174 person-year observations.

The ACL study uses a complex sampling design that oversamples blacks and adults over 60 years of age. A composite weight variable was developed for each wave to adjust sample distributions for the probabilities of selection within households, geographical and race group differences, differential non-response, and poststratification (to match the demographic distributions of the known population as estimated by the 1985 U.S. census). All statistical analyses employed this composite sampling weight to produce unbiased estimates.

We examined three specific health outcomes. *Depressive symptoms* were measured by a standardized index created by the ACL study from an 11-item Center for Epidemiologic Studies Depression Scale (CES-D) of self-reported symptoms of depression, such as feeling depressed, lonely, sad, and so on, the reliability and validity of which have been well established (Radloff 1977). *Physical disability* was indicated by level of difficulty in performing activities of daily living (ADL) and of instrumental ADL (IADL) such as being confined to bed or a chair and climbing a few flights of stairs. A summary index was created by the ACL to indicate an increasing level of functional disability, with 1, 2, 3, and 4 equal to no, least severe, moderately severe, and most severe functional impairment, respectively. Similar to the NHIS data introduced previously, *self-rated health* was measured by a scale indicating perception of general health, with 1 = poor, 2 = fair, 3 = good, 4 = very good, and 5 = excellent.

Age at baseline interview ranged from 25 to 95+. Ages above 95 were recoded as 95 to avoid erratic estimates due to extremely small sample sizes for these ages. Using age and the baseline year of 1986, we grouped respondents into seven 10-year birth cohorts. This operationalization of cohort grouping is conventional in demographic analysis and distinguishes cohorts in a way that is qualitatively meaningful: cohorts 0–6 refer to those born before 1905, young Progressives (1905–1914), Jazz Age Babies (1915–1924), Depression Kids (1925–1934), War Babies (1935–1944), and Baby Boomers (1945–1954, 1955–1964), respectively (Hughes and O'Rand 2004; Yang 2007). We experimented with different cohort groupings, such as 5-year intervals and unequal intervals corresponding to the unequal intervals between each survey, and found largely similar results. The current operationalization, however, was superior in terms of parsimony. At each follow-up, surviving respondents in each cohort aged together, yielding cohort-specific age trajectories. The data structure is exemplified in Table 3.10, where the weighted means and standard deviations of CES-D scores are presented for each cohort, pooled across all surveys by age. The mean CES-D scores tend to decrease with age within cohorts. This is a first rough look at the data, and the evidence does not support a developmental explanation because cohort and age may interact, and other risk factors are not controlled. The lower panel shows that the mean CES-D scores are generally higher in earlier cohorts at each wave.

The analyses adjusted for other key social, demographic, and health behavior variables summarized in Table 3.11. Chronic illnesses are indicated by the number of chronic conditions, including arthritis, lung disease, hypertension, heart attack, diabetes, cancer, foot problems, stroke, broken bones, and urine beyond control. Based on medical guidelines, we recoded the continuous BMI score into categories of underweight (<18.5), overweight ($25 \leq 30$), and obese (30+), with normal ($18.5 \leq 25$) being the reference group.

TABLE 3.10

Summary Statistics of Depressive Symptoms (CES-D) by Age and Birth Cohort: ACLs 1986–2002

					Cohort											
	0 (−1905)		1 (1905–1914)		2 (1915–1925)		3 (1926–1935)		4 (1936–1945)		5 (1946–1955)		6 (1956–1965)		All	
Age	Mean	SD	Mean	SD	Mean	SD	Mean	SD	Mean	SD	Mean	SD	Mean	SD	Mean	SD
25													0.20	0.97	0.20	0.97
26													−0.02	0.93	−0.02	0.93
27													0.45	1.05	0.45	1.05
28													0.04	0.93	0.04	0.93
29													0.09	0.88	0.09	0.88
30													0.03	0.96	0.03	0.96
31													−0.10	1.03	−0.10	1.03
32											−0.10	1.00	−0.03	0.97	−0.08	0.98
33											−0.01	0.90	−0.22	0.87	−0.13	0.88
34											−0.01	0.97	−0.14	1.02	−0.10	1.00
35											−0.11	0.91	0.09	1.11	−0.07	0.96
36											−0.18	1.01	−0.33	0.87	−0.23	0.96
37											0.00	1.09	−0.35	0.82	−0.10	1.03
38											0.00	1.05	−0.44	0.62	−0.14	0.96
39											−0.18	0.97	−0.34	0.99	−0.24	0.97
40											−0.10	1.04	−0.54	0.79	−0.15	1.02
41											−0.18	0.85	−0.28	0.92	−0.20	0.86
42									−0.14	0.97	−0.25	1.10	−0.12	0.89	−0.19	1.02
43									−0.12	1.09	−0.11	1.00	−0.49	0.80	−0.19	1.00

continued

TABLE 3.10 (continued)

Summary Statistics of Depressive Symptoms (CES-D) by Age and Birth Cohort: ACLs 1986–2002

| | Cohort | | | | | | | | | | | | | | | |
| | 0 (–1905) | | 1 (1905–1914) | | 2 (1915–1925) | | 3 (1926–1935) | | 4 (1936–1945) | | 5 (1946–1955) | | 6 (1956–1965) | | All | |
Age	Mean	SD	Mean	SD	Mean	SD	Mean	SD	Mean	SD	Mean	SD	Mean	SD	Mean	SD
44									0.05	0.96	-0.31	0.87	-0.19	0.84	-0.19	0.90
45									-0.19	0.96	-0.26	1.05	-0.15	1.08	-0.19	1.01
46									0.05	1.21	-0.33	0.87	-0.45	0.75	-0.19	1.03
47									-0.09	0.96	-0.13	1.13			-0.11	1.05
48									-0.18	0.87	-0.15	1.03			-0.16	0.95
49									-0.02	1.00	-0.36	0.79			-0.21	0.90
50									-0.15	1.00	-0.21	1.00			-0.17	1.00
51									-0.17	0.95	0.06	1.18			-0.11	1.02
52							0.05	0.99	-0.20	0.85	-0.29	0.81			-0.17	0.88
53							0.07	0.98	-0.36	0.82	-0.19	1.03			-0.22	0.93
54							-0.04	0.94	-0.08	1.12	-0.20	0.94			-0.11	1.03
55							-0.12	0.92	-0.17	1.19	-0.24	0.84			-0.17	0.98
56							0.15	1.10	-0.43	0.70	-0.29	0.70			-0.12	0.94
57							-0.18	0.92	-0.07	0.96					-0.13	0.93
58							-0.22	0.79	-0.40	0.80					-0.31	0.80
59							0.15	1.03	-0.33	0.81					-0.04	0.97
60							-0.08	0.91	-0.40	0.90					-0.15	0.92
61							-0.19	0.95	0.38	1.31					-0.08	1.04

Year												
62					-0.11	1.00	-0.15	1.11	-0.28	1.01	-0.17	1.05
63					-0.13	0.89	-0.34	0.72	-0.43	0.72	-0.30	0.78
64					-0.17	0.77	-0.14	0.93	-0.35	0.95	-0.18	0.88
65					-0.17	0.95	-0.49	0.78	-0.60	0.50	-0.34	0.85
66					-0.18	0.96	-0.40	0.86	-0.44	0.70	-0.27	0.90
67					-0.09	0.91	-0.43	0.90			-0.24	0.92
68					-0.15	0.99	-0.38	0.91			-0.24	0.97
69					-0.19	0.82	-0.26	0.71			-0.21	0.79
70					-0.11	0.97	-0.25	0.59			-0.13	0.92
71					-0.24	0.79	-0.15	1.00			-0.22	0.82
72			-0.18	0.86	-0.16	0.78	-0.51	0.84			-0.24	0.82
73			-0.09	0.94	-0.13	0.95	-0.19	0.64			-0.13	0.89
74			-0.27	0.85	-0.32	0.69	-0.25	0.86			-0.29	0.76
75			0.03	0.94	0.04	0.95	-0.34	0.77			-0.03	0.92
76			0.08	1.01	-0.45	0.78	-0.05	0.94			-0.09	0.96
77			0.03	1.14	-0.19	0.87					-0.09	1.00
78			0.03	0.97	-0.28	0.77					-0.07	0.92
79			0.33	0.90	-0.09	0.91					0.14	0.92
80			0.00	1.00	-0.45	0.86					-0.09	0.99
81			-0.12	0.88	-0.31	0.74					-0.17	0.84
82	0.16	1.27	0.22	1.15	-0.14	0.99					0.13	1.14
83	0.20	0.80	0.02	1.04	-0.14	0.94					0.03	0.97

continued

TABLE 3.10 (continued)

Summary Statistics of Depressive Symptoms (CES-D) by Age and Birth Cohort: ACLs 1986–2002

	Cohort															
	0 (–1905)		1 (1905–1914)		2 (1915–1925)		3 (1926–1935)		4 (1936–1945)		5 (1946–1955)		6 (1956–1965)		All	
Age	Mean	SD	Mean	SD	Mean	SD	Mean	SD	Mean	SD	Mean	SD	Mean	SD	Mean	SD
84	-0.36	0.56	0.15	0.89	-0.58	0.65									-0.18	0.82
85	0.33	1.16	-0.34	0.70	0.17	1.20									0.14	1.11
86	0.26	1.12	-0.36	0.76	-0.27	0.94									-0.02	1.03
87	-0.31	0.67	-0.12	0.71											-0.20	0.69
88	0.52	1.14	-0.16	0.75											0.25	1.04
89	-0.04	0.96	-0.23	0.95											-0.13	0.94
90	0.12	0.90	0.07	0.66											0.10	0.81
91	-0.13	0.85	0.25	1.00											-0.03	0.88
92	-0.28	0.98	-0.49	0.75											-0.34	0.90
93	0.71	1.20	-0.39	0.68											0.30	1.14
94	-0.22	1.25	1.64	.											0.04	1.33
95	-0.02	1.05	-0.54	0.58											-0.17	0.95
All	0.11	1.03	0.00	0.96	-0.18	0.89	-0.16	0.93	-0.18	0.97	-0.15	0.98	-0.11	0.95	-0.14	0.96

TABLE 3.11

Summary Statistics of All Variables in the ACLs Analyses: 1986–2001/2 ($N = 10{,}174$)

	Description and Coding	N	Mean	SD	Min	Max
Dependent Variables						
CES-D	Standardized score of depressive symptoms	10,174	−0.14	0.95	−1.18	4.74
DISABLE	Disability: 1 = most severe; 4 = no impairment	10,174	1.29	0.73	1	4
HEALTH	Self-rated health: 1 = poor; 5 = excellent	10,174	3.61	1.03	1	5
Level 1 Covariates						
AGE	Respondent's age at survey year	10,174	51.25	15.66	25	95
INCOME	Family income: 2001 dollars in thousands	10,174	53.40	60.09	0	2750
NOTMARRIED	Marital status: 1 = not married; 0 = married	10,174	0.31	0.46	0	1
ILLNESS	Number of chronic conditions	10,174	1.06	1.24	0	8
BMI	Body mass index (reference = normal)					
UWEIGHT	Underweight: BMI < 18.5	10,174	0.17	0.38	0	1
OWEIGHT	Overweight: 25 ≤ BMI < 30	10,174	0.30	0.46	0	1
OBESE	BMI ≥ 30	10,174	0.14	0.35	0	1
SMOKE	Number of cigarettes one usually smokes in a day	10,174	4.99	10.22	0	50
Level 2 Covariates						
COHORT	10-year birth cohort	10,174	—	—	0	6
SEX	Respondent's sex: 1 = female; 0 = male	10,174	0.54	0.50	0	1
RACE	Respondent's race: 1 = black; 0 = white	10,174	0.11	0.31	0	1
EDUC	Number of years respondent attended school	10,174	12.62	2.89	0	17
DIED	Attrition type at follow-up: 1 = deceased; 0 = others	10,174	0.15	0.35	0	1
NONRESPONSE	Attrition type at follow-up: 1 = nonresponse; 0 = others	10,174	0.18	0.38	0	1

Respondent's birth cohort membership, sex, race, and education are time-constant or person-level covariates. All other variables are time varying or level 1 within-person covariates with measurements at four interviews. We also controlled for effects of attrition by including dummy variables indicating the deceased and nonrespondents to distinguish those who were lost to follow-ups and those who had complete data from all survey waves.

3.5.2 Health and Retirement Survey (HRS) 1992–2008: Frailty Index

The Health and Retirement Survey (HRS) is a nationally representative longitudinal survey of the U.S. older population conducted every 2 years from 1992–1993 to 2008 that includes several birth cohorts with different entry years. We extended the study by Yang and Lee (2010) to include the most recent wave of data in 2008. We included the maximum number of waves for each of the four birth cohorts for whom data on all health deficits and disorders for the construction of the Frailty Index (FI) are available: the Study of Assets and Health Dynamics Among the Oldest Old (AHEAD) cohort (born before 1924) surveyed in eight waves from 1993 to 2008; Children of Depression (CODA) cohort (born 1924–1930); HRS cohort (born 1931–1941); and War Baby (WB) cohort (born 1942–1947). All were surveyed in seven waves from 1996 to 2008. We did not include the youngest cohort—the Early Baby Boomers (EBB) cohort—because it entered the survey in 2004 and did not have sufficient data on age trajectories over time. Cohort membership is defined by the actual birth year. Age at first entrance into the samples ranges from 70+ for the AHEAD cohort, 66–82 for the CODA cohort, 55–75 for the HRS cohort, and 49–64 for the WB cohort. At each follow-up, respondents in each cohort aged together, yielding age trajectories of health deficits and disorders. The diagram in Figure 3.3 suggests that the observable age trajectories of different cohorts initiate and end at different ages, so comparison of different cohorts potentially can only be based on different segments of the cohort members' life course. As waves of data accumulate, however, age and cohort will become less and less confounded, making it increasingly possible to estimate cohort-specific age trajectories. The HRS data also use the accelerated longitudinal panel design but provide more waves of data, longer age trajectories, and hence higher power than the ACLs to test aging-related hypotheses within the cohort context. The fewer cohorts included, on the other hand, may limit the power of tests for cohort differences.

The analytic samples for all cohorts exclude small numbers of respondents who had missing data on any covariates used in the regression analysis. The numbers of person-year observations pooling all waves for the AHEAD, CODA, HRS, and WB cohorts are 31,565, 17,810, 48,485, and 16,565, respectively, which include observations contributed by respondents who were present at all waves, respondents who died at follow-ups (ranging from 5% for the WB cohort to 57% for the AHEAD cohort), and those who were nonrespondents at follow-ups (14% for all). The HRS oversamples racial minorities such as Hispanics and blacks and provides weighting variables to make these representative of the community-based population. All statistical analyses adjusted for sampling weights (Table 3.12).

Construction of the FI was based on the subset of health deficits most similar to those assessed in the original studies from the Canadian Study of Health and Aging (CSHA) and included symptoms, disabilities or impaired functions, disease classifications, and health attitudes (Mitnitski et al. 2002;

TABLE 3.12

Weighted Summary Statistics of All Variables in the HRS Analysis: 1993–2008

	Description and Coding	All (N = 114,425)		AHEAD (N = 31,565)		CODA (N = 17,810)		HRS (N = 48,485)		WB (N = 16,565)	
		M/%	SD	M/%	SD	M/%	SD	M/%	SD	M/%	SD
Dependent Variable											
FI	Frailty Index	0.16	0.12	0.19	0.13	0.18	0.12	0.15	0.12	0.14	0.12
Level 1 Covariates											
AGE	Respondent's age; centered around cohort median	68.50	10.03	80.83	5.81	74.69	3.94	64.79	5.10	57.74	3.92
POVERTY	Family income is below poverty level: 1 = yes, 0 = no	0.14	0.35	0.20	0.40	0.14	0.35	0.13	0.33	0.12	0.32
NOTMARRIED	Respondent not currently married: 1 = yes; 0 = no	0.37	0.48	0.54	0.50	0.41	0.49	0.30	0.46	0.27	0.45
SMOKED	Respondent ever smoked: 1 = yes, 0 = no	0.62	0.48	0.54	0.50	0.62	0.48	0.66	0.47	0.66	0.47
Level 2 Covariates											
SEX	Respondent's sex: 1 = female; 0 = male	0.55	0.50	0.61	0.49	0.57	0.50	0.53	0.50	0.52	0.50
RACE	Respondent's race: 1 = nonwhite; 0 = white	0.12	0.33	0.09	0.29	0.12	0.32	0.13	0.33	0.15	0.36
EDUC	Years of schooling; 1 = 0–12 yrs; 0 = 13+ yrs	0.60	0.49	0.70	0.46	0.64	0.48	0.60	0.49	0.49	0.50
DIED	Attrition type at follow-up: 1 = Deceased; 0 = other	0.22	0.41	0.57	0.50	0.20	0.40	0.10	0.30	0.05	0.21
NONRESPONSE	Attrition type at follow-up: 1 = Nonresponse; 0 = other	0.14	0.35	0.13	0.34	0.12	0.32	0.15	0.36	0.14	0.35

Notes: The *N*'s presented reflect the unweighted sample size. The weighted sample sizes are slightly smaller due to unavailability of weights for some respondents and are 109,215 for all, 29,665 for AHEAD, 16,155 for CODA, 48,033 for HRS, and 15,362 for WB. AHEAD = Assets and Health Dynamics Among the Oldest Old; CODA = Children of Depression; HRS = Health and Retirement Survey; WB = War Baby.

Mitnitski, Song, and Rockwood 2004). Specifically, we included 30 questions available across waves: eight chronic illnesses respondents ever had (high blood pressure, diabetes, cancer, lung disease, heart problems, stroke, psychological problems, and arthritis); five disabilities in ADL (difficulty in walking across a room, dressing, bathing or showering, eating, getting in/out of bed); seven disabilities in IADL (difficulty in using a map, the toilet, or telephone; managing money; taking medications, shopping for groceries; and preparing hot meals); eight depressive symptoms as measured by the CES-D (felt depressed; everything an effort; sleep was restless; felt lonely, sad; could not get going; enjoyed life; was happy); self-rated health; and obesity (BMI ≥ 30). All but one variable are binary, with 1 indicating the presence and 0 indicating the absence of the deficit. Self-rated health was a 5-point ordinal scale and was mapped into the [0, 1] interval: 0 (excellent), 0.25 (very good), 0.5 (good), 0.75 (fair), and 1 (poor). Because six of seven IADL items and most of the frailty-associated items were not available for the calculation of the FI in the 1992 and 1994 surveys, respectively, we had to omit the data for those two waves for the HRS cohort.

Following extant practice, we defined the FI as a count of deficits for any given person divided by the total number of possible deficits. With no missing data, the denominator would be the theoretical maximum for all individuals (30 in this study). In the HRS data, although the majority of respondents had complete measures of all 30 items, the exclusion of respondents with missing answers would decrease the sample size substantially. We therefore included respondents who had data on at least 25 of the 30 conditions. For instance, if a respondent was administered 30 questions and responded positively (there is a deficit) to 4 and negatively (no deficit) to 23 of them, then the FI for this person is 4/27. Both previous research and preliminary analysis showed no difference in the use of the full set of deficits and a reduced set. We report analysis from the latter sample with a larger number of observations. The FI ranged between 0 and 0.9 and averaged between 0.14 and 0.19 across cohorts. Other key variables and sample characteristics are also summarized in the table. The choice of operational definitions of these variables was based on a test of statistical significance for the regression coefficients and model fit. As mentioned, the analysis accounted for selection due to attrition by controlling for attrition status.

References

Alwin, D. F. 1991. Family of origin and cohort differences in verbal ability. *American Sociological Review* 56:625–638.

Alwin, D. F., and R. J. McCammon. 1999. Aging versus cohort interpretations of intercohort differences in GSS vocabulary scores. *American Sociological Review* 64:272–286.

Arbeev, K. G., S. V. Ukraintseva, L. S. Arbeeva, and A. I. Yashin. 2005. Decline in human cancer incidence rates at old ages: Age-period-cohort considerations. *Demographic Research* 12:273–300.

Bosworth, H. B., I. C. Siegler, B. H. Brummett, et al. 1999. The association between self-rated health and mortality in a well-characterized sample of coronary artery disease patients. *Medical Care* 37:1226–1236.

Burns, D. M., L. Lee, L. Z. Shen, et al. 1998. Cigarette smoking behavior in the United States, In *Changes in cigarette-related disease risk and their implications for prevention and control*, ed. D. M. Burns, L. Garfinkel, and J. M. Samet, 113–304. Bethesda, MD: National Institutes of Health.

Centers for Disease Control and Prevention (CDC). 2010. *National Center for Health Statistics (NCHS). National Health and Nutrition Examination Survey data*. Hyattsville, MD: U.S. Department of Health and Human Services, Centers for Disease Control and Prevention. http://www.cdc.gov/nchs/nhanes/nhanes_questionnaires.htm.

Chen, F., Y. Yang, and G. Liu. 2010. Social change and socioeconomic disparities in health over the life course in China: A cohort analysis. *American Sociological Review* 75:126–150.

Davis, J. A., T. W. Smith, and P. V. Marsden. 2005. *General Social Surveys, 1972–2004* [Cumulative File] [Computer file], 2nd ICPSR version. Chicago: National Opinion Research Center, producer. Storrs, CT: Roper Center for Public Opinion Research, University of Connecticut/Ann Arbor, MI: Inter-university Consortium for Political and Social Research, distributors.

Flegal, K. M., M. D. Carroll, C. L. Ogden, and C. L. Johnson. 2002. Prevalence and trends in obesity among U.S. adults, 1999–2000. *Journal of the American Medical Association* 288:1723–1727.

Frank, S. A. 2007. *Dynamics of cancer: Incidence, inheritance, and evolution*. Princeton, NJ: Princeton University Press.

Freedman, D. S., L. K. Khan, M. K. Serdula, D. A. Galuska, and W. H. Dietz. 2002. Trends and correlates of class 3 obesity in the United States from 1990 through 2000. *Journal of the American Medical Association* 288:1758–1761.

Glass, A. G., J. V. Lacey Jr., J. D. Carreon, and R. N. Hoover. 2007. Breast cancer incidence, 1980–2006: Combined roles of menopausal hormone therapy, screening mammography, and estrogen receptor status. *Journal of the National Cancer Institute* 99:1152–1161.

Glenn, N. D. 1994. Television watching, newspaper reading, and cohort differences in verbal ability. *Sociology of Education* 67:216–230.

Glenn, N. D. 1999. Further discussion of the evidence for an intercohort decline in education-adjusted vocabulary. *American Sociological Review* 64:267–271.

House, J. S., P. M. Lantz, and P. Herd. 2005. Continuity and change in the social stratification of aging and health over the life course: Evidence from a nationally representative longitudinal study from 1986 to 2001/2002 (Americans' Changing Lives Study). *Journal of Gerontology: Social Sciences* 60B:S15–S26.

Hughes, M. E., and A. M. O'Rand. 2004. *The American People Census 2000: The lives and times of the baby boomers*. New York: Russell Sage Foundation.

Idler, E. L., and Y. Benyamini. 1997. Self-rated health and mortality: A review of twenty-seven community studies. *Journal of Health and Social Behavior* 38:21–37.

Jemal, A., M. J. Thun, L. A. G. Ries, et al. 2008. Annual report to the nation on the status of cancer, 1975–2005, featuring trends in lung cancer, tobacco use, and tobacco control. *Journal of the National Cancer Institute* 100:1672–1694.

Jemal, A., E. Ward, Y. Hao, and M. Thun. 2005. Trends in the leading causes of death in the United States, 1970–2002. *Journal of the American Medical Association* 294:1255–1259.

Jemal, A., E. Ward, and M. J. Thun. 2007. Recent trends in breast cancer incidence rates by age and tumor characteristics among U.S. women. *Breast Cancer Research* 9:R28.

Lauderdale, D. S. 2001. Education and survival: Birth cohort, period, and age effects. *Demography* 38:551–561.

Lynch, S. M. 2003. Cohort and life-course patterns in the relationship between education and health: A hierarchical approach. *Demography* 40:309–331.

Manton, K. G., I. Akushevich, and J. Kravchenko. 2009. *Cancer mortality and morbidity patterns in the U.S. population: An interdisciplinary approach*. New York: Springer-Verlag.

Marshall, E. 2011. Cancer research and the $90 billion metaphor. *Science* 331:1540–1541.

Mason, W. M., and S. E. Fienberg, Ed. 1985. *Cohort analysis in social research: Beyond the identification problem*. New York: Springer-Verlag.

Mitnitski, A. B., A. J. Mogilner, C. MacKnight, and K. Rockwood. 2002. The mortality rate as a function of accumulated deficits in a frailty index. *Mechanisms of Ageing and Development* 123:1457–1460.

Mitnitski, A. B., X. Song, and K. Rockwood. 2004. The estimation of relative fitness and frailty in community-dwelling older adults using self-report data. *Journal of Gerontology: Medical Sciences* 59A:M627–M632.

Mokdad, A. H., B. A. Bowman, E. S. Ford, F. Vinicor, J. S. Marks, and J. P. Koplan. 2001. The continuing epidemics of obesity and diabetes in the United States. *Journal of the American Medical Association* 286:1195–1200.

O'Brien, R. M. 2000. Age period cohort characteristic models. *Social Science Research* 29:123–139.

Ogden, C. L., K. M. Flegal, M. D. Carroll, and C. L. Johnson. 2002. Prevalence and trends in overweight among U.S. children and adolescents, 1999–2000. *Journal of the American Medical Association* 288:1728–1732.

Osmond, C. 1985. Using age, period and cohort models to estimate future mortality rates. *International Journal of Epidemiology* 14:124–129.

Pampel, F. C. 2005. Diffusion, cohort change, and social patterns of smoking. *Social Science Research* 34:117–139.

Pompei, F., and R. Wilson. 2001. Age distribution of cancer: The incidence turnover at old age. *Human and Ecological Risk Assessment* 7:1619–1650.

Preston, S. H., P. Heuveline, and M. Guillot. 2001. *Demography: Measuring and modeling population processes*. Malden, MA: Blackwell.

Radloff, L. S. 1977. The CES-D Scale: A self-report depression scale for research in the general population. *Applied Psychological Measurement* 1:385–401.

Reither, E. N., R. M. Hauser, and Y. Yang. 2009. Do birth cohorts matter? Age-period-cohort analyses of the obesity epidemic in the United States. *Social Science & Medicine* 69:1439–1448.

Riley, M. W. 1987. On the significance of age in sociology. *American Sociological Review* 52:1–14.

Schoenfeld, D. E., L. C. Malmrose, D. G. Blazer, D. T. Gold, and T. E. Seeman. 1994. Self-rated health and mortality in the high-functioning elderly—A closer look at health individuals: MacArthur Field Study of Successful Aging. *Journal of Gerontology: Medical Sciences* 49:M109–M115.

Smith, R. A., V. Cokkinides, and O. W. Brawley. 2008. Cancer screening in the United States, 2008: A review of current American Cancer Society guidelines and cancer screening issues. *CA: A Cancer Journal for Clinicians* 58:161–179.

Tonry, M. L., L. E. Ohlin, and D. P. Farrington. 1991. *Human development and criminal behavior: New ways of advancing knowledge.* New York: Springer-Verlag.

Vaupel, J. W., K. G. Manton, and E. Stallard. 1979. The impact of heterogeneity in individual frailty on the dynamics of mortality. *Demography* 16:439–454.

Wang, H., and S. H. Preston. 2009. Forecasting United States mortality using cohort smoking histories. *Proceedings of the National Academy of Sciences of the United States of America* 106:393–398.

Wilson, J. A., and W. R. Gove. 1999. The intercohort decline in verbal ability: Does it exist? *American Sociological Review* 64:253–266.

Yang, Y. 2007. Is old age depressing? Growth trajectories and cohort variations in late-life depression. *Journal of Health and Social Behavior* 48:16–32.

Yang, Y. 2008. Social inequalities in happiness in the United States, 1972 to 2004: An age-period-cohort analysis. *American Sociological Review* 73:204–226.

Yang, Y., and M. Kozloski. 2012. Post-reproductive change in sex gap in total and cause-specific mortality. *Annals of Epidemiology* 22:94–103.

Yang, Y., and L. C. Lee. 2009. Sex and race disparities in health: Cohort variations in life course patterns. *Social Forces* 87:2093–2124.

Yang, Y., and L. C. Lee. 2010. Dynamics and heterogeneity in the process of human frailty and aging: Evidence from the U.S. older adult population. *Journal of Gerontology: Social Sciences* 65B:246–255.

4

Formalities of the Age-Period-Cohort Analysis Conundrum and a Generalized Linear Mixed Models (GLMM) Framework

4.1 Introduction

The three commonly used research designs—tables of rates or proportions, repeated cross-sectional sample surveys, and accelerated longitudinal cohort designs—described in the previous chapter can be used for age-period-cohort (APC) analysis. The use of suitable data alone, however, does not guarantee proper inferences regarding A, P, and C effects. Analytic approaches are often critical in determining the attribution of variations in the outcome of interest to distinct influences of age, period, and birth cohort. As noted previously and discussed in detail in this chapter, the APC underidentification problem, or the "APC conundrum," has been a point of methodological controversy for decades, with little agreement on a systematic set of interrelated models and methods for analysis. Perhaps fueled by this lack of agreed-on methodological procedures for analysis, APC problems and controversies also have flared up from time to time in empirical analyses of substantive questions in the social sciences, some of which have been introduced in Chapter 3.

This chapter first reviews the methods of descriptive and statistical analysis frequently used in previous APC studies. We discuss their utilities and limitations. We then lay out the formal algebra of the APC identification problem and review conventional approaches to model identification. The many pitfalls in empirical APC analysis and lack of major breakthroughs in methodology led to the verdict by Norval Glenn (2005) that statistical APC models cannot be relied on to provide accurate estimates of A, P, and C effects. We reevaluate the state of the field and point out the gap in our knowledge. This is followed by a sketch of a generalized linear mixed model (GLMM) framework that unifies the APC models and methods described in subsequent chapters for the analysis of data from the aforementioned three research designs. This framework and associated analytic and computational tools did not exist when the Mason and Fienberg (1985) synthesis of APC analysis in demographic and social research was articulated. We state the rationale for the GLMM approach to APC analysis in the concluding section of this chapter and provide detailed methodological guidelines on how to conduct APC analysis in the chapters that follow.

4.2 Descriptive APC Analysis

The vast majority of extant APC studies rely on two descriptive techniques to depict the time trends by age, period, and birth cohort. One common practice uses summary measures that are independent of age composition, such as standardized indices (such as *age-standardized death rates*) arrayed by time periods. This is exemplified in Figures 4.1 and 4.2, which show age-standardized lung cancer incidence and death rates, respectively, by sex and race calculated for the period between 1969 and 2008, adjusting the crude rates to the 2000 population age 20 and above. Increases in lung cancer incidence and death rates were evident for men only until the 1980s but continuous for women during the entire period of 40 years.

An alternative device is a *graphical display of the table of age-period-specific rates* or *age-cohort-specific rates*. The former is illustrated in Figures 4.3 and 4.4, which show age-specific rates of lung cancer incidence and mortality for white males and females across time periods. Consistent with previous research, the age-specific rates generally increased steadily from the age of 20 to the 70s and then leveled off at older ages. Both males and females

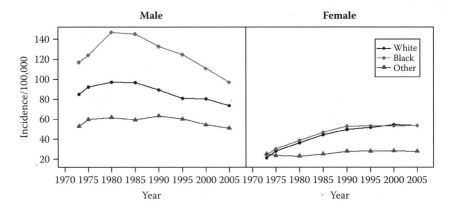

FIGURE 4.1
Age-standardized lung cancer incidence rates by sex and race, United States 1973–2008.

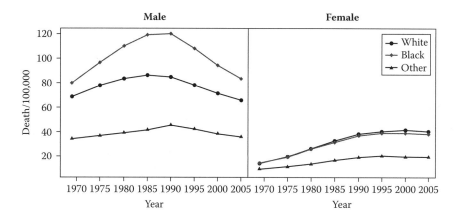

FIGURE 4.2
Age-standardized lung cancer death rates by sex and race, United States 1969–2007.

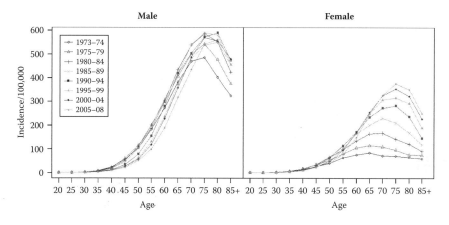

FIGURE 4.3
Age-specific lung cancer incidence rates for white males and females by time period, United States 1973–2008.

showed increases in incidence and death rates from lung cancer for the past 40 years. Whereas the increases for males occurred later in life, were moderate and reversed in more recent years after 1995, female lung cancer incidence and death rates increased substantially and continuously from early adulthood through old age with no marked decrease.

Figures 4.5 and 4.6 present the age-specific rates across birth cohorts. Changes in logarithmic transform of the rates can be interpreted as the proportional increase in rates. The results show leveling and declined incidence and mortality rates across the majority of male cohorts but sharply increasing rates across female cohorts born before 1940.

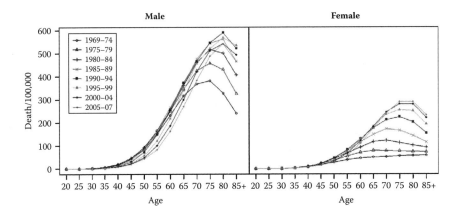

FIGURE 4.4
Age-specific lung cancer death rates for white males and females by time period, United States 1969–2007.

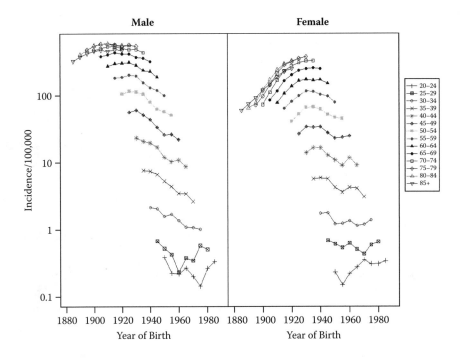

FIGURE 4.5
Age-specific lung cancer incidence rates (logarithmic scale) for white males and females by birth cohort.

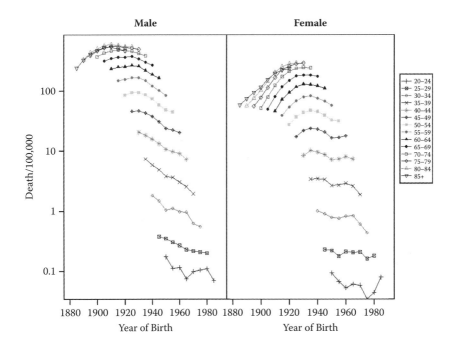

FIGURE 4.6
Age-specific lung cancer death rates (logarithmic scale) for white males and females by birth cohort.

Although comparisons of trends based on these descriptive measures almost always have implications for the A, P, and C effects that underlie the observed data (Hobcraft, Menken, and Preston 1982), neither approach explicitly considers an APC modeling framework. Both are descriptive and share two disadvantages that are evident in the figures. First, they only describe variation in the rates attributable to factors associated with either period of death or birth cohort. Second, neither summarizes the corresponding rate table satisfactorily. Standardization ignores different trends at different ages and graphical methods represent all available rates providing, no summary at all (Osmond 1985). In addition, they each have their own limitations.

Standardized crude rates by definition depend on the age composition of a particular period that is used as the standard and give more weight to the older ages (Preston, Heuveline, and Guillot 2001; Schoen 1970). The resulting rates can be sensitive to the choice of the standard, which changed drastically over the past 50 years (see, e.g., Preston, Heuveline, and Guillot 2001: Table 2.2). Given the substantial population aging that has taken place in many nations in recent decades, including the United States, using the

2000 population as the standard would yield higher rates than using a younger population in a much earlier year. As a result, inconsistent trends may occur from the use of different standard populations.

An example of the possible inconsistencies of graphical summaries of rate data was given by Holford (1991) in his study of lung cancer incidence rates for women aged 20–84 living in Connecticut from 1940 to 1984. He showed that the age-specific rates plotted by time periods tended to increase and then reach a plateau or even decline in the oldest age groups, which contradicts the epidemiologic expectation that lung cancer risk would continue to increase with age. Age-specific rates plotted by birth cohorts, on the other hand, are consistent with such expectation in that the death rates showed a steady increase with age for all birth cohorts. In addition, graphs of rates from two-way age-by-period or age-by-cohort tables are helpful for *qualitative* impressions about temporal patterns, but they provide no *quantitative* assessment of the source of change (Kupper et al. 1985). For example, in Figure 4.7, the curve of age-specific lung cancer death rates for white females in any given time period, say 1995–99, cuts across a number of birth cohort curves, such as 1900, 1905, 1910, and 1920. Therefore, the shape of the period curve is affected by both varying age effects and cohort effects. The understanding of how these effects operate simultaneously to shift period curve requires the use of statistical regression modeling.

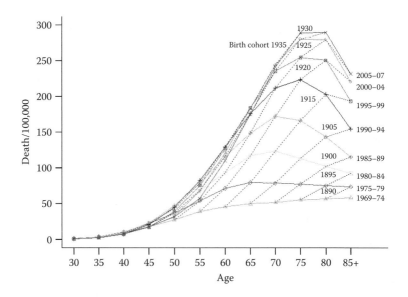

FIGURE 4.7
Age-specific lung cancer death rates for white females by time period and birth cohort.

4.3 Algebra of the APC Model Identification Problem

The essence of APC analysis is the identification and quantification of different sources of variation in an outcome of interest that are associated with age, period, or cohort effects. In addition to descriptive devices, early investigators also developed regression models for situations in which all three factors account for a substantive phenomenon and in which simpler two-factor models (such as an age-period model) are subject to model specification errors and spurious results (Mason et al. 1973). The *conventional linear regression model*, also known as the *APC accounting/multiple classification model*, was introduced to sociologists by Mason and colleagues (1973) and serves as a general methodology for cohort analysis when all three of age, period, and cohort are potentially of interest. This general methodology focuses on the APC analysis of data using the first research design we described in Chapter 3, that is, tables of population rates or proportions, such as occurrence/exposure rates of events such as births, deaths, disease incidence, crimes, and so on, termed "archival data" by Mason et al. (1973). In spite of its theoretical merits and conceptual relevance, APC analysis of tabulated data suffers from the "identification problem" induced by the exact linear dependency between A, P, and C variables defined in this particular data structure. This can be viewed as a special case of collinear regressors that produces, in this case, a singular matrix (of one less than full rank) used in the statistical estimation process. Since a singular matrix produces multiple estimators of the three effects, it is difficult to estimate the unique true separate effects. We now discuss this problem in greater algebraic detail.

We use Table 3.1 of lung cancer incidence rates for U.S. white males as an example to study the algebra of the model identification problem. In this case, this model can be written in *linear regression form* as

$$R_{ij} = I_{ij}/P_{ij} = \mu + \alpha_i + \beta_j + \gamma_k + \varepsilon_{ij} \tag{4.1}$$

where R_{ij} denotes the observed incidence rate for the ith age group for $i = 1$, ..., a age groups at the jth time period for $j = 1$, ..., p time periods of observed data; I_{ij} denotes the number of incidences or newly diagnosed cases of lung cancer in the ijth group; P_{ij} denotes the size of the estimated population in the ijth group, the population at risk; μ denotes the intercept or adjusted mean incidence rate; α_i denotes the ith row age effect or the coefficient for the ith age group; β_j denotes the jth column period effect or the coefficient for the jth time period; γ_k denotes the kth diagonal cohort effect or the coefficient for the kth cohort for $k = 1$, ..., $(a + p - 1)$ cohorts, with $k = a - i + j$; and ε_{ij} denotes a random error with expectation $E(\varepsilon_{ij}) = 0$ and var$(\varepsilon_{ij}^2) = \sigma^2$.

Conventional APC models as represented in Equation (4.1) fall into the class of *generalized linear models* (GLIM or GLM) (see McCullagh and Nelder

1989 or McCulloch and Searle 2001 for expositions) that can take various alternative forms. First, model (4.1) can take a *log-linear regression form* via a *log link* as

$$\log(E_{ij}) = \log(P_{ij}) + \mu + \alpha_i + \beta_j + \gamma_k \qquad (4.2)$$

where E_{ij} denotes the expected number of incidences in cell (i, j) that is assumed to be distributed as a Poisson variate, and $\log(P_{ij})$ is the log of the exposure P_{ij} in (4.1) and is called the "offset" or adjustment for the log-linear contingency table model. Models of this type are widely used in demography and epidemiology, for which the counts of demographic events such as deaths or the incidence of diseases generally follow Poisson distributions, and the rates are estimated through log-linear models (Agresti 1996). A second alternative formulation of the model, often used in studies of mortality, is to treat the underlying number of events (deaths) as a binomial variate. The canonical link changes from a log link to a *logit link*, which yields a *logistic model*:

$$\theta_{ij} = \log\left(\frac{m_{ij}}{1 - m_{ij}}\right) = \mu + \alpha_i + \beta_j + \gamma_k \qquad (4.3)$$

where θ_{ij} is the log odds of death, and m_{ij} is the probability of death in cell (i, j). This model has been implemented more widely in demographic research (e.g., Mason and Smith 1985).

Regression models (4.1)–(4.3) can be treated as *fixed effects generalized linear models* (GLMs) after a reparametrization to center the parameters:

$$\sum_i \alpha_i = \sum_j \beta_j = \sum_k \gamma_k = 0 \qquad (4.4)$$

After reparameterization, model (4.1) can be written in the conventional matrix form of a least-squares regression:

$$Y = Xb + \varepsilon \qquad (4.5)$$

where Y is a vector of incidence rates or log-transformed rates, X is the regression design matrix consisting of "dummy variable" column vectors for the vector [of dimension $m = 1 + (a - 1) + (p - 1) + (a + p - 2)$] of model parameters b:

$$b = (\mu, \alpha_1, \ldots, \alpha_{a-1}, \beta_1, \ldots, \beta_{p-1}, \gamma_1, \ldots, \gamma_{a+p-2})^T \qquad (4.6)$$

with the T superscript denoting vector transposition, and where ε in (4.5) is a vector of random errors with mean 0. Note that the parameters α_a, β_p, and

γ_{a+p-1} are not included in the parameter vector b because of the constraints (4.4) and can be uniquely determined by use of (4.4) in conjunction with each estimator of b. In Chapter 5, Section 5.6, we illustrate by an empirical example that the use of reference categories is equivalent to the translation by a constant of the parameter estimates produced by the constraints (4.4) and thus of no substantive importance.

The ordinary least squares (OLS) estimator of the matrix regression model (4.5), which also is the maximum likelihood estimator (MLE) when the error term is specified as to be normally distributed with expected value of zero and constant variance, is the solution b of the normal equations:

$$\hat{b} = (X^T X)^{-1} X^T Y \tag{4.7}$$

However, $(X^T X)^{-1}$ does not exist. This is due to the fact that the design matrix X is singular with one less than full column rank (Kupper et al. 1985) as a result of the perfect linear relationship between the A, P, and C variables defined in the data design shown in Figure 3.1:

Period − Age = Cohort.

Therefore, this OLS/MLE estimator *does not exist* (i.e., there is no uniquely defined vector of coefficient estimates). This is the *model identification problem* of APC analysis. It implies that there are an infinite number of possible solutions of the matrix equation (4.7) [i.e., estimators of model (4.5)], one for each possible linear combination of column vectors that results in a vector identical to one of the columns of X. Therefore, it is not possible to estimate the effects of A, P, and C separately without imposing at least one constraint on the coefficients in addition to the reparameterization (4.4).

4.4 Conventional Approaches to the APC Identification Problem

The Mason et al. (1973) article spawned a large body of methodological literature in the social sciences, beginning with Glenn's critique (1976) and Mason, Mason, and Winsborough's reply (1976), followed by Fienberg and Mason's work (1979). Similar debates occurred between Rodgers (1982) and Smith, Mason, and Fienberg (1982). These early investigations culminated in an edited volume on cohort analysis by Mason and Fienberg in 1985. Limitations of those analytic strategies propelled the search for other approaches. For example, a Bayesian approach developed by Nakamura (1982, 1986) and introduced to American social scientists by Saski and Suzuki (1987) was again critiqued by

Glenn, who cautioned against "mechanical solutions" to the problem (1989). The challenge of developing statistical models for APC analysis also stimulated considerable methodological work in biostatistics and epidemiology (e.g., Clayton and Schifflers 1987a,b; Holford 1991; Kupper et al. 1985). Several recent reviews provided useful additional material on these and related contributions to cohort analysis (Mason and Wolfinger 2002; Yang 2007, 2009).

We discuss next several conventional solutions to the identification problem, all of which are confined to conventional linear regression models and their GLM extensions. The purpose is often to break the linear dependency between the three APC dimensions and thus to identify the model, but each solution has limitations.

4.4.1 Reduced Two-Factor Models

One possible solution is just to estimate a reduced age and period two-factor model that contains no cohort effects and can be written as

$$\log(E_{ij}) = \log(I_{ij}/P_{ij}) + \mu + \alpha_i + \beta_j \qquad (4.8)$$

An example of this approach can be found in empirical investigations of sources of recent mortality reductions (Yang 2008a). Studies are usually confined to changes in one or two demographic components. The cohort effect is less frequently tested, but its presence implies that certain assumptions currently used by demographers and other social scientists to analyze factors contributing to mortality declines can be misleading. For instance, it is frequently assumed that rates of mortality declines over time are equal across birth cohorts. It is also assumed that these declines depend on rates of changes in period-specific conditions such as economic advance and health care technology that are independent of the year of birth.

Because cohort effects can be interpreted as a special form of interaction effect between the categorical age and period variables, model (4.8) rests on the assumption of no interaction effect (Fienberg and Mason 1985). In particular, the expected rate in age-by-period cell (i, j) is modeled a function of the *marginal* or gross effects of age i and period j only, and not also of the cell-specific effect, such as γ_{a-i+j}, which is a function of both i and j. Violation of this assumption can be detected by plots of age-specific rates by time period, and *a lack of parallelism* among these curves suggests that birth cohort effects are operating (Kupper et al. 1985). The same applies to the period effect as a particular type of age-cohort interaction (Holford 1991). Figures 4.3 and 4.4 show evidence of nonparallelism among age curves by period in plots of lung cancer incidence and mortality, confirming the existence of birth cohort effects. Age curves of incidence and mortality rates by cohort also show nonparallelism. Therefore, it is likely that period effects are also important in lung cancer trends. So, results of preliminary graphical analyses argue against the plausibility of reduced two-factor models. Regression analyses comparing model fit of reduced and full APC models further suggest that the latter is favored by taking all three variables into account (see Chapter 5). Therefore, the assumption of no cohort effects greatly simplifies estimations but can lead to model misspecification and is inconsistent with accumulating evidence of cohort changes in a variety of health outcomes and mortality.

4.4.2 Constrained Generalized Linear Models (CGLIMs)

Since the work of Fienberg and Mason (1979, 1985), the conventional and most widely used approach to estimating model (4.5) has been a *coefficients-constraints approach*, which takes the form of placing one or more additional identifying constraint (usually an equality constraint) on the parameter vector (4.6) to just identify or overidentify the model. For example, in an APC analysis of U.S. tuberculosis mortality, Mason and Smith (1985) placed the constraint by equating the contrasts for ages 0–9 and 10–19. This yielded unique estimates for age, period, and cohort effects. Similarly, in the case of lung cancer incidence rates shown in Table 3.1, one can constrain the effect coefficients of the first two age groups (20–24 and 25–29) to be equal, $\alpha_1 = \alpha_2$ or two periods (1973–74 and 1980–84) to be equal, $\beta_1 = \beta_2$. With this one additional constraint, the model (4.5) is just identified, the matrix $(X^T X)$ becomes nonsingular, and the least squares estimator (4.7) exists [as do related MLEs for models (4.1), (4.2), and (4.3)]. Overidentified constraints extend the just-identified constraints and in essence group multiple ages, periods, or cohorts into less-refined categories. A common CGLIM (constrained generalized linear model) approach is to create time periods or birth cohorts in longer intervals than the interval for age. The use of differential temporal grouping of the three variables breaks their exact linear dependency.

The main problems with this CGLIM approach have been discussed in a large methodological literature in demography, epidemiology, and statistics. First, the analyst needs to rely on *external or side information* to find constraints, but such information often does not exist or cannot easily be verified (Mason and Wolfinger 2002). Second, different choices of identifying constraints can produce widely different estimates of patterns of change across the A, P, and C categories. As demonstrated, for example, by Mason and Smith (1985) and Yang, Fu, and Land (2004), estimates of model effect coefficients are sensitive to the choice of the equality coefficient constraint. In the case of overidentification, the estimates may also be sensitive to the choice of interval widths that the analysts used to group variables because longer widths allow higher degrees of overidentification. There are additional problems with differential interval grouping of A, P, and C variables that we discuss in Chapter 7. Third, all just-identified models will produce the same levels of goodness of fit to the data, making model fit a useless criterion for selecting the best just-identified constrained model.[*]

4.4.3 Nonlinear Parametric Transformation

The nonlinear parametric transformation approach defines a nonlinear parametric or algebraic function for at least one of the A, P, and C variables so that its relationship to others is nonlinear (Mason and Fienberg 1985). This requires reformulating model (4.1) in terms of continuous A, P, and C variables, which allows for the specification of polynomials in one or more of these variables. For example, one can specify a quadratic or cubic function of age to model the nonlinear rate of change in verbal ability (Yang and Land 2006) or happiness with age (Yang 2008b). The drawbacks of this approach are that (1) specifying a functional form of one or more of the APC dimensions makes the model less flexible, and (2) it may not be clear what nonlinear function should be defined. More important, the use of this approach alone does not completely solve the identification problem. As Mason and Fienberg (1985) pointed out, the linear terms of these variables are still unidentified.

4.4.4 Proxy Variables

The proxy variables approach uses one or more proxy variables to replace the A, P, or C variable. This is a popular approach because of its substantive appeal. After all, the indicator variables of A, P, and C serve as surrogates for different sets of unmeasured structural correlates (Hobcraft, Menken, and Preston 1982). Examples of proxy variables for cohort effects include relative cohort size (O'Brien, Stockard, and Isaacson 1999) and cohort mean years of smoking before age 40 (Preston and Wang 2006). The use of cohort characteristics to replace cohort effects in APC accounting models has recently been labeled the age-period-cohort characteristics (APCC) model by O'Brien, Stockard, and Isaacson (1999). Unemployment rates, labor force size, and gender role attitudes have been used as proxy variables for period effects (Pavalko, Gong, and Long 2007).

But, there are caveats in the employment of this approach: (1) Proxy variables should not be linearly related to cohort or period. Otherwise, they will be highly collinear with the other two effects, just as the effects they are intended to replace in the models (Glenn 2005). (2) As Mason and Smith (1985) clearly indicated, in the case of interactions that can affect the coefficients of the included variables, the use of measured variables will not provide any

---

[*] Goodness-of-Fit and model selection statistics can be useful for ascertaining the extent to which one, two, or three of the age, period, and cohort temporal dimensions are necessary to model the patterns in a given age-by-time period array of rates or proportions; see Chapter 5, Section 5.5.2.

particular advantage. (3) Even in the absence of any interaction, the use of measured variables is not necessarily superior, as Smith, Mason, and Fienberg (1982: 792) cautioned, for replacing cohort or period with a measured variable "leaves open the question of whether all of the right measured variables have been included in an appropriately wrought specification. Although replacing an accounting dimension with measured variables solves an identification problem, it makes room for others." Thus, assuming that *all* of the variation associated with the A, P, or C dimension is fully accounted for by the chosen proxy variables may be unwarranted. We test this assumption empirically in subsequent analyses using the GLMM approach by entering key cohort and period-level covariates previously hypothesized to proxy cohort and period effects in studies of mortality and health. For example, we compare estimates of cohort variance before and after the adjustment of the cohort mean years of smoking, which was used to replace the cohort effects in a recent study of U.S. adult mortality (Preston and Wang 2006). The lack of reduction or elimination of cohort variance then suggests that the proxy cohort effect variable chosen does not completely account for the cohort effect as expected and hence is not sufficient to represent the cohort effect.

Winship and Harding (2008) recently proposed a mechanism-based approach that accommodates a more general set of models. Using the framework of causal modeling described by Pearl (2000) to achieve identification, the approach allows any given measured variable to be associated with more than one of the age, period, and cohort dimensions and provides statistical tests for the plausibility of alternative restrictions. If a rich set of mechanism variables is available and the original age, period, and cohort categories can be conceived as the exogenous elements of a causal chain (Smith 2008), this is an enriched and sophisticated alternative to the proxy variable approach.

4.4.5 Other Approaches in Biostatistics

In biostatistics and epidemiology, a number of solutions to the identification problems have also been proposed over the past two decades. There are three broad classes of solutions other than the CGLIM approach. The first and most widely adopted one uses estimable functions, deviations, curvatures, and drift and focuses solely on the nonlinear rather than linear components of trends (Clayton and Shifflers 1987a,b; Holford 1983; Tarone and Chu 1992). The second is based on the use of minimizing a penalty function to derive the necessary extra linear constraint (Osmond and Gardner 1989; Fu 1998, 2000). The third includes the methods that use individual records of cases to construct a three-way APC table (Robertson and Boyle 1998). Reviews and comparisons of these methodologies are available in the work of Robertson, Gandini, and Boyle (1999) and Holford (2005). Generally, the statistical limitations of existing approaches have been acknowledged, and analysts have been advised that any statistical modeling of APC data should be carried out in conjunction with a detailed descriptive analysis (Kupper et al. 1985; Mason and Smith 1985). Substantively, there also is a lack of clear and parsimonious interpretation of the results in terms of the A, P, and C trends. And, similar to the conventional solutions in social sciences, they are confined to linear models (LMs) that necessarily incur the identification problem.

4.5   Generalized Linear Mixed Models (GLMM) Framework

Various analytic approaches mentioned have produced ambiguous and inconsistent results. Researchers do not agree on methodological solutions to these problems and conclude that APC analysis is still in its infancy (Kupper et al. 1985; Mason and Wolfinger 2002). Where does the early literature on APC analysis leave us today? If a researcher has a temporally ordered dataset and wants to tease out its age, period, and cohort components, how should the researcher proceed? Can any methodological guidelines be recommended? A problem with much of the extant literature is that there is a lack of useful guidelines on how to conduct an APC analysis. Instead, one is left with the impression that either it is *impossible* to obtain meaningful estimates of the distinct contributions of age, cohort, and time period (Glenn 2005) or that the conduct of an APC analysis is an *esoteric art* that is best left to a few skilled methodologists (Yang 2009).

We seek to redress this situation by focusing on recent developments in APC analysis for three common research designs. The guiding principle of our work is a famous quotation from the celebrated statistician George E. P. Box (1979: 202), emeritus professor of statistics at the University of Wisconsin at Madison: "All models are wrong but some are useful." Our version of this statement is that all statistical models are wrong but, some models in APC analysis have better statistical properties and employ constraints that are more reasonable and substantively relevant than others, which may make them useful. We show that the models we present in this volume possess good statistical properties in comparison to alternatives. Nonetheless, as stated by Yang et al. (2008: 1733), every statistical model has its limits and will break down under some conditions. The models presented here are no exception, and we try to identify some conditions under which they become unstable and break down.

We begin with the observation that the identification problem is not inevitable in all settings—only in the case of conventional linear accounting models that assume additive fixed effects of A, P, and C. In a special issue of *Sociological Methods & Research* (2008) that collected recent new developments of APC models in sociology, Wenjiang Fu (2008) insightfully pointed out that the APC conundrum is "not data specific but model specific." Second, LMs also have conceptual limitations. Assuming additivity of A, P, and C effects, such models may be poor approximations of the processes generating changes. Third, LMs in the form of an APC accounting model cannot include explanatory variables and test substantive hypotheses. These major limitations suggest the use of different families of models that are not subject to the identification problem and are perhaps suitable for addressing many new research questions. A useful alternative is a family of models that (1) do not assume fixed A, P, and C effects that are additive and therefore avoid the identification problem; (2) can statistically characterize contextual effects of historical time and cohort membership; and (3) can accommodate covariates to aid better conceptualization of specific social and biological processes generating observed patterns in the data.

Over three decades ago, Nelder and Wedderburn (1972; see also McCullagh and Nelder 1989) introduced the class of GLMs as a generalization of classical LMs. GLMs facilitate the regression analysis of data that are not normally distributed and the specification of nonlinear link functions that specify the expected value of the dependent or response variable as a nonlinear function of parameters; the class of GLMs thus includes LMs as a special case. As noted by McCulloch and Searle (2001: 2), the last 30 years have seen GLMs extended further to the class of GLMMs, which permit the specification of some parameters as fixed and others as random.

It is clear that the conventional models for APC analysis that we have discussed thus far are LMs or GLMs that are members of the GLMM class. In the next chapter, we describe a recently developed method of estimation, the Intrinsic Estimator (IE), to identify APC accounting models of the form of (4.5) utilizing Moore-Penrose generalized inverse and an estimable function (Fu 2000; Yang, Fu, and Land 2004; Yang et al. 2008). The IE, then, is a pure fixed effects LM if the response variable is treated as normally distributed; if the response variable is an incidence or occurrence/exposure rate and therefore more properly modeled by a log-linear or logit link, then the IE becomes a GLM.

A mixed effects modeling framework has been developed for APC analysis. Specifically, the class of *hierarchical APC (HAPC) models* initially was proposed (Yang and Land 2006, 2008; Yang 2006) to examine microdata in the second research design described in Chapter 3, repeated cross-sectional surveys, but they also can be readily applied to aggregate data in the first research design (O'Brien, Hudson, and Stockard 2008). The HAPC approach conceptualizes time periods and cohort memberships as social historical contexts within which individuals are embedded and ordered by age and models them as random as opposed to fixed effects additive to that of age. This contextual approach broadens the theoretical foundation of

APC analysis, helps to deal with (actually completely avoids) the identification problem, and accounts for potentially correlated errors. As shown in Chapter 7, HAPC models are mixed effects models. While HAPC models applied to the General Social Survey (GSS) verbal test score data described in Chapter 3 use a linear link and Gaussian errors, we can extend such applications within the GLMM framework to allow a flexible choice of nonlinear link functions and nonlinear errors, as are the cases for GSS happiness data, NHANES (National Health and Nutrition Examination Survey) obesity data, and SEER (Surveillance, Epidemiology, and End Results) cancer registry and mortality data of Chapter 3. In this way, HAPC models can accommodate a larger range of data used in social behavioral, life, and animal sciences.

Hierarchical or mixed effects models are similarly useful tools for modeling accelerated longitudinal data in the third research design. Because repeated observations over time (level 1 units) can be viewed as nested within individuals (level 2 units) from multiple cohorts followed over time, one can use the individual growth/change model, a specification of mixed effects models, to assess simultaneously the intracohort age changes and intercohort differences. We show in Chapter 7 the estimation of individual growth curve models for cohort analysis of data on physical and mental health.

In sum, it is clear that the class of GLMMs is an appropriate context within which to view the newly developed APC models and within which to further explore APC analysis. Using the GLMMs framework, we seek to develop an integrated statistical methodology for APC analysis that yields consistent models and substantive inferences across all three research designs. We make comparisons of inferences and conduct model validation/robustness analyses across different research designs and model specifications. There is, of course, no reason that APC inferences must be identical across all data and modeling formats. But, as statistical methodologists, it is incumbent on us to identify and explore the similarities, differences, and constraints that may apply to the more limited data designs (repeated cross sections, tabulated rates) as compared to a full accelerated longitudinal cohort design. This should yield some guidelines that may be useful in empirical studies.

Indeed, since the APC underidentification problem is an instance of a larger family of structural underidentification problems, the potential range of application of the GLMM may be even larger. *Structural underidentification* problems occur when a conceptualization of effects of structural arrangements leads to an exact linear dependency among the effects. An example is the classical problem in mobility analysis of distinguishing the effects of socioeconomic mobility or distance moved on an outcome variable from the effects of origin and destination statuses (see, e.g., Duncan 1966). Another example pertains to the estimation of the effects of years of labor force experience separately from current age and age at labor force entry. These and similar problems of structural underidentification occur frequently in sociology and related disciplines. The new models and methods may thus prove useful in applications to such problems.

References

Agresti, A. 1996. *An introduction to categorical data analysis*. New York: Wiley-Interscience.

Box, G. E. P. 1979. Robustness in the strategy of scientific model building. In *Robustness in statistics: Proceedings of a workshop*, ed. R. L. Launer and G. N. Wilkinson, 201–236. New York: Academic Press.

Clayton, D., and E. Schifflers. 1987a. Models for temporal variation in cancer rates. I: Age-period and age-cohort models. *Statistics in Medicine* 6:449–467.

Clayton, D., and E. Schifflers. 1987b. Models for temporal variation in cancer rates. II: Age-period-cohort models. *Statistics in Medicine* 6:469–481.

Duncan, O. D. 1966. Path analysis: Sociological examples. *American Journal of Sociology* 72:1–16.

Fienberg, S. E., and W. M. Mason. 1979. Identification and estimation of age-period-cohort models in the analysis of discrete archival data. *Sociological Methodology* 10:1–67.

Fienberg, S. E., and W. M. Mason. 1985. Specification and implementation of age, period, and cohort models. In *Cohort analysis in social research: Beyond the identification problem*, ed. W. M. Mason and S. E. Fienberg, 45–88. New York: Springer-Verlag.

Fu, W. J. 1998. Penalized regressions: The bridge versus the lasso. *Journal of Computational and Graphical Statistics* 7:397–416.

Fu, W. J. 2000. Ridge estimator in singular design with application to age-period-cohort analysis of disease rates. *Communications in Statistics—Theory and Methods* 29:263–278.

Fu, W. J. 2008. A smoothing cohort model in age-period-cohort analysis with applications to homicide arrest rates and lung cancer mortality rates. *Sociological Methods & Research* 36:327–361.

Glenn, N. D. 1976. Cohort analysts' futile quest: Statistical attempts to separate age, period and cohort effects. *American Sociological Review* 41:900–904.

Glenn, N. D. 1989. A caution about mechanical solutions to the identification problem in cohort analysis: Comment on Sasaki and Suzuki. *American Journal of Sociology* 95:754–761.

Glenn, N. D. 2005. *Cohort Analysis*. 2nd ed. Thousand Oaks, CA: Sage.

Hobcraft, J., J. Menken, and S. Preston. 1982. Age, period, and cohort effects in demography: A review. *Population Index* 48:4–43.

Holford, T. R. 1983. The estimation of age, period and cohort effects for vital rates. *Biometrics* 39:311–324.

Holford, T. R. 1991. Understanding the effects of age, period, and cohort on incidence and mortality rates. *Annual Review of Public Health* 12:425–457.

Holford, T. R. 2005. Age-period-cohort analysis. In *Encyclopedia of biostatistics*, ed. P. Armitage and T. Colton, 82–99. Hoboken: Wiley.

Kupper, L. L., J. M. Janis, A. Karmous, and B. G. Greenberg. 1985. Statistical age-period-cohort analysis: A review and critique. *Journal of Chronic Diseases* 38:811–830.

Mason, K. O., W. M. Mason, H. H. Winsborough, and W. K. Poole. 1973. Some methodological issues in cohort analysis of archival data. *American Sociological Review* 38:242–258.

Mason, W. M., and S. E. Fienberg, ed. 1985. *Cohort analysis in social research: Beyond the identification problem*. New York: Springer-Verlag.

Mason, W. M., K. O. Mason, and H. H. Winsborough. 1976. Reply to Glenn. *American Sociological Review* 41:904–905.

Mason, W. M., and H. L. Smith. 1985. Age-period-cohort analysis and the study of deaths from pulmonary tuberculosis. In *Cohort analysis in social research: Beyond the identification problem*, ed. W. M. Mason and S. E. Fienberg, 151–228. New York: Springer-Verlag.

Mason, W. M., and N. H. Wolfinger. 2002. Cohort analysis. In *International encyclopedia of the social and behavioral sciences*, ed. N. J. Smelser and P. B. Baltes, 2189–2194. New York: Elsevier.

McCullagh, P., and J. A. Nelder. 1989. *Generalized linear models*. 2nd ed. Boca Raton, FL: CRC Press.

McCulloch, C. E., and S. R. Searle. 2001. *Generalized, linear, and mixed models*. New York: Wiley-Interscience.

Nakamura, T. 1982. A Bayesian cohort model for standard cohort table analysis. *Proceedings of the Institute of Statistical Mathematics* 29:77–97.

Nakamura, T. 1986. Bayesian cohort models for general cohort table analyses. *Annals of the Institute of Statistical Mathematics* 38:353–370.

Nelder, J. A., and R. W. M. Wedderburn. 1972. Generalized linear models. *Journal of the Royal Statistical Society. Series A (General)* 135:370–384.

O'Brien, R. M., K. Hudson, and J. Stockard. 2008. A mixed model estimation of age, period, and cohort effects. *Sociological Methods & Research* 36:402–428.

O'Brien, R. M., J. Stockard, and L. Isaacson. 1999. The enduring effects of cohort characteristics on age-specific homicide rates, 1960–1995. *American Journal of Sociology* 104:1061–1095.

Osmond, C. 1985. Using age, period and cohort models to estimate future mortality rates. *International Journal of Epidemiology* 14:124–129.

Osmond, C., and M. J. Gardner. 1989. Age, period, and cohort models: Non-overlapping cohorts don't resolve the identification problem. *American Journal of Epidemiology* 129:31–35.

Pavalko, E. K., F. Gong, and J. S. Long. 2007. Women's work, cohort change, and health. *Journal of Health and Social Behavior* 48:352–368.

Pearl, J. 2000. *Causality: Models, reasoning, and inference*. New York: Cambridge University Press.

Preston, S. H., P. Heuveline, and M. Guillot. 2001. *Demography: Measuring and modeling population processes*. Malden, MA: Blackwell.

Preston, S., and H. Wang. 2006. Sex mortality differences in the United States: The role of cohort smoking patterns. *Demography* 43:631–646.

Robertson, C., and P. Boyle. 1998. Age-period-cohort analysis of chronic disease rates. I: Modelling approach. *Statistics in Medicine* 17:1305–1323.

Robertson, C., S. Gandini, and P. Boyle. 1999. Age-period-cohort models: A comparative study of available methodologies. *Journal of Clinical Epidemiology* 52:569–583.

Rodgers, W. L. 1982. Estimable functions of age, period, and cohort effects. *American Sociological Review* 47:774–787.

Sasaki, M., and T. Suzuki. 1987. Changes in religious commitment in the United States, Holland, and Japan. *American Journal of Sociology* 92:1055–1076.

Schoen, R. 1970. The geometric mean of the age-specific death rates as a summary index of mortality. *Demography* 7:317–324.

Smith, H. L. 2008. Advances in age-period-cohort analysis. *Sociological Methods & Research* 36:287–296.

Smith, H. L., W. M. Mason, and S. E. Fienberg. 1982. Estimable functions of age, period, and cohort effects: More chimeras of the age-period-cohort accounting framework: Comment on Rodgers. *American Sociological Review* 47:787–793.

Tarone, R. E., and K. C. Chu. 1992. Implications of birth cohort patterns in interpreting trends in breast cancer rates. *Journal of the National Cancer Institute* 84:1402–1410.

Winship, C., and D. J. Harding. 2008. A mechanism-based approach to the identification of age-period-cohort models. *Sociological Methods & Research* 36:362–401.

Yang, Y. 2006. Bayesian inference for hierarchical age-period-cohort models of repeated cross-section survey data. *Sociological Methodology* 36:39–74.

Yang, Y. 2007. Age/period/cohort distinctions. In *Encyclopedia of health and aging*, ed. K. S. Markides, 20–22. Los Angeles: Sage.

Yang, Y. 2008a. Trends in U.S. adult chronic disease mortality: Age, period, and cohort variations. *Demography* 45:387–416

Yang, Y. 2008b. Social inequalities in happiness in the United States, 1972 to 2004: An age-period-cohort analysis. *American Sociological Review* 73:204–226.

Yang, Y. 2009. Age, period, cohort effects. In *Encyclopedia of the life course and human development*, ed. D. Carr, R. Crosnoe, M. E. Hughes, and A. M. Pienta, 6–10. New York: Gale.

Yang, Y., W. J. Fu, and K. C. Land. 2004. A methodological comparison of age-period-cohort models: The intrinsic estimator and conventional generalized linear models. *Sociological Methodology* 34:75–110.

Yang, Y., and K. C. Land. 2006. A mixed models approach to the age-period-cohort analysis of repeated cross-section surveys, with an application to data on trends in verbal test scores. *Sociological Methodology* 36:75–97.

Yang, Y., and K. C. Land. 2008. Age-period-cohort analysis of repeated cross-section surveys: Fixed or random effects? *Sociological Methods & Research* 36:297–326.

Yang, Y., S. Schulhofer-Wohl, W. J. Fu, and K. C. Land. 2008. The intrinsic estimator for age-period-cohort analysis: What it is and how to use it. *American Journal of Sociology* 113:1697–1736.

5

APC Accounting/Multiple Classification Model, Part I: Model Identification and Estimation Using the Intrinsic Estimator

5.1 Introduction

The age-period-cohort (APC) accounting/multiple classification model described by Mason et al. (1973) has served for over three decades as a general methodology for estimating A, P, and C effects in demographic and social research. It also has been the dominant model for APC analysis in biostatistics and epidemiology (Holford 2005). This methodology uses conventional linear regression models or their generalized linear model extensions and has mainly been applied to data using the first research design (tables of rates or proportions) described in Chapter 3. The problem of identification of model parameters that arises from the application of APC accounting models to tables of rates or proportions has long been recognized and deemed difficult to deal with, if not completely unsolvable, in the social science and biostatistics communities.

This chapter focuses on an innovative approach to the model identification and estimation problem within the linear/generalized linear models framework for APC analysis. Incorporating recent methodological developments on estimable functions in the context of the APC accounting model, a new method of estimation termed the *Intrinsic Estimator* (IE) was proposed and compared to the conventional coefficient constraints approach, that is, the constrained generalized linear model (CGLIM) (Fu 2000; Yang, Fu, and Land 2004). Subsequent methodological studies have further shown that the IE performs well compared to conventional approaches to the identification problem (Yang et al. 2008; Fu, Land, and Yang 2011). To date, the IE has been empirically applied in a number of studies in epidemiology, demography, and social science and has yielded sensible results on age, period, and cohort trends in a variety of topics, including both general and cause-specific mortality rates (Yang 2008); homicide arrest rates (Fu, Land, and Yang 2011); rates of religious activities and beliefs (Schwadel 2011); mortality due to accidental poisoning (Miech, Koester, and Dorsey-Holliman 2011); trends in marijuana use (Miech and Koester 2012); and deer hunter demography (Winkler, Huck, and Warnke 2009). This chapter defines the IE, describes its statistical properties, and illustrates the robustness of these properties via a comparison with age, period, and cohort estimates from an empirical application of a hierarchical APC (HAPC) model (the subject of Chapters 7, 8, and 9) and from a numerical simulation analysis. Because an understanding of the numerical and statistical properties of the new estimator requires a substantial mathematical background, readers who are not familiar with linear algebra or have no formal training in advanced statistical methods can choose to skip the latter part of the chapter and turn to the empirical applications of the APC accounting model described in Chapter 6.

5.2 Algebraic, Geometric, and Verbal Definitions of the Intrinsic Estimator

The objectives of this section are to formally define the IE, review properties of the IE as a statistical estimator, report results of model validation assessments of the IE both from an empirical example and from a simulation exercise, give some "how-to-use" advice, and show how to relate the coefficients of the IE to those of conventional constrained APC models with applications to U.S. female cancer mortality rates, 1969–2007.[*] Since the IE is a general-purpose method of APC analysis of potentially wide applicability in the social, life, and animal sciences, it merits recalling the criteria for acceptability of such a general-purpose method articulated by Norval Glenn, a long-time critic of attempts to provide general solutions to the APC analysis problem. Glenn (2005: 20) stated that such a method "may prove to be useful … if it yields approximately correct estimates 'more often than not,' if researchers carefully assess the credibility of the estimates by using theory and side information, and if they keep their conclusions about the effects tentative." These are strong criteria, and we agree with them. The purpose of this and the next sections is to assess the extent to which the IE satisfies them.

The consensus has been that the key problem for APC analysis in the conventional linear accounting/multiple classification model is to identify an estimable function that uniquely determines the parameter estimates. But, the controversy continues whether there exists such an estimable function that solves the identification problem. The conventional wisdom is that only the nonlinear, but not the linear, components of APC models are estimable

---

[*] Parts of Section 5.2 and Appendices 5.1 and 5.2 are adapted from Yang, Y., W. J. Fu, and K. C. Land. 2004. *Sociological Methodology* 34:75–110 and Yang, Y., S. Schulhofer-Wohl, W. J. Fu, and K. C. Land. 2008. *American Journal of Sociology* 113:1697–1736.

(Holford 1985; Rodgers 1982a). As noted by Fu (2008), however, there have been only numeric demonstrations, but no rigorous proofs, to support the idea that no estimable function exists. It should also be noted that Kupper et al. (1985) provided a condition for estimable functions (see Section 5.2) and suggested that an estimable function satisfying this condition resolves the identification problem. Subsequent publications have shown that the IE satisfies this condition and estimates the unique estimable function, including both the linear and the nonlinear components of the parameter vector of the multiple classification model (Fu 2000; Fu and Hall 2006; Fu, Hall, and Rohan 2004; Yang, Fu, and Land 2004).

Within the context of the description of the algebra of the APC identification problem in conventional linear/generalized linear APC regression models given in Chapter 4, we next describe the IE in three ways: algebraically, geometrically, and verbally.

5.2.1 Algebraic Definition

First, as concerns the algebraic definition of the IE, Yang, Fu, and Land (2004) showed that, because the design matrix X is one less than full-column rank, the parameter space of the unconstrained APC regression model (4.5) of Chapter 4 can be decomposed into the direct sum of two linear subspaces that are perpendicular to each other. One subspace corresponds to the unique zero eigenvalue of the matrix X^TX of equation (4.7) and is of dimension one; it is termed the *null subspace* of the design matrix X. The *nonnull subspace* is the complement subspace orthogonal to the null space.

Due to this orthogonal decomposition of the parameter space, each of the infinite number of solutions of the unconstrained APC accounting model (4.5) can be written as

$$\hat{b} = B + sB_0 \tag{5.1}$$

where s is a scalar corresponding to a specific solution, and B_0 is a unique eigenvector of Euclidean norm or length one. The eigenvector B_0 does not depend on the observed rates Y, only on the design matrix X, and thus is completely determined by the number of age groups and period groups—regardless of the event rates. In other words, B_0 has a specific form that is a function of the design matrix. To give an explicit representation of B_0, note that the exact linear dependency between A, P, and C variables in model (4.5) is mathematically equivalent to

$$XB_0 = 0 \tag{5.2}$$

This equation expresses the property that X is singular, that is, there exists a linear combination of the columns of the design matrix X that equals a zero vector. Kupper et al. (1985) showed that B_0 has the algebraic form

$$B_0 = \frac{\tilde{B}_0}{\left\| \tilde{B}_0 \right\|} \tag{5.3}$$

that is, B_0 is the normalized vector of $\tilde{B}_0$:

$$\tilde{B}_0 = (0, A, P, C)^T \tag{5.4}$$

where

$$A = \left(1 - \frac{a+1}{2}, \cdots, (a-1) - \frac{a+1}{2} \right)$$

$$P = \left(\frac{p+1}{2} - 1, \cdots, \frac{p+1}{2} - (p-1) \right)$$

and

$$C = \left(1 - \frac{a+p}{2}, \cdots, (a+p-2) - \frac{a+p}{2} \right)$$

with a, p, and c denoting the number of age categories, time periods, and cohorts, respectively, in the age-by-time period array of rates.

It is important to note that the vector B_0 is fixed or nonrandom because it is a function solely of the dimension of the design matrix X or the number of age groups a and periods p and independent of the response variable Y. This can be illustrated with a specific numerical example. Suppose $a = 3$ and $p = 3$. Then from Equation (5.4) we have

$$A = \left(1 - \tfrac{3+1}{2}, 2 - \tfrac{3+1}{2} \right) = (-1, 0)$$

$$P = \left(\tfrac{3+1}{2} - 1, \tfrac{3+1}{2} - 2 \right) = (1, 0)$$

$$C = \left(1 - \tfrac{3+3}{2}, 2 - \tfrac{3+3}{2}, 3 - \tfrac{3+3}{2}, 4 - \tfrac{3+3}{2} \right) = (-2, -1, 0, 1)$$

which yields

$$\tilde{B}_0 = (0, -1, 0, 1, 0, -2, -1, 0, 1)^T$$

We can then compute the vector B_0 as follows:

$$B_0 = \frac{\tilde{B}_0}{\left\| \tilde{B}_0 \right\|} = \frac{\tilde{B}_0}{(\tilde{B}_0^T \tilde{B}_0)^{1/2}} = (0, -0.354, 0, 0.354, 0, -0.707, -0.354, 0, 0.354)^T$$

where $(\tilde{B}_0{}^T \tilde{B}_0)^{1/2} = 8^{1/2}$. The fact that the trend estimates should be completely determined by Y (and hence are intrinsic to Y) means that arbitrary terms associated with B_0 should be removed. But the conventional CGLIM approach violates this principle if the scalar s in Equation (5.1) is nonzero.

The idea that B_0 should not affect the results is a key point, as intuition suggests that the eigenvector corresponding to the zero eigenvalue would be an arbitrary vector. And, indeed, sB_0 is arbitrary. But B_0 is not arbitrary; it is fixed by the design matrix. Furthermore, by Equation (5.1), any APC estimator, obtained by placing any identifying constraints on the design matrix, can be written as a linear combination $B + sB_0$, where B is the special estimator termed the IE that lies in the nonnull parameter subspace and is determined by the Moore-Penrose generalized inverse.[*]

One computational algorithm for the IE is based on an *orthonormal transformation of a principal components regression* (PCR) (for a standard exposition of PCR, see, e.g., Sen and Srivastava 1990) consisting of the following steps:

(a) Compute the eigenvectors $u_1, \ldots, u_m$ of matrix $X^T X$, where m denotes the number of rows (columns) of the $X^T X$ matrix [i.e., $m = 1 + (a - 1) + (p - 1) + (a + p - 2)$]. Normalize them to have unit length with $\|u_m\|$ and denote the orthonormal matrix as $U = (u_1, \ldots, u_m)^T$;

(b) Identify the special eigenvector B_0 corresponding to eigenvalue 0. Denote $u_1 = B_0$ without loss of generality;

(c) Select the principal components to be the remaining eigenvectors $u_2, \ldots, u_m$ with nonzero eigenvalues;

(d) Estimate a PCR model with the outcome variable of interest (e.g., logged death rates) as the response using a design matrix V whose column vectors are the principal components $u_2, \ldots, u_m$, that is, $V = (u_2, \ldots, u_m)$, to obtain the coefficients $(w_2, \ldots, w_m)$; and

(e) Set coefficient $w_1 = 0$ and transform the coefficients vector $w = (w_1, \ldots, w_m)^T$ by the orthonormal matrix of all eigenvectors $U = (u_1, \ldots, u_m)$ to transform the coefficients of the PCR model to obtain the Intrinsic Estimator $B = Uw$.

Instead of using reference categories, the IE uses the "usual analysis-of-variance (ANOVA) type constraints" that the sums of the respective A, P, and C coefficients equal zero, termed *effect coding*. The computational algorithm used by the IE estimates the resulting *effect coefficients* for each of the $a - 1$, $p - 1$, and $a + p - 2$ A, P, and C categories, respectively, which is consistent with the definition of the parameter vector in Equation (4.1) of Chapter 4. Then, the IE uses the zero-sum constraints to obtain the numerical values of the deleted A, P, and C categories.

---

[*] See, for example, Searle (1971: 16–19) for a definition of the Moore-Penrose generalized inverse and its properties.

Two other perspectives on the IE are useful for interpreting and calculating the IE. First, Fu (2000) showed that the IE can be viewed as a special case of the classical ridge estimator (Hoerl and Kennard 1970a, 1970b) for the conventional linear regression model that is used when the regressors are highly collinear. The ridge estimator shrinks the least squares estimator toward zero as a function of a shrinkage parameter $\lambda \geq 0$, thus correcting the tendency of ordinary least squares (OLS) to produce estimated coefficients that are too far from zero when the regressors are highly correlated. This produces a biased estimator that has a smaller mean-squared error than the least squares estimator, thus trading off bias for a reduction of error of estimation. Fu (2000) studied the ridge estimator in the singular design case, where the design matrix X has one less than full rank, of which the design matrix for the APC accounting model is an example, and showed (1) that the ridge estimator lies in a subparameter space orthogonal to the null space of the design matrix generated by the eigenvector of the zero eigenvalue, and (2) that the ridge estimator converges to the IE as the shrinkage parameter λ tends to 0. In other words, the IE can be interpreted as the limit of the ridge estimator as its shrinkage penalty goes to zero. Second, Tu, Kramer, and Lee (2012) studied the application of partial least squares (PLS) to the APC accounting model. Whereas PCR extracts the components independently of the outcome variable, PLS maximizes the covariance of the components with the outcome variable Y, extracting the components by order of this covariance from the highest to the lowest. Tu, Kramer, and Lee (2012) showed with an empirical application that, as the number of components extracted by PLS approaches the maximum number possible for a design matrix, the numerical values of the PLS estimates of the age, period, and cohort effect coefficients approach, and are within sampling error of, the corresponding coefficients estimated by the IE (which uses the maximum possible number of components). They also showed that an estimator based on the first three PLS components is a numerically reasonable approximation to the PLS effect coefficients estimated by using the maximum possible number of components.

5.2.2 Geometric Representation

The parameter space of the unconstrained vector b of model (4.5) of Chapter 4, **P**, can be decomposed into the direct sum of two linear subspaces that are orthogonal (independent) to each other: $\mathbf{P} = \mathbf{N} \oplus \mathbf{\Theta}$, where one subspace (**N**) denotes the null space of X defined by the vector B_0, corresponding to the unique zero eigenvalue of the matrix $X^T X$, and the other subspace (**Θ**) denotes the complement nonnull space orthogonal to **N**. The parameter vector decomposition is

$$b = b_0 + sB_0 \tag{5.5}$$

where $b_0 = P_{proj}b$ is a special parameter vector that is a linear function of b corresponding to the projection of the unconstrained parameter vector b to

Θ, the nonnull space of X. Specifically, the special parameter vector b_0 corresponding to $s = 0$ satisfies the geometric projection

$$b_0 = (I - B_0 B_0^T)b \tag{5.6}$$

Note that B_0 represents a special direction in the parameter space defined by the difference between two arbitrary CGLIM estimators, $\hat{b}_1 - \hat{b}_2$:

$$X(\hat{b}_1 - \hat{b}_2) = X(sB_0) = 0 \tag{5.7}$$

where s is an arbitrary real number, and sB_0 represents arbitrary linear trends. Thus, the difference of any two arbitrary CGLIM estimators must be in the null space of X, that is, the space defined by B_0. Different equality constraints of the CGLIM estimators assign different values to s. And, the arbitrary term in these parameter vectors, sB_0, leads to different estimates.

This projection is illustrated in Figure 5.1, which shows the projection of two parameterizations, b_1 and b_2, onto the nonnull parameter space (the vertical axis in Figure 5.1), which is independent of the real number s. The

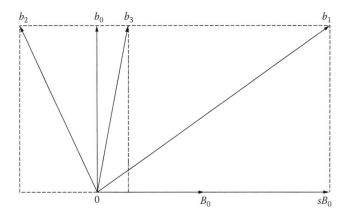

FIGURE 5.1
Geometric projection of parameters vectors. Note: b_1, b_2, and b_3 are three sets of parameter vectors that result from the imposition of various constraints on elements of the parameter vector b of the APC regression model; sB_0 represents the arbitrary term in these parameter vectors; projection of b_1, b_2, and b_3 to the vertical axis yields the same estimable function b_0. The vertical vector b_0 is orthogonal to and therefore independent of the component sB_0, which represents effects of the design matrix. While b_0 is estimable, none of the others, $b_0 + sB_0$ with $s \neq 0$ such as b_1, b_2, or b_3 are. As shown later in the text, however, a statistical test can be defined and applied to ascertain whether or not the vector of estimated coefficients corresponding to constraints such as those used to define, say, b_3 are sufficiently close to those corresponding to b_0, as estimated by the IE, that it can be concluded that the scalar $s = 0$, and, correspondingly, that the equality constraint that produced b_3 also produces an estimable function in a statistical sense. (Adapted from Yang, et al. 2008: Figure 1.)

geometric representation in Figure 5.1 can be thought of as either a simple parameter space of dimension two or as multidimensional with the vertical axis representing a direction in a multidimensional nonnull space. In either case, since the projection of any parameterization of b yields the same parameter vector b_0, the latter is estimable. Figure 5.1 provides only a geometric illustration.

Figure 5.1 also helps to illustrate geometrically that the IE may, in fact, also be viewed as a constrained estimator. But, in contrast to the equality constraints on two or more *coefficients* of the parameter vector b that are imposed in conventional approaches to the estimation of APC accounting models, *the constraint imposed by the IE to identify model (4.5) is a constraint on the geometric orientation of the parameter vector* b *in parameter space.* Specifically, the IE imposes the constraint that the direction in parameter space defined by the eigenvector B_0 in the null space of the design matrix X has zero influence on the parameter vector b_0, that is, on the specific parameterization of the vector b that is estimated by the IE. Since B_0 is a fixed vector that is a function solely of the design matrix and does not depend on the observed event rates or frequencies being analyzed, this would seem to be a reasonable constraint.

Corresponding to the projection of the parameter vector b onto b_0, we have the following projection of the estimators of Equation (5.6) onto the IE:

$$B = (I - B_0 B_0^T)\hat{b} \qquad (5.8)$$

This equation provides another algorithm for computing the IE, namely, compute an initial estimator $\hat{b}$ of model (4.5), say by an equality constraint on two of the A, P, and C parameters, and then geometrically project $\hat{b}$ to the IE B by removing the component in the B_0 direction. Combining Equations (5.1) and (5.2) yields

$$X\hat{b} = X(B + tB_0) = XB + tXB_0 = XB + 0 = XB \qquad (5.9)$$

This shows that the IE B is the estimator that uniquely estimates the A, P, and C effects in the constrained parameter vector $b_0 = P_{proj}b$. This point merits emphasis: The IE does not estimate the unconstrained coefficient vector b of the APC accounting/multiple classification model; rather, it estimates the projection of the unconstrained vector onto the nonnull space of the design matrix X.

5.2.3 Verbal Description

Statisticians have known since the work of Kupper et al. (1985) that the dimension of the design matrix in the APC accounting model (i.e., the number of age groups and time periods) may affect the estimates obtained from

CGLIM estimators. Put in the simplest possible terms, the basic idea of the IE is to remove the influence of the design matrix on coefficient estimates. As noted in Section 5.3 and the appendices of this chapter, this approach produces an estimator that has desirable statistical properties.

The IE also can be viewed as a special form of PCR estimator, as shown in Section 5.2.1, that removes the influence of the null space of the design matrix X on the estimator. It specifically estimates a constrained parameter vector b_0 that is a linear function of the parameter vector b of the unconstrained APC accounting model (4.5). This constrained parameter vector b_0 corresponds to the projection of the unconstrained parameter vector b onto the nonnull subspace of the design matrix X.

Since the IE is a principal components estimator, one might well ask: Why not just calculate the eigenvectors of the matrix $X^T X$ by application of principal components, regress the observed rates on the subspace spanned by these eigenvectors, and leave it at that? The answer is that regression coefficient estimates in this subspace are not directly interpretable in terms of A, P, and C effects. Therefore, the IE uses the extra step of inverse orthonormal transformation of the coefficient estimates of the PCR back to the original space of A, P, and C coordinates. The inverse transformation is what makes the IE a special form of principal components estimator. It yields coefficients, as illustrated by empirical applications in Section 5.4 and in Chapter 6, that are directly interpretable as A, P, and C effects and that can be compared to corresponding effects estimated by the conventional imposition of equality constraints on parameters.

5.2.4 Computational Tools

Programs for estimating the IE can be written as add-on files to commercially available software packages. We provide codes for users of R and Stata in the online computational guides that accompany some analyses shown in the following material. An ado-file to calculate the IE in Stata may be obtained by typing "ssc install apc" on the Stata command line on any computer connected to the Internet or by downloading from the Statistical Software Components archive at http://ideas.repec.org/s/boc/bocode.html. The program uses much the same syntax as Stata's glm command for generalized linear models. For example, a user whose dataset contains a dependent variable y, an exposure variable x, an age variable a, and a period variable t can fit a Poisson model with age, period, and cohort effects by typing

```
apc_ie y, exposure(x) family(poisson) link(log) age(a) period(t)
```

Stata will then display statistics usually seen in applications of glm command. The program is documented more fully in a help file that can be read by typing "help apc_ie" in Stata after installation. The package also includes a command "apc_cglim" for calculating CGLIM estimators for readers interested in methodological comparisons. This package, however, is not recommended for use in empirical analysis due to its previously stated problems.

5.3 Statistical Properties

In general, it is desirable that estimators of coefficients in statistical models be unbiased in finite samples, relatively efficient in finite-time-period analyses in the sense of having a sample variance, or mean squared error (MSE) in the case of biased estimators, that is smaller than that of other estimators and consistent or asymptotically unbiased in large samples (Casella and Berger 2001). We now examine the IE with respect to these properties. While readers are encouraged to read and understand all of these properties, those who are not familiar with linear algebra and asymptotics can skip the proofs in the appendices and Section 5.3.2.

5.3.1 Estimability, Unbiasedness, and Relative Efficiency

To study the statistical properties of the IE, consider first the role of estimable functions in APC accounting models. A large part of discussions of the parameter identification issue in APC models, which may be confusing and bewildering, pertains to what exactly it is for which an estimator is sought. The long methodological discussion of the APC identification and estimation problem in the 1970s and 1980s was centered on the presumption that the objective was to estimate the unidentified coefficient vector of the APC multiple classification model—the b vector in Equation (4.5)—and, as noted in Appendix 5.4, this search continues to motivate some methodological work today. But, the unconstrained b vector is not estimable in the sense of satisfying the estimability condition described in the following discussion. Accordingly, a concluding point of those methodological discussions was that only estimable functions of the model parameters should be recommended as they provide unbiased estimation of special characteristics of the model parameters (Kupper et al. 1983, 1985). Nonlinear estimable functions also have been applied to the APC accounting model (Clayton and Schifflers 1987; Holford 1985, 1991; Robertson, Gandini, and Boyle 1999), although the classical definition of estimable functions is limited to only linear combinations of the model parameters that are independent of parameter constraint (Searle 1971: 180).

For context, consider first the analysis of an APC dataset for a finite number of time periods p. That is, suppose that an APC analysis is to be conducted for a fixed matrix of observed rates or event counts. This implies that the corresponding design matrix X is fixed (i.e., X has a fixed number of age groups and time periods). The randomness in the error term ε of model (4.5) of Chapter 4 then corresponds to measurement errors in the rates or in the event counts or to intrinsic randomness in the rates or counts. In this context of an age-by-time period table of population rates with a fixed number of time period p of data, Yang, Fu, and Land (2004: 101) showed that the IE satisfies a condition for estimability of linear functions of the parameter vector b that was established by Kupper et al. (1985: Appendix B) and recently further elaborated by Fu (2008). *Estimable functions*[*] are invariant with respect to whatever solution [see Equation (4.7) in Chapter 4] is obtained to the normal equations; estimable functions are desirable as statistical estimators because they are linear functions of the unidentified parameter vector that can be estimated without bias, that is, they have unbiased estimators.[†] Specifically, the condition for estimability of a constraint on the parameter vector of the APC accounting/multiple classification model that was established algebraically by Kupper et al. (1985) is, in the notation defined previously, that $l^T B_0 = 0$, where l^T is a constraint vector (of appropriate dimension) that defines a linear function $l^T b$ of b. In words, the Kupper et al. (1985) condition for estimability of a constraint on the parameter vector of the APC accounting model is that the constraint must be orthogonal or perpendicular to the null vector of the model.

Applying this condition to the IE, note first that, since the IE imposes the constraint that $s = 0$, that is, that the arbitrary vector B_0 has zero influence, the constraint vector is $l^T = (I - B_0 B_0^T)$. Since $B_0^T B_0 = 1$, it follows that $l^T B_0 = (I - B_0 B_0^T) B_0 = B_0 - B_0 B_0^T B_0 = B_0 - B_0 = 0$; that is, the Kupper et al. condition holds for the IE. Note also that the Kupper et al. condition implies that any constrained estimator that is obtained by imposing an equality constraint on the parameter vector b and that contains any nonzero component due to the vector B_0 defined by the design matrix is not estimable; that is, such a constrained estimator produces biased estimates of the A, P, and C effect coefficients in the projection of the unconstrained parameter vector b to Θ, the nonnull space of X, that is, $b_0 = P_{proj} b$, the coefficient vector estimated by the IE.

Because the IE B satisfies the Kupper et al. estimability condition for APC models, it follows from properties of estimable functions (Searle 1971: 181) that, for a fixed number of time periods of data, the IE B is an unbiased estimator of the special parameterization (or linear function) b_0 of b defined in

---

[*] See Searle (1971: 180–188) or McCulloch and Searle (2001: 120–121) for expositions of this concept.

[†] In the history of methodological discussions of the APC accounting model in sociology, Rodgers (1982a) was early to argue that analysts should seek estimable functions of the unidentified parameter vector; see also the comment by Smith, Mason, and Fienberg (1982) and the response by Rodgers (1982b). In some respects, the IE can be regarded as providing a practical, easily applicable method to produce estimates of estimable functions from data in the form of age-by-time period tables of rates, as called for by Rodgers over two decades ago. The estimability referred by Rodgers, however, essentially means identifiability that can be achieved by any linear constraints, which differs from statistical estimability defined for APC models by Kupper et al. (1985).

Equation (5.6).* Thus, *a first statistical property of the IE in the context of a linear model for APC analysis with a fixed number of time periods of data is that it produces unbiased estimates of the regression coefficients of the projected coefficient vector* b_0.

In addition, Yang, Fu, and Land (2004: 108) also showed that, *for a fixed number of time periods of data, the IE is more statistically efficient (has a smaller variance) as an estimator of* b_0 *than any CGLIM estimator that is obtained from a nontrivial equality constraint on the unconstrained regression coefficient estimator* b, *that is, any equality constraint that does not produce a projection of* b *onto* b_0. The proof of this property is given in Appendix 5.2. In sum, the IE then has desirable finite-time period properties.

In brief, given an age-by-time period table of population rates or proportion for a finite number of time periods p of data—the classical context of the APC accounting/multiple classification model of demography, epidemiology, and the social sciences—the IE possesses the desirable statistical properties of unbiasedness and relative efficiency as an estimator of the $b_0 = P_{proj}b$ constrained APC coefficient vector. When these statistical properties are taken together, and noting that the choice of the variance of estimators as a criterion for assessing their relative desirability is equivalent to choosing a quadratic norm in the underlying Euclidean vector space, *the IE can be characterized as a minimum norm quadratic unbiased estimator (MINQUE).*† This property is explained in Appendix 5.3.

5.3.2 Asymptotic Properties

The asymptotic properties of APC estimators, including the CGLIM and the IE, as the number of time periods p of data increase have been studied by Fu, Hall, and Rohan (2004), Fu and Hall (2006), and Yang, Fu, and Land (2004). These properties derive largely from the fact that the eigenvector B_0 converges elementwise to zero with increasing numbers of time periods of data. The vector can converge elementwise to zero even though its length is fixed at one because the number of elements of the vector grows as we add time periods. Therefore, for any two estimators $\hat{b}_1 = B + s_1 B_0$ and $\hat{b}_2 = B + s_2 B_0$, where s_1 and s_2 are nonzero and correspond to different identifying constraints placed on model (4.5), as the number of time periods in an APC analysis increases, the difference between these two estimators $\hat{b}_1$ and $\hat{b}_2$ decreases toward zero, and in fact, the estimators converge toward the IE B. Suffice it to say that the proof proceeds by demonstrating that the coordinates of B_0 are bounded by a quantity that is a function of the

---

* A direct proof of the unbiased property of the IE also is given in Appendix 5.1.

† No other estimator of the coefficient vector of the APC accounting model can be a minimum quadratic norm estimator unless it is identical to that of the IE. This is due to the fact that the IE is estimated by the Moore-Penrose generalized inverse matrix, whose projection on the null space of the APC model is zero. This projection defines the minimum of the quadratic norm; see Appendix 5.3.

number of age groups and periods, and this function converges to 0 as $p \rightarrow \infty$ (Yang, Fu, and Land 2004: Lemma 1).

In addition, for the conventional linear regression specification of the APC accounting model [Equation (4.5) of Chapter 4] with a zero expected value and finite variance specification on the error term, Fu and Hall (2006) proved that (1) as the number of time periods of data increases, there are definite bounds on the absolute values of the differences between the effect coefficients estimated by the IE and the age effect coefficients of the projected coefficient vector $b_0 = P_{proj}b$ and on the expected values of the maximum differences between the estimated values of the period and cohort coefficients and their b_0 counterparts; (2) these bounds decrease with an increasing number of time periods of data, and the convergence/rate of decrease is most rapid for the age effect coefficients. In other words, in many circumstances likely to apply to empirical data, the IE gives consistent estimators of the components of b_0.

5.3.3 Implications

These statistical properties are not trivial and merit comment. Both the IE B and any other estimator $\hat{b} = B + sB_0$ with $s \neq 0$ obtained from an equality constraint are asymptotically consistent as the number of time periods of data increases without bound. Therefore, with a large number (e.g., 30 or 40) of periods of data, differences among estimators decline, and it makes little difference which identifying constraint is employed. In most empirical APC analyses, however, there usually are a small number (e.g., 4 or 5) of time periods of observations available for analysis. In these cases, the differences can be substantial.

As just noted, the IE, by its very definition and construction, satisfies the estimability condition of Kupper et al. (1985). Other estimators using equality or other linear constraints on the parameter vector b derived from theory or prior research or information on side conditions on a specific process being studied may satisfy this condition either exactly or statistically (in a sense defined further in the chapter). If other estimators do indeed satisfy the estimability condition, then they also produce unbiased estimates of the A, P, and C effect coefficients in the b_0 vector. If not, then the estimates they produce are biased.

These properties provide a means for differentiating among estimators. That is, for tables of rates with a finite number of time periods of data, especially a small number (e.g., 4 or 5), an unbiased estimator of b_0 should be preferred to a biased estimator as the latter can be misleading with respect to the estimated trends across the age, time period, and cohort categories.

In contrast, as has been noted many times over the years in discussions of the APC accounting model (see, e.g., Pullum 1978, 1980; Rogers 1982; Smith, Mason, and Fienberg 1982), different just-identified models will generate the same data and yield exactly the same model fit. In particular, linear transformations of the estimated A, P, and C coefficients obtained by the IE (i.e., linear transformations of the elements of the B vector) will fit the observed data just as well as the IE.[*] If, however, such a linear transformation of the A, P, and C coefficients (or any subset thereof) results in coefficients that depart sufficiently far from the coefficients in B that the transformed coefficients contain a significant component of the B_0 vector (i.e., contain a significantly nonzero s coefficient), then the resulting transformed vector may not be estimable in the sense of satisfying the Kupper et al. (1985) condition for estimability; that is, it may not be unbiased.[†] Thus, even though the transformed coefficients will reproduce the data just as well as those obtained by the IE, they will be biased and will give poor indications of the patterns of change across the A, P, and C categories used in the analysis. Therefore, goodness of fit to the data (as measured, e.g., by log-likelihood functions or deviance statistics) *cannot* be used as a criterion for selecting among estimators. But, estimability can be so used.

Because of its estimability and unbiasedness properties, the IE may provide a means of accumulating reliable estimates of the trends of coefficients across the A, P, and C categories of the APC accounting model. To provide intuition for this statement, recall the distinction between the *steady-state* and *general solutions* to the ordinary differential equation for, say, Hooke's law for the motion of a displaced spring-mass system subject to an additional forcing motion in classical mechanics. This law has the algebraic form $F = -kx + a\cos(\omega t)$, where F denotes acceleration (second derivative of the motion with respect to time) of the mass, x denotes the distance of displacement, k is a constant unique to the particular spring under study, and $a\cos(\omega t)$ is the forcing term (see, e.g., Marion and Thornton 1995: 125). If the mass is displaced by, say, a distance of 2 feet, it will oscillate back and forth with some influence of the length of the initial displacement, but it eventually will settle down to a characteristic pattern of oscillations that depends only on the driving force. By comparison, if the initial displacement of the mass is a distance of 4 feet, then the mass will display an initial set of larger oscillations back and forth that are different from the pattern observed for the 2-foot displacement. But, after an initial series of oscillations, the driving force will cause the mass to settle into the same set of oscillations back and forth as those found after an initial 2-foot displacement. Mathematically, the pattern of oscillations observed after the impacts of the initial lengths of displacements have worn off are termed the *steady-state solution* of the Hooke's law differential equation, whereas the *general solution* of the equation consists of the steady-state solution plus a factor that takes into account the initial conditions or displacement of the spring. Because initial conditions can vary from

---

[*] Geometrically, a linear transformation of the coefficient vector corresponds to a rotation of the vector in parameter space. Such a rotation will produce distorted and misleading indications of patterns of change across the age, period, and cohort categories used in the analysis.

[†] This point is illustrated in our empirical analyses and is the basis of the statistical test for estimability derived in Section 5.6.

application to application, the general solution of the differential equation can be quite unique to the application. On the other hand, the steady-state part of the solution is invariant and generalizable to the motion of the system regardless of the initial conditions.

Analogously, the IE is essentially a steady-state solution to the APC accounting model estimation problem that factors out the initial conditions of the dimension of the matrix of observed data, namely, the number of age and time period categories that define the design matrix. Because the IE does not allow these "initial conditions" to influence the estimates it produces of the A, P, and C effect coefficients, they will be more invariant to changes in the design matrix, such as an additional time period of data, than estimates produced by estimators that incorporate such influences. In this sense, the IE removes the part of the subjectivity in the estimator that is due to the shape of the data. We illustrate this feature of the IE in our model validation analyses that follow.

5.4 Model Validation: Empirical Example

In brief, the IE possesses some valuable properties as a statistical estimator. But, given the long history of problems and pitfalls in proposed methods of APC analysis, it is reasonable to question whether this estimator gives numerical estimates of A, P, and C effect coefficients that are valid. This is a question of *model validation*—that is, does the identifying constraint imposed by the IE produce estimated coefficients that meaningfully capture the underlying age, time period, and cohort trends.

One approach to the question of validity is to compare results from an APC analysis of empirical data by application of the IE with results from an analysis of the same empirical data by application of a different family of models that do not use the same or any identifying constraint. As an instance of such an empirical comparison, we next analyze the General Social Survey (GSS) verbal ability data described in Chapter 3. We apply the IE to data on verbal test scores grouped into five-year age groups and time periods shown in Table 5.1. In this age-by-period array, there are 12 five-year age categories from ages 20 to 75+,[*] 6 five-year period groups from 1976 to 2006, and 12 + 6 − 1 = 17 ten-year birth cohorts born in 1901 to 1981. This yields 72 degrees of freedom. The event/exposure rates (sample mean proportions) of correct answers to GSS vocabulary questions can be transformed by a log

---

[*] We excluded ages 18–19 to obtain age groups of equal interval length so that the diagonal elements of the age-by-period matrix refers to cohort members. The results are not influenced by omitting this cell of small sample size. Ages 75–89 are grouped into a last 75+ category to combine small population exposures for more stable estimates.

TABLE 5.1

Verbal Test Correct Rates (Exposure): GSS 1976–2006

Age	Period					
	1976–1980	**1981–1985**	**1986–1990**	**1991–1995**	**1996–2000**	**2001–2006**
20–24	0.55	0.53	0.53	0.55	0.56	0.57
	(8203)	(8489)	(5553)	(4950)	(5045)	(1930)
25–29	0.60	0.55	0.58	0.59	0.58	0.57
	(9671)	(10653)	(6946)	(6480)	(6780)	(2730)
30–34	0.63	0.63	0.59	0.61	0.60	0.60
	(8643)	(8718)	(7096)	(7745)	(7740)	(2660)
35–39	0.63	0.64	0.63	0.63	0.62	0.61
	(6776)	(7946)	(6799)	(7960)	(8400)	(2780)
40–44	0.62	0.63	0.64	0.65	0.63	0.62
	(5619)	(5996)	(5548)	(6940)	(8355)	(2870)
45–49	0.61	0.60	0.62	0.68	0.65	0.61
	(4595)	(5112)	(4734)	(5915)	(6775)	(3150)
50–54	0.62	0.60	0.59	0.62	0.66	0.65
	(5454)	(4560)	(3314)	(4770)	(5885)	(2570)
55–59	0.63	0.59	0.59	0.62	0.68	0.65
	(5439)	(5335)	(3207)	(3590)	(4225)	(2400)
60–64	0.63	0.60	0.56	0.61	0.60	0.66
	(4915)	(4830)	(3319)	(3430)	(3700)	(1950)
65–69	0.61	0.58	0.58	0.65	0.60	0.65
	(4010)	(498)	(3633)	(3000)	(2985)	(1520)
70–74	0.52	0.57	0.61	0.59	0.61	0.63
	(3420)	(3333)	(2789)	(3305)	(3540)	(1290)
75+	0.55	0.53	0.54	0.57	0.56	0.63
	(4394)	(4264)	(3584)	(4890)	(4970)	(1890)

link and modeled by a log-linear regression for which the IE can be obtained. The events are the total number of correct answers for every age and period group and are calculated by multiplying the mean verbal scores by the number of individuals. These are nonnegative counts and can be considered to be distributed as Poisson variates. The population exposure is calculated as the product of number of people in each cell and the total number of possible correct answers (10). Vocabulary knowledge test scores are not available every year from 1974 to 2006. Therefore, for those missing years before 2000, we interpolated the mean verbal scores and numbers of individuals at risk based on the data of neighboring years. There are only two waves of data after 2000, so we just combined them in the calculation.

Figure 5.2 shows the results from application of the IE to these data (accompanying codes available online). Estimated coefficients and their 95% confidence intervals are plotted for successive categories within the age, period, and cohort classifications. Since they indicate changes in correct answer rates from one age group to the next, from one time period to the next,

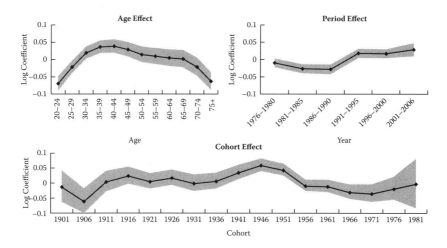

FIGURE 5.2
Intrinsic Estimator coefficient estimates of the age, period, and cohort effects on vocabulary scores.

and from one cohort to the next, the estimated coefficients represent the temporal trends of vocabulary knowledge along each of these three dimensions, net of the effects of the other two. The same data were analyzed by a series of studies by Yang and colleagues (Yang and Land 2006; Yang 2006; Frenk, Yang, and Land 2012) using a completely different approach to APC analysis—namely, a HAPC analysis in the form of a cross-classified random effects regression model. We more fully describe the HAPC model in Chapter 7. Briefly, this approach proceeds by building a level 1 fixed effects regression model at the individual level of analysis and then a random effects model for cohort and time period effects at level 2. The mixed effects model does not rest on the assumption of additive and fixed A, P, and C effects used by the conventional linear regression models that cause the identification problem. The resulting estimates of cohort and period effect coefficients are average residual effects of the cohorts and period across all time periods and cohorts, respectively, and are not constrained in any way to conform to the IE or any other identifying constraint required by the accounting model. The results of the HAPC analysis are shown in Figure 5.3, which is reproduced from Frenk, Yang, and Land (2012) and Table 7.2 in Chapter 7.

While the coefficient metrics and timescales for the age and period effects in Figure 5.3 are in single years rather than the 5-year groups of those from the IE analysis in Figure 5.2, the trends of estimated effect coefficients are quite similar across the graphs. That is, both exhibit age effect curves that are quadratic and concave, corroborating those found by Wilson and Gove (1999): low at youth, rising to a peak in the 40s, staying largely flat until the mid-50s, and declining gradually into late life. Both produced period effect curves that showed slight declines from the 1970s into the 1980s followed by

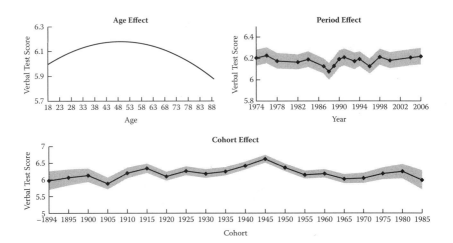

FIGURE 5.3
HAPC model estimates of age, period, and cohort effects on verbal scores.

slight rises into the 1990s that flattened out until 2006. Variation in vocabulary knowledge over time for the past 30 years, therefore, is quite revealing given the absence of direct estimates of period effects in the previous studies. And, the cohort effect curves are also quite similar—showing two peaks for cohorts born in the early and middle twentieth century followed by declines in recent cohorts. These results give mixed support for the hypothesis of intercohort declines in previous studies (Alwin and McCammon 1999; Glenn 1999) as there are also increases in vocabulary knowledge for cohorts born before World War I and between 1930 and 1950.

In sum, this comparison shows that the independent estimates from a mixed effects models analysis corroborate the estimated patterns of change across age, period, and cohort categories that are obtained by the IE analysis. Of course, this is only one example of comparative analysis, and additional empirical studies are needed before it can be concluded that the IE produces substantively meaningful and empirically valid results under various circumstances. Additional model validation assessments of the intrinsic estimation through empirical analyses of various datasets are reported in Section 5.6 and in Chapter 8, Section 8.3.

5.5 Model Validation: Monte Carlo Simulation Analyses

Given the long history of debates over the existence of any solution to the APC model identification problem (Glenn 2005), questions can be raised regarding whether the IE method is based on a constraint of purely algebraic convenience and just as arbitrary as previous methods of constraints. Is the IE just another way to go wrong? The conceptual foundations of the IE have been established statistically using mathematical proofs and thus remain abstract and potentially difficult to understand. Further exposition and analysis, especially using straightforward and replicable numerical illustrations, is needed to directly address this question.

This section reports the numerical results from Monte Carlo (MC) simulations that are conducted systematically for the IE and CGLIM estimators of APC multiple classification models based on the work of Yang, Schuholfer-Wohl, and Land (2007). We compare results from application of an estimator such as the IE or CGLIM to artificial data wherein we know the form of the underlying model that generated the data.[*] The simulation analyses can help to determine conditions under which the IE indeed recovers the parameters of the underlying model while CGLIM estimators do not. Then, we revisit a recent critique of the utility of APC models in social research through numerical examples (Glenn 2005), show the consequence of misuse and misinterpretation of such models, and make suggestions to future research in light of findings from the simulation analysis.

We first investigate whether the IE is unbiased and is relatively efficient in samples with a fixed number of age groups and time periods. The asymptotic results for the IE summarized in Section 5.3.2 apply as the number of periods in the dataset goes to infinity. But, any dataset used in practice has only a finite number of periods. Thus, we explore whether the asymptotic results give good approximations to the behavior of the IE in finite samples by simulating datasets with 5, 10, and 50 periods. Based on the mathematical analyses of Fu and Hall (2006) summarized in Section 5.3.2, the basic asymptotic result we investigate in these numerical simulations is that, as the number of periods increases, estimated age effects should converge to the underlying age effects when using the IE but not necessarily when using other estimators. If the underlying processes generating the period-to-period changes in the observed outcomes are constant throughout the periods of the simulation, then we also expect the estimated period and cohort effects converge to the underlying period and cohort effects when using the IE.[†] We conduct the simulation analyses systematically to examine the performance of the IE and alternative methods, CGLIM in particular,

---

[*] In the methodological literature on estimation of the APC accounting model, a distinction often is made between the "true (but unknown) population parameter values" (Kupper et al. 1985: 822) or "the population parameters that generated the data ... the 'true' solution (the one representing the process that generated the outcomes)" (O'Brien 2011a: 436). In the exposition that follows, we use the term *underlying model* for this referent.

[†] If such processes are not constant and change from period to period, regardless of the estimator used, estimated period and cohort effects cannot be expected to converge to their underlying values as the number of periods increases because adding a period to the dataset does not add information about the previous periods or about cohorts not present in the period just added.

in reproducing the underlying models when the underlying model is a full APC model in which all three of the A, P, and C effects are present.

5.5.1 Results for APC Models: True Effects of A, P, and C All Present

We fix the number of age categories in all simulations given the fact that humans have a relatively fixed life span. We let the number of age categories be 10 without loss of generality. For a given number of periods *P*, we generate 1,000 datasets by MC simulation in which the entries in the 10 × *P* outcome matrix are distributed according to

$$y_{ij} \sim N(\mu, \sigma^2)$$

where[*]

$$\mu = 0.3 + 0.1(age_{ij} - 5.5)^2 + 0.1\sin(period_{ij}) + 0.1\cos(cohort_{ij}) + 0.1\sin(10 \cdot cohort_{ij})$$

$$\sigma^2 = 25$$

This equation for the data-generating process tells us what the underlying age, period, and cohort effects are:

Age effect at age *a*	$0.1(a - 5.5)^2$
Period effect in period *p*	$0.1\sin(p)$
Cohort effect in cohort *c*	$0.1\cos(c) + 0.1\sin(10c)$

So that the underlying effects have mean zero in each category in accord with the constraints on the effect coefficient specified previously, we subtract constants from the effects listed, where the constants are calculated as the mean effects for each temporal category. To explore the finite time period properties of various estimators, we then estimate A, P, and C effects in each simulated dataset for a given *P* using the IE and using three different CGLIM estimators: one with the first two age effects constrained to be equal (CGLIM_a), one with the first two period effects constrained to be equal (CGLIM_p), and one with the first two cohort effects constrained to be equal (CGLIM_c). To explore the large-sample properties of these estimators, we let P increase from 5 to 10 and to 50 and repeated the simulations for each number of P.

Table 5.2 reports the results for A, P, and C effects estimated from data simulated with five time periods (*P* = 5). For each effect in the model, we show the underlying value and, for each estimator, the mean, standard deviation,

---

[*] We chose the variance of 25 so that the sampling variability of the estimator would be visible in our graphs. Experiments with smaller and larger variances produced qualitatively similar results.

TABLE 5.2

Simulation Results (n = 1000) of the IE and CGLIM Estimators of APC Models: P = 5

Variable	True Effect	IE			CGLIM_a (a1=a2)			CGLIM_p (p11=p12)			CGLIM_c (c1=c2)		
		Mean	sd	MSE	Mean	sd	MSE	Mean	sd	MSE	Mean	sd	MSE
Age													
a1	1.20	1.17	2.41	5.80	-2.20	14.04	208.30	1.36	10.88	118.20	1.86	30.11	906.20
a2	0.40	0.42	2.22	4.92	-2.20	14.04	203.50	0.57	8.57	73.40	0.96	23.46	549.90
a3	-0.20	-0.25	2.28	5.21	-2.12	9.22	88.60	-0.15	6.27	39.30	0.13	16.70	278.70
a4	-0.60	-0.60	2.36	5.57	-1.73	5.95	36.70	-0.54	4.29	18.30	-0.37	10.25	105.00
a5	-0.80	-0.80	2.33	5.43	-1.17	3.05	9.40	-0.78	2.61	6.80	-0.72	3.98	15.80
a6	-0.80	-0.78	2.31	5.34	-0.41	2.84	8.20	-0.80	2.64	6.90	-0.86	4.09	16.70
a7	-0.60	-0.64	2.26	5.10	0.48	5.80	34.80	-0.71	4.14	17.10	-0.87	10.27	105.50
a8	-0.20	-0.15	2.33	5.43	1.73	9.20	88.40	-0.25	6.44	41.40	-0.53	16.77	281.10
a9	0.40	0.36	2.24	4.99	2.98	12.59	165.00	0.21	8.60	73.90	-0.18	23.81	566.50
a10	1.20	1.27	2.30	5.30	4.64	16.17	273.20	1.08	10.68	113.90	0.58	29.48	868.70
Period													
p11	-0.11	-0.02	1.43	2.05	1.48	7.16	53.70	-0.11	3.61	13.00	-0.33	13.11	171.90
p12	-0.06	-0.06	1.48	2.17	0.68	3.78	14.90	-0.11	3.61	13.00	-0.22	7.02	49.20
p13	0.03	0.00	1.46	2.14	0.00	1.46	2.10	0.00	1.46	2.10	0.00	1.46	2.10
p14	0.09	0.11	1.39	1.93	-0.64	3.79	14.90	0.15	2.74	7.50	0.26	6.79	46.10
p15	0.06	-0.02	1.49	2.21	-1.52	7.10	52.80	0.06	4.92	24.10	0.28	13.42	179.90

continued

TABLE 5.2 (continued)

Simulation Results (n = 1000) of the IE and CGLIM Estimators of APC Models: P = 5

Variable	True Effect	IE			CGLIM_a (a1=a2)			CGLIM_p (p11=p12)			CGLIM_c (c1=c2)		
		Mean	sd	MSE	Mean	sd	MSE	Mean	sd	MSE	Mean	sd	MSE
Cohort													
c1	−0.01	−0.25	4.71	22.26	−14.32	65.59	4502.30	0.07	41.57	1726.50	2.09	108.60	11786.30
c2	0.04	−0.09	3.47	12.02	−13.41	61.13	3913.60	0.18	39.21	1535.90	2.09	108.60	11786.10
c3	−0.20	−0.24	3.00	8.96	−12.81	57.62	3475.80	−0.01	36.90	1360.00	1.79	99.86	9965.50
c4	0.00	−0.03	2.53	6.38	−11.86	54.20	3074.90	0.15	34.49	1188.20	1.84	93.31	8700.90
c5	0.00	0.02	2.32	5.38	−11.06	50.75	2695.40	0.16	32.54	1057.80	1.74	86.49	7475.20
c6	0.06	−0.05	2.54	6.45	−10.37	47.12	2326.90	0.06	30.08	903.90	1.53	79.84	6370.50
c7	0.15	0.17	2.65	7.00	−9.41	43.72	2000.90	0.23	27.76	769.60	1.59	73.44	5390.10
c8	−0.12	−0.11	2.45	5.97	−8.93	40.14	1687.30	−0.08	25.57	653.00	1.16	66.66	4440.20
c9	−0.01	0.06	2.46	6.03	−8.02	36.36	1385.00	0.04	23.61	557.00	1.17	60.07	3605.70
c10	−0.14	−0.11	2.27	5.14	−7.44	33.49	1173.50	−0.17	21.56	464.50	0.85	53.46	2856.60
c11	−0.01	0.05	2.54	6.44	−6.53	30.17	951.70	−0.05	19.62	384.60	0.86	46.77	2185.80
c12	0.14	0.26	2.77	7.69	−5.57	26.90	755.50	0.12	17.40	302.30	0.92	40.30	1623.20
c13	−0.01	0.21	3.27	10.72	−4.88	24.05	601.50	0.02	15.56	241.80	0.71	33.97	1153.30
c14	0.11	0.11	5.33	28.40	−4.23	22.50	524.60	−0.12	14.13	199.60	0.46	27.65	763.90

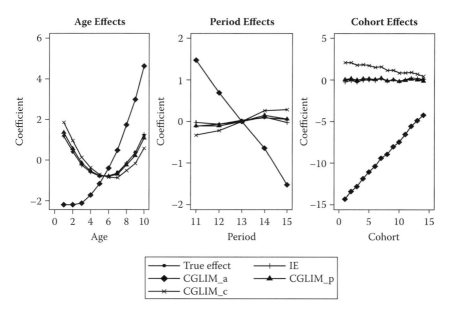

FIGURE 5.4
Means of estimates from 1,000 simulations of APC models with P = 5: IE versus CGLIM.

and MSE of the estimated effect across 1,000 simulations. By comparing the mean of the simulated estimates to the underlying values, we can assess the degree of *unbiasedness* for each estimator. The standard deviation of the simulated estimates shows how much the estimated parameters vary from sample to sample. Smaller variance relates to *relative efficiency*. MSE is the average squared difference between the estimated parameter and the truth; this measure of accuracy takes into account both bias and variance.

Figure 5.4 compares the means of IE and CGLIM estimates shown in Table 5.2. Two of the four estimators recover the profile of the A, P, and C effects qualitatively: the IE and the CGLIM_p. The other two sets of CGLIM estimates that constrain the first two age and first two cohort effects to be equal clearly fail to recover underlying forms of these effects because the constraints are incorrect: The first two underlying age effects and the first two underlying cohort effects are not equal, and the differences between the two underlying effects are large. The CGLIM_p estimator recovers the qualitative shapes of underlying effects more closely because constraining the coefficients of the first two period effects to be equal more closely approximates the fact that the difference between the two underlying effects is much smaller. Scrutiny of the numerical results in Table 5.2, however, suggests that CGLIM_p estimates are far off the mark in quantitative terms; only for the IE is the mean of each estimated effect close to the underlying value and hence unbiased. This is a direct result of the nonestimability of the CGLIM estimator. That is, any substantial departure from the IE constraint, which incorporates a large nonzero s, will not yield an estimable function and thus will induce bias in the estimates.

5.5.1.1 Property of Estimable Constraints

This constrained model illustrates numerically a property established algebraically by Kupper et al. (1985: 830), namely, that, if the constraint used to just identify the APC accounting model is, indeed, satisfied by the underlying model, then the Kupper et al. orthogonality condition stated previously ($I^T B_0 = 0$) will hold, and the corresponding constrained coefficient vector is estimable.

With respect to the specification of the underlying effect coefficients in the present simulation, for example, it can be noted that, due to the periodicity of the age coefficients, several are equal. Thus, for example, if the analyst were to impose the identifying constraint a1 = a10, then the corresponding constrained coefficient vector will be estimable, and in fact, the resulting CGLIM estimated coefficients will be within sampling and rounding error of those estimated by the IE. We make this point more precise through a statistical test described in Section 5.6.

Table 5.2 also shows that the IE exhibits substantially less sampling variation than the CGLIM estimators. The IE estimates of A, P, and C effects have standard deviations that range between 0 and 1. CGLIM_p estimates have the smallest standard deviations among all CGLIM estimates, but their standard deviations are still at least 10 times larger than those of the IE. The IE also has much smaller MSEs. The MSEs of the IE estimates are close to 0, whereas those of the CGLIM_p estimators can be as large as 6. All estimators have larger MSEs for the youngest and oldest cohorts because these cohorts are located at the upper and lower corners of the age-by-period table and have the smallest sample sizes. The CGLIM_a and CGLIM_c estimates have MSEs too large to provide reliable findings. It is noteworthy that the cohort effects are particularly poorly estimated by the CGLIM models.

To see what happens to these estimators in cases when analysts have access to more data, we next increased the number of time periods to 10 and 50. Because the CGLIM_p continued to yield the best estimates among all CGLIM estimators, the following analysis focuses on the comparison of the CGLIM_p and the IE and is summarized in Table 5.3.

The means and standard errors of age effects estimates are shown in Figure 5.5 for the IE and the CGLIM_p by number of periods P. The means of the IE are extremely close to the underlying age effects for all P and rapidly approach the underlying age effects as P increases from 5 to 50. The means of the CGLIM_p also recover the underlying age effects well and do better with increasing P, but to a less extent than the IE. Although the means of the IE and CGLIM_p are close, the CGLIM_p shows much larger standard errors and thus a statistically significant difference between the mean and the truth. Comparison of the standard errors across P (not shown) suggests decreasing sampling variations for both estimators with increasing P but much smaller

TABLE 5.3

Simulation Results (n = 1000) of the IE and CGLIM Estimators: Age Effects by Number of Time Periods

Age	TRUE Effect	IE			CGLIM_a (a1=a2)			CGLIM_p (p11=p12)			CGLIM_c (c1=c2)		
		P = 5	P = 10	P = 50	P = 5	P = 10	P = 50	P = 5	P = 10	P = 50	P = 5	P = 10	P = 50
a1	1.20												
	mean	1.17	1.22	1.23	−2.20	−2.39	−2.60	1.36	1.41	1.52	1.86	1.76	3.16
	sd	2.41	1.59	0.69	14.04	9.53	4.18	10.88	10.66	10.36	30.11	28.48	27.69
	MSE	5.80	2.53	0.48	208.34	103.66	31.90	118.22	113.55	107.35	906.21	810.66	769.65
a2	0.40												
	mean	0.42	0.41	0.38	−2.20	−2.39	−2.60	0.57	0.56	0.61	0.96	0.84	1.88
	sd	2.22	1.56	0.70	14.04	9.53	4.18	8.57	8.34	8.09	23.46	22.25	21.53
	MSE	4.92	2.43	0.49	203.54	98.55	26.46	73.44	69.57	65.48	549.88	494.56	465.31
a3	−0.20												
	mean	−0.25	−0.19	−0.19	−2.12	−2.19	−2.31	−0.15	−0.09	−0.03	0.13	0.11	0.88
	sd	2.28	1.55	0.67	9.22	6.09	2.68	6.27	6.09	5.78	16.70	15.82	15.38
	MSE	5.21	2.41	0.45	88.65	41.08	11.66	39.26	37.04	33.34	278.72	250.15	237.40
a4	−0.60												
	mean	−0.60	−0.61	−0.59	−1.73	−1.81	−1.86	−0.54	−0.55	−0.49	−0.37	−0.43	0.06
	sd	2.36	1.57	0.69	5.95	3.93	1.69	4.29	3.84	3.49	10.25	9.54	9.24
	MSE	5.57	2.46	0.47	36.68	16.87	4.46	18.34	14.75	12.18	105.00	90.89	85.67
a5	−0.80												
	mean	−0.80	−0.85	−0.79	−1.17	−1.25	−1.22	−0.78	−0.83	−0.76	−0.72	−0.79	−0.58
	sd	2.33	1.61	0.72	3.05	2.01	0.90	2.61	1.98	1.36	3.98	3.51	3.15
	MSE	5.43	2.59	0.51	9.45	4.25	0.99	6.79	3.92	1.85	15.81	12.33	9.93

continued

TABLE 5.3 (continued)

Simulation Results (n = 1000) of the IE and CGLIM Estimators: Age Effects by Number of Time Periods

Age		TRUE Effect	IE			CGLIM_a (a1=a2)			CGLIM_p (p11=p12)			CGLIM_c (c1=c2)		
			P = 5	P = 10	P = 50	P = 5	P = 10	P = 50	P = 5	P = 10	P = 50	P = 5	P = 10	P = 50
a6	mean	-0.80	-0.78	-0.83	-0.80	-0.41	-0.43	-0.37	-0.80	-0.85	-0.83	-0.86	-0.89	-1.01
	sd		2.31	1.59	0.67	2.84	2.01	0.88	2.64	1.96	1.30	4.09	3.57	3.10
	MSE		5.34	2.53	0.45	8.19	4.17	0.96	6.94	3.83	1.69	16.73	12.74	9.62
a7	mean	-0.60	-0.64	-0.68	-0.62	0.48	0.52	0.66	-0.71	-0.75	-0.72	-0.87	-0.86	-1.26
	sd		2.26	1.55	0.67	5.80	3.79	1.72	4.14	3.95	3.52	10.27	9.64	9.25
	MSE		5.10	2.41	0.45	34.78	15.60	4.53	17.10	15.61	12.37	105.46	92.86	85.97
a8	mean	-0.20	-0.15	-0.16	-0.20	1.73	1.84	1.93	-0.25	-0.27	-0.36	-0.53	-0.46	-1.27
	sd		2.33	1.61	0.67	9.20	6.18	2.65	6.44	6.00	5.77	16.77	15.80	15.39
	MSE		5.43	2.59	0.45	88.35	42.25	11.55	41.43	35.98	33.24	281.06	249.43	237.74
a9	mean	0.40	0.36	0.42	0.39	2.98	3.23	3.37	0.21	0.27	0.17	-0.18	0.00	-1.11
	sd		2.24	1.57	0.68	12.59	8.49	3.71	8.60	8.47	8.12	23.81	22.34	21.59
	MSE		4.99	2.46	0.47	165.02	79.93	22.57	73.90	71.72	65.85	566.53	498.86	467.90
a10	mean	1.20	1.27	1.27	1.18	4.64	4.88	5.01	1.08	1.08	0.89	0.58	0.73	-0.74
	sd		2.30	1.55	0.68	16.17	10.78	4.69	10.68	10.52	10.36	29.48	28.23	27.64
	MSE		5.30	2.40	0.47	273.16	129.65	36.50	113.86	110.58	107.24	868.68	796.56	766.96

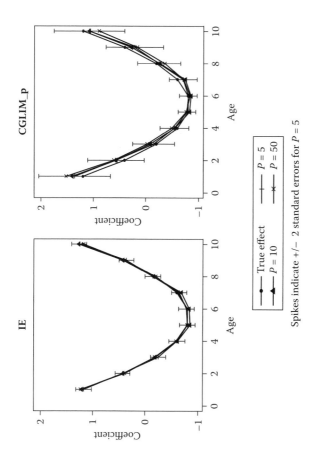

FIGURE 5.5
Means and standard errors of age effects estimates from 1,000 simulations of APC models by P: IE versus CGLIM.

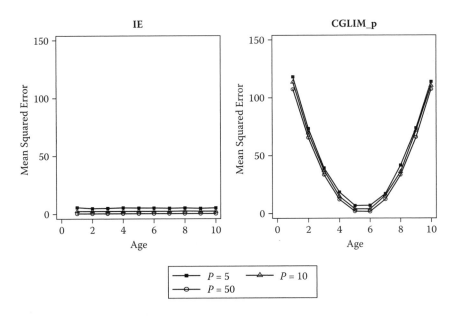

FIGURE 5.6
Mean squared errors of the age effects estimates from 1,000 simulations of APC models by P:
IE versus CGLIM.

variability for the IE for all P. Figure 5.6 further shows the advantage of the
IE in terms of MSE. The IE has MSE much closer to 0 than the CGLIM_p
for all P. Whereas the MSE of the IE approaches 0 as P increases, that of the
CGLIM_p, although decreasing, is far above 0.

Figures 5.7 and 5.8 present the mean and MSE of the IE and CGLIM_p estimates of period effects and cohort effects, respectively. Similar to the results
shown previously, the IE recovers the underlying effects much better for all
P, increases in precision, and decreases in MSE with increasing P. In contrast,
the CGLIM_p shows much larger departures from the underlying effects that
do not decrease with increasing P. The first two time period coefficients were
constrained to be equal by the CGLIM_p, whereas in fact they increase slightly
from time 1 to time 2. As a result, the period effects estimated by the CGLIM_p
rotate the underlying period effects (horizontal oscillations) upward. And, the
cohort effects estimates are rotated downward. While the IE has a MSE close
to 0, the CGLIM_p also produces much larger and in some cases increasing
MSE with increasing P. This illustrates that linear constraints with even small
deviations from the truth can result in coefficient estimates with large bias in
unknown directions that will not lessen with more periods of data.

Several insights follow from this analysis. First, the IE produces estimates
of the A, P, and C effects that are more invariant to changes in the design
matrix, such as additional time periods of data, than estimates produced
by estimators that incorporate functions of the design matrix (such as the

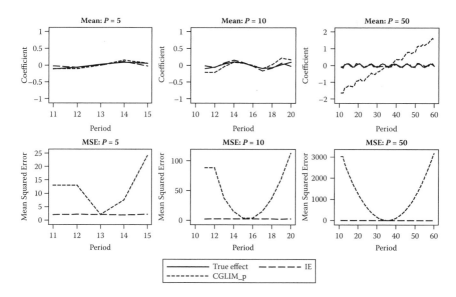

FIGURE 5.7
Period effects estimates from 1,000 simulations of APC models by P: IE and CGLIM.

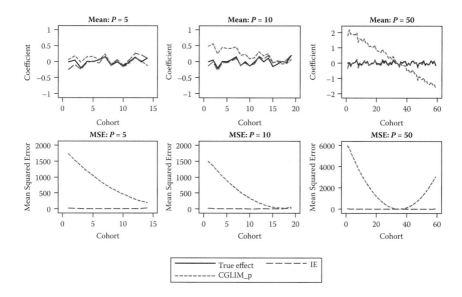

FIGURE 5.8
Cohort effects estimates from 1,000 simulations of APC Models by P: IE and CGLIM.

CGLIM). This precisely is because of its estimability and unbiasedness property. In this sense, the IE reduces the part of the subjectivity in the estimator that is due to the influence of fixed component determined by the shape of the data by removing it.

Second, both the IE B and any other estimator $\hat{b} = B + sB_0$ with $s \neq 0$ obtained from an equality constraint produce asymptotically consistent age effects as the number of time periods of data increase without bound. Therefore, with a large number (e.g., 50) of periods of data, differences among estimators decline, and it makes little difference which identifying constraint is employed. In most empirical APC analyses, however, there usually are a small number (e.g., 5) time periods of observations available for analysis. In these cases, the differences can be substantial, and an unbiased estimator should be preferred to a biased estimator as the latter can be misleading with respect to the estimated trends.

Third, as a result of the properties mentioned, the IE may provide a means for the accumulation of reliable estimates of the A, P, and C trends when more data become available over time, whereas the other estimators may not.

5.5.2 Misuse of APC Models: Revisiting a Numerical Example

The previous exposition and numerical simulation analyses suggested that (1) the IE indeed yields unbiased and relatively efficient estimates of A, P, and C effects in the IE constrained coefficient vector $b_0 = P_{proj} b$; (2) the estimates of the elements of b_0 will be relatively close to the estimate of any equality constrained coefficient vector that satisfies the estimability condition of Kupper et al. (1985); and (3) the elements of b_0 converge to the underlying effects with an increasing number of time periods. This seems to be in conflict with the age-old notion that there is no solution to the APC accounting model identification problem because there can be any number of estimates that fit the data equally well. This notion is best represented in a recent critique of the utility of APC models in social research raised by Norval Glenn (2005). We next revisit the numerical example given by Glenn to evaluate this critique.

Glenn based his analysis on some hypothetical datasets that are cited here as Table 5.4 for purpose of illustration. These data potentially show very different A, P, and C effects. The values of the dependent variable in Table 5.4a show obvious age variations with an increment of 5 for each successive age, but seemingly no period or cohort effects. Glenn correctly pointed out that there could be some combination of A and offsetting P and C effects, and an infinite number of combinations of such effects can produce the pattern of variation. Data in Tables 5.4b and 5.4c show stronger period and cohort variations, respectively, and similarly can arise from many different combinations of underlying effects. In subsequent analyses, Glenn used the CGLIM approach to estimate APC models of these data. We present these results of the CGLIM analyses for data of Table 5.4a in Table 5.5. Corresponding analyses were conducted for the other two datasets but are not reported because

TABLE 5.4

Patterns of Data Showing Age, Period, and Cohort Effects (Glenn 2005)

Table 5.4a "Pure Age Effects" (Glenn 2005: Table 1.2)

Age	Year					
	1950	1960	1970	1980	1990	2000
20-29	50	50	50	50	50	50
30-39	55	55	55	55	55	55
40-49	60	60	60	60	60	60
50-59	65	65	65	65	65	65
60-69	70	70	70	70	70	70
70-79	75	75	75	75	75	75

Table 5.4b "Pure Period Effects" (Glenn 2005: Table 1.3)

Age	Year					
	1950	1960	1970	1980	1990	2000
20-29	30	35	40	45	50	55
30-39	30	35	40	45	50	55
40-49	30	35	40	45	50	55
50-59	30	35	40	45	50	55
60-69	30	35	40	45	50	55
70-79	30	35	40	45	50	55

Table 5.4c "Pure Cohort Effects" (Glenn 2005: Table 1.4)

Age	Year					
	1950	1960	1970	1980	1990	2000
20-29	50	55	60	65	70	75
30-39	45	50	55	60	65	70
40-49	40	45	50	55	60	65
50-59	35	40	45	50	55	60
60-69	30	35	40	45	50	55
70-79	25	30	35	40	45	50

Source: Norval Glenn (2005). *Cohort Analysis*, 2nd ed. Thousand Oaks, CA: Sage Publications.

Note: As noted by Glenn, data in Table 5.4a could show alternative true effects such as pure A effects, offsetting P and C effects, or a combination of A effects and offsetting P and C effects. The same applies to data in Tables 5.4b and 5.4c.

TABLE 5.5

Regression Coefficients of Linear APC Models of
Data in Table 5.4 Estimated by the CGLIM and IE

Variable	CGLIM				IE
	1	**2**	**3**	**4**	**5**
Intercept	62.5	62.5	62.5	62.5	62.5
Age					
20-29	−12.5	−12.5	0	**−5**	−11.2
30-39	−7.5	−7.5	0	**−5**	−6.7
40-49	−2.5	−2.5	**0**	−1.4	−2.2
50-59	2.5	2.5	**0**	1.1	2.2
60-69	7.5	7.5	0	3.6	6.7
70-79	12.5	12.5	0	6.4	11.2
Period					
1950	0	0	−12.5	**−5**	−1.3
1960	0	0	−7.5	**−5**	−0.8
1970	0	0	−2.5	−1.4	−0.3
1980	0	**0**	2.5	1.1	0.3
1990	0	**0**	7.5	3.6	0.8
2000	0	0	12.5	6.4	1.3
Birth Cohort					
1880	0	0	25	11.2	2.6
1890	0	0	20	10.1	2.1
1900	0	0	15	7.7	1.6
1910	0	0	10	5.3	1.1
1920	0	0	5	3	0.5
1930	0	0	0	0.2	0
1940	0	0	−5	−2	−0.5
1950	0	0	−10	−4.7	−1
1960	**0**	0	−15	−7.3	−1.6
1970	**0**	0	−20	−9.9	−2.1
1980	0	0	−25	−13.8	−2.6

Source: Data Modified from Norval Glenn, *Cohort Analysis*, 2nd ed (2005: Table 2.1) Thousand Oaks, CA: Sage Publications; the regression coefficients are centered to sum to zero within age, period, and cohort categories based on the usual constraint, $\Sigma_i \alpha_i = \Sigma_j \beta_j = \Sigma_k \gamma_k = 0$; coefficients highlighted in bold are constrained to be equal.

the results are similar. Models 1 to 4 using Glenn's results confirm the point made previously that different equality constraints result in drastically different estimates of A, P, and C effects. For purpose of comparison, we add model 5 estimated by the IE. There are two fallacies in Glenn's interpretation of results of models 1 to 4.

First, patterns in hypothetical data do not suggest underlying effects generating them. Glenn acknowledged that the data in Table 5.4a can be generated by any forms of the underlying A, P, and C effects, but then contradicted himself in the discussion of the modeling results, claiming that "For this simulation experiment, I know what the effects are and can apply the Mason et al. method to the data to see how well it performs" (p. 12). Specifically, he assigned the underlying effects to be the age effects shown in models 1 and 2 (although this guess is not wrong, as will be shown in the following by model selection analysis). Based on this logically incorrect starting point of what the underlying effects are, he went on to conclude that the APC accounting model gives "grossly incorrect results" using certain constraints (like those in models 3 and 4) and hence it is impossible to estimate APC effects with this model in practice when one cannot know what the right constraints are. It is clear from our simulation analysis shown previously that one can only examine the performance of model estimators by specifying the underlying effects that generate the data rather than using certain data to speculate what the underlying effects are.

Second, the assumption that the age, period, and cohort trends in any given set of data can be best estimated by full APC models rather than by reduced models of one or two of the three effects needs to be tested. If the effects of one or two of the three factors are null, the full model will statistically overfit the data and produce biased and inaccurate estimates. In addition, the full model has the model identification problem. The results shown in models 1 to 5 reflect precisely this problem. And, different linear constraints used to estimate the full model are bound to produce different estimates that are inaccurate in different degrees. Unlike in the simulation exercise, analysts cannot know which underlying effects are present and which are not, given only observed data.

One way to select among alternative models is to conduct model fit tests of whether all three of the A, P, and C effects are present and should be simultaneously estimated (see, e.g., Mason and Smith 1985). That is, analysts should successively estimate model with the A, P, C, AP, AC, PC, and APC sets of effect coefficients and examine the corresponding model fit statistics for improvement as additional sets and combinations of coefficients are added. This gives a sense of the relative importance of A, P, and C effects and the best model that summarizes the trends in the observed data. Accordingly, for the data of Table 5.4, we estimated nested models and computed model selection statistics—log likelihoods and Bayesian information criteria statistics (which are log likelihoods penalized for numbers of parameters estimated; see Raftery 1986, 1995)—for the three

TABLE 5.6

Model Fit Statistics for Data in Table 5.4

	For Data in Table 5.4a		
Models	Log-Likelihood	DF	BIC
A	**121.8**	**6**	**−107.5**
P	−128.3	6	2516.4
C	−115.8	11	1222.5
AP	123.1	11	−89.6
AC	132.6	16	−71.7
PC	132.6	16	−71.7
APC	122.6	20	−57.3

	For Data in Table 5.4b		
Models	Log-Likelihood	DF	BIC
A	−128.3	6	2517.6
P	**119.9**	**6**	**−107.5**
C	−115.8	11	1223.5
AP	121.4	11	−89.6
AC	125.4	16	−71.7
PC	129.8	16	−71.7
APC	131.5	20	−57.3

	For Data in Table 5.4c		
Models	Log-Likelihood	DF	BIC
A	−128.3	6	2518.3
P	−128.3	6	2516.9
C	**126.1**	**11**	**−89.6**
AP	**125.4**	**11**	**−89.6**
AC	133.5	16	−71.7
PC	129.4	16	−71.7
APC	124.4	20	−57.3

Source: Data Modified from Norval Glenn, *Cohort Analysis*, 2nd ed. Thousand Oaks, CA: Sage Publications.

Note: Model fit statistics, BIC (Bayesian Information Criterion), are calculated by Stata GLM; the smaller the AIC and BIC, the better the model fit. The best fitting models for each dataset are highlighted in bold.

sets of data. Applying the usual criterion of selecting models with the smallest values of these statistics, the results shown in Table 5.6 suggest that the best-fitting models for data sets one, two, and three are A effects only, P effects only, and C effects only or AP effects models, respectively. Because the full APC models are not the preferred models, the discussion of which identifying constraint gives the correct estimates is not productive and can be avoided given the findings from the model selection analysis.

Every statistical model has its limits and will break down under some conditions. We have found in additional simulation analysis one such condition in which the IE produces larger bias in small samples when the underlying effects are zero than when they are not zero. We also have shown, by analyses of Glenn's (2005) numerical example, that researchers should conduct careful model selection tests before using full APC models. *We reiterate that imposition of a full APC model on data when a reduced model fits the data equally well or better constitutes a model misspecification and should be avoided.* On the other hand, when model selection tests indicate that all three of the A, P, and C dimensions are operative in producing a given set of data, application of the IE may be quite useful in producing meaningful and stable estimates. And, MC simulation analysis is one important avenue for model validation. In conclusion, the IE performs well as a statistical estimator under most conditions. The results of the simulation studies can inform empirical studies that use the IE for statistical estimation of linear APC models.

5.6 Interpretation and Use of the Intrinsic Estimator

Given the desirable properties the IE possesses as a statistical estimator and its evident ability to produce valid estimates of A, P, and C effect coefficients from an underlying generating model, the question becomes one of how to interpret and use this estimator. The question of interpretation arises because the identifying constraint imposed by the IE on the unidentified APC accounting model parameter vector *b*—namely, projection onto the nonnull (column) space of the design matrix *X*—appears to be a constraint of purely algebraic convenience, devoid of substantive meaning. By contrast, conventional equality-constrained estimators of APC accounting models often are motivated by substantive hypotheses derived from theory or prior studies or side information about a process under study that indicate that certain coefficients are, say, of the same magnitude and hence can be constrained to be equal. Thus, the question becomes: How can the IE and

---

* An extreme example of patterns of effects of the A, P, and C temporal dimensions is the case in which the effect coefficients of two of the dimensions change linearly across their respective domains. In this case, one of the linear trends can be written as an exact algebraic linear transformation of the other. Accordingly, a reduced model with fewer than three dimensions is sufficient to model the data and the estimation of the full model with the IE will result in biased and inaccurate estimates of the effect coefficients. APC accounting model applications to real empirical data rarely would exhibit such an extreme linear degeneracy, but nonetheless can produce approximations thereto.

the Moore-Penrose generalized inverse matrix be used to statistically evaluate equality-constrained APC accounting models in which the identifying equality constraint is based on prior substantive theory or empirical research findings? Recall the *property of estimable constraints* stated in Section 5.5 to the effect that if the equality constraints are, in fact, valid, the constraints produce an estimable function of the unconstrained coefficient vector *b* and produce CGLIM estimates of APC effect coefficients that will be numerically close to those estimated by the IE. This property provides the key to how the IE can be used to evaluate substantively motivated equality constraints.

To focus the discussion, consider the effect coefficient estimates reported in Table 5.7 for the cancer mortality data for U.S. females, 1969–2007 (computational codes available online). The first two columns of the table report the coefficient estimates, standard errors (SEs), model deviance, and overdispersion coefficient for the IE applied to these data.[*] The next two columns report the corresponding estimates for a CGLIM model, which (for reasons that will be made clear) we label CGLIM 3, wherein the coefficients for the respective first categories of the A, P, and C groups are taken as the reference categories and have effects set to zero, and the identifying constraint is that the second birth cohort (C2) is constrained to have the same effect coefficient (zero) as the first cohort (C1). The bottom rows give the overall model fit statistics for these two models. Previously, we noted that all just-identified models that incorporate effect coefficients for the full array of A, P, and C categories will fit the data equally well. This is evident in the model fit statistics here. Thus, to reiterate the point made previously, one cannot use fit statistics to discriminate among just-identified models. Rather, some other criterion must be employed. The criterion applied here is that the constrained vector must be estimable.

The numerical estimates in Table 5.7 show that the constraint imposed in the CGLIM 3 model (the effect of the 1885 cohort constrained to equal that of the 1890 cohort) is statistically valid (i.e., within sampling variability) for these data—the estimated effect for the 1885 cohort in the IE column is .69, and that for the 1890 cohort is .66 with the SE of the former being .08 and latter being .04. Thus, the difference of these two effect coefficient estimates, .04, is well within sampling error and therefore effectively zero. In other words, this equality constraint produces an estimable function in a statistical sense (to be made precise in the following discussion). The consequence is that the coefficient vectors for these two models are statistically identical up to a centering or normalizing transformation. This equivalence is demonstrated numerically in the last column of Table 5.7, which gives the numerical values of the corresponding recentered/renormalized CGLIM 3

---

[*] For both the CGLIM and IE models, the Poisson variance function is assumed to have a multiplicative overdispersion factor that can be estimated by dividing the deviance by degrees of freedom (DF). And, the overdispersion coefficients were estimated using the quasi-likelihood method (McCullagh and Nelder 1989).

TABLE 5.7

IE and CGLIM Estimates, U.S. Female Cancer Mortality, 1969–2007

	Intrinsic Estimator		CGLIM 3 (C1 = C2)		Centered CGLIM 1 (A1 = A2)	Centered CGLIM 2 (P1 = P2)	Centered CGLIM 3 (C1 = C2)
	Effect	S.E.	Effect	S.E.	Effect	Effect	Effect
Intercept	−5.39	0.01	−7.74	1.23	−5.39	−5.39	−5.39
Age							
20-24	−2.21	0.04	0.00	0.00	1.05	−2.73	−2.45
25-29	−1.70	0.03	0.54	0.11	1.05	−2.15	−1.91
30-34	−1.25	0.02	1.03	0.21	1.00	−1.62	−1.42
35-39	−0.80	0.02	1.52	0.31	0.95	−1.09	−0.93
40-44	−0.36	0.02	2.00	0.41	0.90	−0.56	−0.45
45-49	0.01	0.01	2.41	0.51	0.77	−0.11	−0.04
50-54	0.28	0.01	2.71	0.61	0.53	0.23	0.26
55-59	0.50	0.01	2.97	0.71	0.25	0.54	0.52
60-64	0.70	0.01	3.21	0.81	−0.05	0.83	0.76
65-69	0.87	0.01	3.41	0.91	−0.39	1.07	0.96
70-74	0.97	0.01	3.55	1.01	−0.79	1.25	1.10
75-79	1.03	0.01	3.65	1.11	−1.23	1.39	1.20
80-84	1.03	0.01	3.69	1.21	−1.73	1.47	1.23
85+	0.94	0.01	3.63	1.31	−2.32	1.47	1.18
Period							
1969-74	−0.34	0.02	0.00	0.00	−2.10	−0.06	−0.21
1975-79	−0.26	0.01	0.04	0.10	−1.51	−0.06	−0.16
1980-84	−0.15	0.01	0.12	0.20	−0.90	−0.02	−0.09
1985-89	0.00	0.01	0.22	0.30	−0.25	0.04	0.02
1990-94	0.07	0.01	0.26	0.40	0.32	0.03	0.05
1995-99	0.16	0.01	0.31	0.50	0.91	0.04	0.11
2000-04	0.23	0.01	0.34	0.60	1.48	0.03	0.14
2005-07	0.28	0.01	0.36	0.70	2.04	0.00	0.15
Cohort							
1885	0.69	0.08	0.00	0.00	5.71	−0.12	0.32
1890	0.66	0.04	0.00	0.00	5.17	−0.08	0.32
1895	0.56	0.03	−0.06	0.13	4.57	−0.09	0.26
1900	0.47	0.02	−0.11	0.22	3.98	−0.10	0.20
1905	0.39	0.02	−0.15	0.32	3.40	−0.09	0.17
1910	0.34	0.02	−0.16	0.42	2.85	−0.06	0.15
1915	0.29	0.01	−0.18	0.52	2.30	−0.04	0.14
1920	0.22	0.01	−0.21	0.62	1.73	−0.02	0.11
1925	0.16	0.01	−0.23	0.72	1.17	0.00	0.09

continued

TABLE 5.7 (continued)

IE and CGLIM Estimates, U.S. Female Cancer Mortality, 1969–2007

	Intrinsic Estimator		CGLIM 3 (C1 = C2)		Centered CGLIM 1 (A1 = A2)	Centered CGLIM 2 (P1 = P2)	Centered CGLIM 3 (C1 = C2)
	Effect	S.E.	Effect	S.E.	Effect	Effect	Effect
1930	0.08	0.01	−0.28	0.83	0.58	−0.01	0.04
1935	0.00	0.01	−0.31	0.93	0.00	0.00	0.00
1940	−0.08	0.02	−0.35	1.03	−0.58	0.01	−0.04
1945	−0.16	0.02	−0.40	1.13	−1.17	0.00	−0.09
1950	−0.27	0.02	−0.47	1.23	−1.77	−0.02	−0.15
1955	−0.34	0.02	−0.50	1.33	−2.34	−0.01	−0.19
1960	−0.40	0.02	−0.52	1.43	−2.90	0.01	−0.21
1965	−0.45	0.02	−0.54	1.53	−3.46	0.04	−0.22
1970	−0.48	0.03	−0.53	1.63	−3.99	0.09	−0.21
1975	−0.52	0.03	−0.54	1.74	−4.53	0.13	−0.22
1980	−0.60	0.05	−0.58	1.84	−5.11	0.13	−0.26
1985	−0.58	0.09	−0.52	1.94	−5.60	0.23	−0.21
t ratio					9.59	−3.26	−0.37
p-value					<0.001	0.001	0.712
Deviance		1825.3		1825.3	1825.3	1825.3	1825.3
DF		72		72	72	72	72
Overdispersion		25.4		25.4	25.4	25.4	25.4

effect coefficients, that is, the CGLIM 3 APC coefficients transformed by subtracting their respective group means so that the transformed coefficients sum to zero.* The resulting recentered CGLIM 3 and IE effect coefficients generally agree. The patterns of the respective groups of A and P effects are more similar than those of C effects, which reflects the impact of the cohort equality constraint used by CGLIM3.

---

* CGLIM coefficients can be transformed as follows to the normalization that A, P, C effects each sum to zero: Under the original CGLIM normalization, let a_i be the estimated A effect for age i, let p_j be the estimated P effect for period j, and let c_k be the estimated C effect for cohort k. Let d be the estimated intercept. Transforming these estimates to a different normalization means that we want to find new coefficients a_i', p_j', c_k' and d' such that (1) the predicted values for each data point do not change, and (2) the A, P, and C effects each sum to zero. The solution is to subtract the mean of the original A effects from each a_i to obtain a new A effect a_i' (this guarantees that the new A effects sum to zero) but then add the mean of the original A effects to the intercept, which guarantees that the predicted values do not change. The same process can be used to transform the P and C effects. One can compute the corresponding standard errors using the variance of the new or transformed design matrix, $X^{new} = LX$, where L is the normalizing matrix that transforms the X into the form introduced previously. That is, the standard errors for $X^{new} = sqrt(diag(cov(X^{new})))$, where $cov(X^{new}) = Lcov(X)L^T$ is the new variance-covariance matrix. As stated elsewhere in the chapter, such linear transformation only makes the coefficients more comparable across different methods of estimation in this methodological exposition and does not change the substantive findings of A, P, and C trends.

Other possible identifying assumptions do not fare so well. To illustrate this, Table 5.7 reports comparable effect coefficient estimates for two alternative CGLIM models: CGLIM 1, which identifies the model by constraining the effect coefficient for the group aged 25–29 (A2) to be the same as that of the group aged 20–24 (A1); and CGLIM 2, which achieves identification by constraining the effect coefficient for the 1975–79 time period (P2) to equal that for the 1969–74 period (P1). Comparing all three sets of CGLIM coefficients, it can be seen that those produced by the CGLIM 1 and CGLIM 2 models bracket those produced by the CGLIM 3 model; that is, the two alternative CGLIM models yield effect coefficients that diverge substantially from those given by the CGLIM 3 model. The divergences are substantial and dramatic. The reason for this behavior is that the equality constraints used to produce the CGLIM 1 and CGLIM 2 models do not produce statistically estimable functions and corresponding coefficient estimates.

To proceed more systematically, a method is needed for assessing whether two estimated coefficient vectors are within sampling error of being equal. There are several ways to accomplish this. We have compared centered CGLIM effect coefficients *element by element* with the corresponding estimated IE effect coefficients to determine whether the former are within, say, two standard errors of the latter. Using the estimated standard errors of the IE coefficients, it can be seen that most of the centered effect coefficients for the CGLIM 3 model are within two standard errors of the IE coefficients. By comparison, this is not the case for the centered effect coefficients of the CGLIM 1 and CGLIM 2 models. An alternative procedure is to define a test statistic based on the *entire vector of coefficients*. To do so, we state the null hypothesis:

$$H_0 : E(\hat{b}^T B_0) = (B + sB_0)^T B_0 = s = 0 \tag{5.10}$$

In words, this null hypothesis is that the expected value of the product of the estimated and re-normalized CGLIM vector $\hat{b}$ and the eigenvector B_0 that is fixed by the design matrix is zero. Due to the orthogonality of the vectors B and B_0, this, as Equation (5.10) indicates, is equivalent to the hypothesis that the expected valued of the scalar s is equal to zero. Using the geometric projection illustrated in Figure 5.1, this test is equivalent to testing whether the estimated parameter vector under a given set of constraints (b_1 or b_2 or b_3) lies significantly far away from the estimable function b_0 so that one can infer that its horizontal projection results from a nonzero s.

To specify a test for this null hypothesis, we build on a well-known asymptotic distribution property of the maximum likelihood estimator (MLE) used to estimate $\hat{b}$: Under broad regularity conditions, as sample size or the number of time periods of data increases, the MLE of $\hat{b}$ is consistent and asymptotically normally distributed (McCulloch and Searle 2001: 306). This property facilitates the definition of an asymptotic t test for the null hypothesis (15) as

$$t = \frac{s-0}{se(s)} = \frac{s}{se(s)} \tag{5.11}$$

where $se(s)$ denotes the estimated (asymptotic) standard error of the scalar s. To obtain this test statistic, note that the numerator s can be computed, as indicated in Equation (5.10), by calculating the product of the vectors $\hat{b}$ and B_0. Then, the denominator can be computed by transforming the asymptotic variance-covariance matrix $\hat{\Sigma}^{-1}$ (i.e., the inverse of the Fisher information matrix) that is obtained in the process of estimating b by maximum likelihood to obtain $\hat{b}$. Because the vector B_0 is orthogonal to the design matrix [see Equation (5.2)], the variance of the scalar s can be computed by imposition of the usual quadratic form transformation applied to obtain the variance of the restricted MLE from the MLE, namely, $B_0^T \hat{\Sigma}^{-1} B_0$.* Taking the square root of this transformation yields an estimate of the standard deviation of $s = \hat{b}^T B_0$, which, when divided by the degrees of freedom of the model, $df = ap - (1 + (a-1) + (p-1) + (a+p-2))$, produces an estimate of the standard error of s in the denominator of Equation (5.11).

Applied to the three alternative CGLIM models for which the renormalized coefficients are given in Table 5.7, we obtain the following t ratios: 9.59 ($p < 0.001$) for CGLIM 1, -3.26 ($p = 0.001$) for CGLIM 2, and -0.37 ($p = 0.712$) for CGLIM 3. The results show that only the CGLIM 3 models estimates are not statistically different from those of the IE, but the CGLIM1 and CGLIM 2 model estimates are. Reiterating the point made above, the results of applying the asymptotic t test to the CGLIM1, CGlIM2, and CGLIM3 models show that estimators that incorporate significant components of the B_0 vector (and thus the design matrix) may give misleading results concerning the patterns of change of the estimated effect coefficients across the age, period, and cohort categories. Significant t ratios indicate statistically significant departures from the IE coefficients and patterns of change that depart substantially from those of the IE. The much larger t ratio produced by CGLIM1 means that equating age effects is a particularly inadequate constraint to use. This statistical result is perfectly consistent with the strong age gradient in cancer mortality rates documented in biodemographic and epidemiologic studies.

In brief, the foregoing test for statistical estimability leads to new avenues for using and interpreting the IE as applied to a particular dataset. On the one hand, an analyst can apply the IE in an *exploratory data analysis* manner, in

---

* Formally, the asymptotic normality property of the MLE yields $\hat{b} \stackrel{asy}{\sim} N(b, \Sigma^{-1})$, that is, the sampling distribution of the estimated coefficient vector of a model that is identified by imposition of a theoretically motivated equality constraint on two or more coefficients is asymptotically multivariate normal with a mean (expected) parameter vector b and a variance-covariance matrix Σ^{-1}, where Σ^{-1} is estimated by $\hat{\Sigma}^{-1}$, the inverse of the Fisher information matrix. By properties of the MLE (see, e.g., McCulloch and Searle 2001: 309), a linear transformation of $\hat{b}$ using B_0 then yields $\hat{b}_0^T B_0 \stackrel{asy}{\sim} N(B_0^T \hat{\Sigma}^{-1} B_0)$, where $\hat{b}^T B_0 = s$. From this, the asymptotic standard error of s can be computed as the square root of $B_0^T \hat{\Sigma}^{-1} B_0$ divided by degrees of freedom as indicated in the text.

which the objective is to ascertain good estimates of the patterns of A, P, and C effects in a table or set of tables of rates.[*] In such an exercise, the analyst does not approach the data with strong prior notions about particular patterns of effects that should be evident in the data, but rather seeks to let the intrinsic patterns emerge from application of the IE, taking advantage of the fact that the IE is an estimable function and thus has desirable statistical properties.

On the other hand, using the definition of statistical estimability and the test described, an analyst can use the IE in a more *confirmatory data analysis* manner, in which a vector of effect coefficients estimated from application of the IE is used as a benchmark to assess whether a corresponding vector of coefficients estimated from the imposition of one or more theoretically or substantively motivated constraints to achieve model identification is acceptable. In this type of exercise, an analyst approaches a table or tables of rates with a definite hypothesis or set of hypotheses about the underlying A, P, and C effects that generated the data. The analyst can use the vector of effect coefficients estimated by the IE to assess the empirical plausibility of the hypotheses. In this way, the definition of statistical estimability and the test described directly address a criticism that often has been lodged against general-purpose methods of APC analysis, namely, that they provide no avenue for testing specific substantive hypotheses and thus are mere devices of algebraic convenience that may be misleading.

Appendix 5.1: Proof of Unbiasedness of the IE as an Estimator of the $b_0 = P_{proj}b$ Constrained APC Coefficient Vector

This proof is for the linear model form of the APC accounting model. The same results apply, however, for generalized linear models through the Fisher information matrix as the linear part of the models remains the same.

We begin by citing the PCR formulation of the IE noted previously as one method of estimation of the IE. Denote by m the number of eigenvalues (including the zero eigenvalue) of the matrix $X^T X$ in the solution to normal equations (4.7) in Chapter 4 and denote by r the rank of the $X^T X$ matrix. A general property of the principal components estimator (Sen and Srivastava 1990: 256) in the linear model is that the bias in the principal components estimator of the regression coefficient vector β induced by the deletion of $m - r$ variables in the regression model (corresponding to eigenvalues equal to zero or nearly equal to zero) is $Q_{(m-r)}\beta_{(m-r)}$. In this expression, $Q_{(m-r)}$ denotes a $(m - r) \times (m - r)$ matrix, the columns of which are the eigenvectors corresponding to the $m - r$ eigenvalues that are zero or near zero, and $\beta_{(m-r)}$ denotes

---

[*] The distinction between "exploratory" and "confirmatory" data analysis dates in statistics at least to the classic work of Mosteller and Tukey (1977).

the regression coefficients of these eigenvectors. In the present case of the APC regression model (4.5) in Chapter 4, $m - r = 1$, $Q_{(m-r)} = B_0$, and $\beta_{(m-r)} = s$. Since $s = 0$ in the IE, the unbiased property applies here. Furthermore, since an orthonormal linear transformation of an unbiased estimator (which, as described in the text, is used in the principal components approach to estimation to transform the PCR estimator to the special parameterization or linear function b_0 of the parameter vector b that is estimated by the IE) remains unbiased, it follows that the IE estimate obtained by this algorithm is unbiased.

Appendix 5.2: Proof of Relative Efficiency of the IE as an Estimator of the $b_0 = P_{proj}b$ Constrained APC Coefficient Vector

We prove the theorem only for linear models as the same results can be derived for generalized linear models through the Fisher information matrix. Recall that X denotes the design matrix of the APC multiple classification model (4.5) in Chapter 4. Let x denote the row vector with elements $-1, 0,$ or 1 corresponding to the identifying constraint on the parameters in β necessary to achieve a unique estimator. Let W denote the augmented design matrix in which X has been expanded by the constraint vector x, so that we can write $W^T = [X^T | x^T]$. Thus, $W^T W$ is invertible, and by linear model theory, the variance-covariance matrix for the constraint estimator is $var(\hat{b}) = \sigma^2 (W^T W)^{-1}$, where σ^2 denotes the variance of the model random error term.

Let U be the orthonormal matrix such that the matrix $X^T X$ is diagonalized, that is, $X^T X = U \Lambda U^T$, where $\Lambda = diag\ [\lambda_1, \ldots, \lambda_{m-1}, \lambda_m]$ with $\lambda_1, \geq \ldots \geq \lambda_{m-1} > 0$ and $\lambda_m = 0$ being the eigenvalues of matrix $X^T X$. Let $\Lambda_1 = diag\ [\lambda_1, \ldots, \lambda_{m-1}]$. Since the IE is a special principal component estimator with the principal components $\lambda_1 \geq \ldots \geq \lambda_{m-1} > 0$, by linear model theory

$$var(B) = \sigma^2 U \Lambda^G U^T = \sigma^2 U \begin{pmatrix} \Lambda_1^{-1} & \\ & 0 \end{pmatrix} U^T \tag{5.12}$$

where Λ^G denotes the generalized inverse. Note also that $W^T W = X^T X + x^T x = U(\Lambda + z^T z)U^T$, where $z = xU$.

By the principal component decomposition of the CGLIM estimator $\hat{b} = B + B_1$, B and B_1 are orthogonal in the parameter space. Since Λ_1 corresponds to the variance of B, $z^T z$ corresponds to the variance of B_1. Therefore, letting $z^T z = diag\ [0, \ldots, 0, c]$ with a constant $c > 0$, we have

$$var(\hat{b}) - var(B) = \sigma^2 \left[(W^T W)^{-1} - U \begin{pmatrix} \Lambda^{-1} & \\ & 0 \end{pmatrix} U^T \right] = \sigma^2 c^{-1} U_m U_m^T \tag{5.13}$$

is positive definite for nontrivial constraint z with $c > 0$, where U_m is the mth column vector of matrix U, which is the eigenvector of matrix X^TX with eigenvalue 0.

Appendix 5.3: IE as a Minimum Norm Quadratic Unbiased Estimator of the $b_0 = P_{proj}b$ Constrained APC Coefficient Vector

In the context of the estimation of the $b_0 = P_{proj}b$ constrained coefficient vector of the APC accounting/multiple classification model for a fixed number of time periods p of data, the unbiasedness and relative efficiency of the IE can be combined with the numerical properties of the IE in the following two properties.

Property 1: The IE is a well-defined numerical solution to problem of matrix inversion in the rank deficient linear equation system defined by the APC accounting/ multiple classification model. As noted, for example, by Girosi and King (2008: 237) in discussing the Moore-Penrose generalized inverse solution to rank deficient linear systems: "There is always a well-defined solution that lies in [the nonnull space of a rank deficient matrix A], that is, a solution whose projection on the null space is zero. We take this as the 'representative' solution." This is the solution that defines the IE.

It follows from this property that any solution of the rank deficient linear equation system of the APC accounting model that incorporates elements of the null space of the system is not a well-defined numerical solution. Does this mean that no estimator of the APC accounting model coefficient vector b obtained by placing an equality constraint on two coefficients is a well-defined solution to the system of equations defined by the model? No. The reason is that such an estimated coefficient vector can be estimable in a statistical sense; see the definition of statistical estimability in Section 5.6.

Property 2: The IE is a MINQUE of the projection $b_0 = Proj\ b$ of the unconstrained APC multiple accounting model coefficient vector b onto the nonnull space of the design matrix X. The MINQUE method of statistical estimation was proposed and developed by the esteemed statistician C. R. Rao (1970, 1971, 1972; see also Jiang 2007: 28–29) in a series of papers published at about the time the Mason et al. (1973) article articulated the APC accounting/multiple classification model and its identification problem. Concerning the IE as a MINQUE, note that we have proven in Appendices 5.1 and 5.2 that for a fixed number of time periods of data, the IE is unbiased and has minimum variance as an estimator of the coefficient vector b_0. The definition of the variance of estimators is equivalent to choosing a quadratic norm in the underlying Euclidean vector space. And, as shown by Girosi and King (2008: 238), "The solution [to the rank deficient system of linear equations $Ax = y$] whose projection in the null space of A is 0 and the solution of minimum [quadratic] norm coincide." Since the projection of the IE onto the null space of the design matrix X in the APC accounting/multiple classification model is 0, it follows that it also is a MINQUE estimator.

Appendix 5.4: Interpreting the Intrinsic Estimator, Its Relationship to Other Constrained Estimators in APC Accounting Models, and Limits on Its Empirical Applicability

Recently, O'Brien (2011a) discussed constrained estimators in the APC accounting model and the IE in particular. Fu, Land, and Yang (2011) similarly discussed these topics and commented on the work of O'Brien (2011a), to which O'Brien (2011b) responded. Because of the relevance of this exchange to the contents of this chapter, we summarize and comment on these articles.

O'Brien (2011a) showed algebraically how each just-identifying equality constraint on the coefficient vector of the APC accounting model is associated with a specific generalized inverse matrix that produces the least squares estimator of the resulting constrained coefficient vector. He also discussed the geometry of constrained estimators in terms of solutions to the unidentified equation system corresponding to various constraints and generalized inverses being orthogonal to constraints, solutions to various constraints all lying on a single line in multidimensional space, the distance on that line between various solutions, and the crucial role of the null vector. This then is used to characterize properties of all constrained estimators and those that are unique to the IE. Among other things, O'Brien (2011a) showed that the identifying constraint used by the IE, which results in the Moore-Penrose generalized inverse solution and the various desirable statistical properties cited in this chapter and the appendices, produces the "most representative" among all possible solutions corresponding to all possible just-identifying equality constraints. As Fu, Land, and Yang (2011) and O'Brien (2011b) pointed out, these are useful illustrations and derivations of a number of properties of all constrained estimators and the IE on which there is much agreement. O'Brien (2011b) also identified some disagreements, on which we now comment.

First, as noted in Section 5.3.1, many of the articles on the identification and estimation problem published in the 1970s and 1980s were based on the presumption that the objective was to estimate the "true" or "generating" parameters, that is, those in the unidentified and unconstrained regression coefficient vector of the APC multiple classification model that generated the observed table of rates or proportions. This objective also motivated O'Brien (2011a: 435): "The crucial question, from our perspective, is whether the parameter estimates are unbiased in the sense that their expected values equal the values of the parameters that generated *Y*." As pointed out by Fu, Land, and Yang (2011), however, data on an outcome variable from a regression model with linearly dependent covariates can be generated by multiple sets of generating parameters. Algebraically, in the case of the APC accounting model, this corresponds to the existence of an infinite set of generalized inverses that solve the normal equations for the unconstrained coefficient vector of the APC accounting model—geometrically illustrated by the "line of solutions" in O'Brien's (2011a) figures.

In face of this, as stated in the various articles by Yang, Fu, and Land and their associates, and as stated and emphasized in this chapter, the objective of the IE is not to estimate the unidentifiable regression coefficient vector but rather its projection to the nonnull space of the design matrix, an estimable function. As Tu, Kramer, and Lee (2012: 592) stated, the IE and related estimators are based on the mathematical relation within rows/columns in $X^T X$ or $X^T y$, and "this relation is a natural consequence of the mathematical relations among" the A, P, and C temporal dimensions in the classical APC accounting model. After this projected coefficient vector is estimated, it then can be used, as shown in Section 5.5, to statistically assess the probability that substantively motivated coefficient equality constraints result in an estimated APC regression coefficient vector that also is estimable. This then achieves the long-standing search, and that of O'Brien (2011a), for estimates of a vector of estimated effect coefficients of the APC accounting model that are based on prior research or theory. In addition, the properties of the IE and the Moore-Penrose generalized inverse matrix associated therewith that have been described in this chapter and appendices will more often than not produce useful and replicable empirical estimates of the trends across the age, period, and cohort temporal dimensions of the APC accounting model.

Second, in Section 5.3.1, we showed that the identifying constraint of the IE satisfies the estimability condition derived by Kupper et al. (1985) for the APC accounting model and thus conclude that the IE is an estimable function of the unconstrained coefficient vector *b*. Yet, O'Brien (2011b: 468) stated that the IE "is not an estimable function in the traditional sense, since it does not meet the necessary and sufficient condition for an estimable function described by Searle (1971: 185)." This statement disregarded the fact that the Kupper et al. estimability condition was derived by application of the traditional definition and criteria for estimable functions to the APC accounting model. It also disregarded Searle's (1971: 180) characterization of an estimable function as "a linear function of the parameters for which an estimator can be found from [any generalized inverse solution to the normal equations] that is invariant to whatever solution of the normal equations is used." As the exposition in Section 5.3 clearly showed, the IE has this invariance property.

Third, O'Brien (2011a: 440) was "skeptical" of the asymptotic properties of the IE described in Section 5.3.2 that imply that the expected value of any just-identified constrained estimator converges in value to the expected value of the IE as the number of time periods or age groups increases without bound.

In support of this skepticism, O'Brien (2011a) reported results of an analysis of a simulation of a specific numerical specification of an APC accounting model. As stated by Fu, Land, and Yang (2011: 464), however, even this simulation illustrates the convergence properties proven mathematically in the articles cited in Section 5.3.2. Note also that the numerical simulations reported in Section 5.5 exhibited these asymptotic convergence properties. In addition, Fu, Land, and Yang (2011: 463) noted that the nonconstancy in the scalar coefficient that O'Brien (2011a) found to vary with differing numbers of time periods of data is due to a different parameterization from that used to mathematically prove the asymptotic properties of the IE.

In the end, as emphasized in this chapter and elsewhere in this book, the utility of the constraint imposed on the APC accounting model by the IE and the corresponding Moore-Penrose generalized inverse matrix solution of the deficient normal equations of the model will be determined by their ability to meet Glenn's (2005) criteria as cited in Section 5.2, that is, to yield approximately correct estimates more often than not in empirical applications. As concerns this criterion, note again the empirical analyses reported in Sections 5.4 and 5.5 and in Chapter 8 that show similarities of inferences regarding the temporal patterns of age, period, and cohort effect when those estimated by the IE are compared with those estimated with the very different HAPC model that is the subject of Chapters 7 and 8. In addition, using data on U.S. homicide arrest rates, Fu, Land, and Yang (2011: 461–462) reported estimates of the temporal trends of the age, period, and cohort effect coefficients by the IE that are quite consistent with those estimated by O'Brien (2000) using an APC characteristics (APCC) model analysis (the APCC model is a variant of the proxy variable approach to identification and estimation of the APC accounting model; see Section 4.4.4 of Chapter 4). In brief, the prior empirical findings regarding these temporal patterns by O'Brien using a proxy-variable-based APC model he previously developed are replicable by the IE. Again, however, we do not claim that the IE is a universal solution to the APC accounting model identification and estimation problem. As emphasized in Section 5.5.2, every statistical model will break down under some circumstances, and the IE is no exception.

References

Alwin, D. F., and R. J. McCammon. 1999. Aging versus cohort interpretations of intercohort differences in GSS vocabulary scores. *American Sociological Review* 64:272–286.

Casella, G., and R. L. Berger. 2001. *Statistical inference*. 2nd ed. Pacific Grove, CA: Duxbury Press.

Clayton, D., and E. Schifflers. 1987. Models for temporal variation in cancer rates. II: Age-period-cohort models. *Statistics in Medicine* 6:469–481.

Frenk, S. M., Y. Yang, and K. C. Land. 2012. *Assessing the significance of cohort and period effects in hierarchical age-period-cohort models with applications to verbal test scores and voter turnout in U.S. Presidential elections.* Under review.

Fu, W. J. 2000. Ridge estimator in singular design with application to age-period-cohort analysis of disease rates. *Communications in Statistics—Theory and Methods* 29:263–278.

Fu, W. J. 2008. A smoothing cohort model in age-period-cohort analysis with applications to homicide arrest rates and lung cancer mortality rates. *Sociological Methods and Research* 36:327–361.

Fu, W. J., and P. Hall. 2006. Asymptotic properties of estimators in age-period-cohort analysis. *Statistics & Probability Letters* 76:1925–1929.

Fu, W. J., P. Hall, and T. Rohan. 2004. Age-period-cohort analysis: Structure of estimators, estimability, sensitivity, and asymptotics. Michigan State University, Department of Epidemiology. Research paper.

Fu, W. J., K. C. Land, and Y. Yang. 2011. On the intrinsic estimator and constrained estimators in age-period-cohort models. *Sociological Methods & Research* 40:453–466.

Girosi, F., and G. King. 2008. *Demographic forecasting.* Princeton, NJ: Princeton University Press.

Glenn, N. D. 1999. Further discussion of the evidence for an intercohort decline in education-adjusted vocabulary. *American Sociological Review* 64:267–271.

Glenn, N. D. 2005. *Cohort analysis.* 2nd ed. Thousand Oaks, CA: Sage.

Hoerl, A. E., and Kennard, R. W. 1970a. Ridge regression: Applications to nonorthogonal problems. *Technometrics* 12:69–82.

Hoerl, A. E., and Kennard, R. W. 1970b. Ridge regression: Biased estimation for nonorthogonal problems. *Technometrics* 12:55–67.

Holford, T. R. 1985. An alternative approach to statistical age-period-cohort analysis. *Journal of Chronic Diseases* 38:831–836.

Holford, T. R. 1991. Understanding the effects of age, period, and cohort on incidence and mortality rates. *Annual Review of Public Health* 12:425–457.

Holford, T. R. 2005. Age-period-cohort analysis. In *Encyclopedia of biostatistics*, ed. P. Armitage and T. Colton, 82–99. Hoboken, NJ: Wiley.

Jiang, J. 2007. *Linear and generalized linear mixed models and their applications.* New York: Springer-Verlag.

Kupper, L. L., J. M. Janis, A. Karmous, and B. G. Greenberg. 1985. Statistical age-period-cohort analysis: A review and critique. *Journal of Chronic Diseases* 38:811–830.

Kupper, L. L., J. M. Janis, I. A. Salama, C. N. Yoshizawa, B. G. Greenberg, and H. H. Winsborough. 1983. Age-period-cohort analysis: An illustration of the problems in assessing interaction in one observation per cell data. *Communications in Statistics—Theory and Methods* 12:201–217.

Marion, J. B., and S. T. Thornton. 1995. *Classical dynamics of particles and systems.* 4th ed. Fort Worth, TX: Harcourt Brace College.

Mason, K. O., W. M. Mason, H. H. Winsborough, and W. Kenneth Poole. 1973. Some methodological issues in cohort analysis of archival data. *American Sociological Review* 38:242–258.

Mason, W. M., and H. L. Smith. 1985. Age-period-cohort analysis and the study of deaths from pulmonary tuberculosis. In *Cohort analysis in social research: Beyond the identification problem,* ed. W. M. Mason and S. E. Fienberg, 151–228. New York: Springer-Verlag.

McCullagh, P., and J. A. Nelder. 1989. *Generalized linear models.* 2nd ed. Boca Raton, FL: CRC Press.

McCulloch, C. E., and S. R. Searle. 2001. *Generalized, linear, and mixed models.* New York: Wiley-Interscience.

Miech, R., and S. Koester. 2012. Trends in U.S. past-year marijuana use from 1985–2009: An age-period-cohort analysis. *Drug and Alcohol Dependence* 124:259–267.

Miech, R., S. Koester, and B. Dorsey-Holliman. 2011. Increasing U.S. mortality due to accidental poisoning: The role of the baby boom cohort. *Addiction* 106:806–815.

Mosteller, F., and J. W. Tukey. 1977. *Exploratory data analysis.* Reading, MA: Addison-Wesley.

O'Brien, R. M. 2000. Age Period cohort characteristic models. *Social Science Research* 29:123–139.

O'Brien, R. M. 2011a. Constrained estimators and age-period-cohort models. *Sociological Methods and Research* 40:419–452.

O'Brien, R. M. 2011b. Intrinsic estimators as constrained estimators in age-period-cohort accounting models. *Sociological Methods and Research* 40:467–470.

Pullum, T. W. 1978. Parametrizing age, period, and cohort effects: An application to U.S. delinquency rates, 1964–1973. *Sociological Methodology* 9:116–140.

Pullum, T. W. 1980. Separating age, period, and cohort effects in white U.S. fertility, 1920–1970. *Social Science Research* 9:225–244.

Raftery, A. E. 1986. Choosing models for cross-classifications. *American Sociological Review* 51:145–146.

Raftery, A. E. 1995. Bayesian model selection in social research. *Sociological Methodology* 25:111–164.

Rao, C. R. 1970. Estimation of heteroscedastic variances in linear models. *Journal of the American Statistical Association* 65:161–172.

Rao, C. R. 1971. Estimation of variance and covariance components—MINQUE theory. *Journal of Multivariate Analysis* 1:257–275.

Rao, C. R. 1972. Estimation of variance and covariance components in linear models. *Journal of the American Statistical Association* 67:112–115.

Robertson, C., S. Gandini, and P. Boyle. 1999. Age-period-cohort models: A comparative study of available methodologies. *Journal of Clinical Epidemiology* 52:569–583.

Rodgers, W. L. 1982a. Estimable functions of age, period, and cohort effects. *American Sociological Review* 47:774–787.

Rodgers, W. L. 1982b. Reply to comment by Smith, Mason, and Fienberg. *American Sociological Review* 47:793–796.

Schwadel, P. 2011. Age, period, and cohort effects on religious activities and beliefs. *Social Science Research* 40:181–192.

Searle, S. R. 1971. *Linear models.* New York: Wiley.

Sen, A., and M. Srivastava. 1990. *Regression analysis: Theory, methods and applications.* New York: Springer-Verlag.

Smith, H. L., W. M. Mason, and S. E. Fienberg. 1982. Estimable functions of age, period, and cohort effects: More chimeras of the age-period-cohort accounting framework: Comment on Rodgers. *American Sociological Review* 47:787–793.

Tu, Y.-K., N. Krämer, and W.-C. Lee. 2012. Addressing the identification problem in age-period-cohort analysis: A tutorial on the use of partial least squares and principal components analysis. *Epidemiology* 23:583–593.

Wilson, J. A., and W. R. Gove. 1999. The intercohort decline in verbal ability: Does it exist? *American Sociological Review* 64:253–266.

Winkler, R., J. Huck, and K. Warnke. 2009. Deer hunter demography: Age, period, and cohort analysis of trends in hunter participation in Wisconsin. University of Wisconsin-Madison: Applied Population Lab. Retrieved from http://paa2009.princeton.edu/papers/91178

Yang, Y. 2006. Bayesian inference for hierarchical age-period-cohort models of repeated cross-section survey data. *Sociological Methodology* 36:39–74.

Yang, Y. 2008. Trends in U.S. adult chronic disease mortality, 1960–1999: Age, period, and cohort variations. *Demography* 45:387–416.

Yang, Y., W. J. Fu, and K. C. Land. 2004. A methodological comparison of age-period-cohort models: The intrinsic estimator and conventional generalized linear models. *Sociological Methodology* 34:75–110.

Yang, Y., and K. C. Land. 2006. A mixed models approach to the age-period-cohort analysis of repeated cross-section surveys, with an application to data on trends in verbal test scores. *Sociological Methodology* 36:75–97.

Yang, Y., S. Schulhofer-Wohl, W. J. Fu, and K. C. Land. 2008. The intrinsic estimator for age-period-cohort analysis: What it is and how to use it. *American Journal of Sociology* 113:1697–1736.

Yang, Y., S. Schulhofer-Wohl, and K. C. Land. 2007. A simulation study of intrinsic estimator for age-period-cohort analysis. Presented at the annual meetings of the *American Sociological Association*, New York, August.

6

APC Accounting/Multiple Classification Model, Part II: Empirical Applications

6.1 Introduction

This chapter continues the discussion of the conventional age-period-cohort (APC) accounting model but focuses on empirical applications. We first describe a three-step procedure that combines descriptive and statistical APC analysis using the Intrinsic Estimator (IE) through studies of recent cancer incidence and mortality trends by sex and race in the United States.

The utilities of linear models (LMs)/generalized linear models (GLMs) of APC extend beyond the identification and estimation of individual A, P, and C coefficients. Long-term trends estimated from such models also can be used to make projections/forecasts of the future. The special advantage of using APC models relative to conventional techniques for forecasting lies in the fact that they take into account cohort effects as an additional source of variation. We build on a previous study of APC model-based projection and forecasting (Osmond 1985) and illustrate the utility of linear APC models in forecasting future trends of mortality from four leading cancers.

6.2 Recent U.S. Cancer Incidence and Mortality Trends by Sex and Race: A Three-Step Procedure

We now turn to empirical analyses utilizing the new methodology of the IE to reveal the age, period, and cohort trends in cancer incidence and mortality since the launching of the SEER (Surveillance, Epidemiology, and End Results) Program by the National Cancer Institute (NCI) in 1973. Previous epidemiologic and demographic studies of trends in cancer mostly focused on period changes for specific age groups (e.g., Jemal et al. 2008; Manton, Akushevich, and Kravchenko 2009). Those that explicitly took cohort variation into account used either descriptive analyses (e.g., Manton 2000) or conventional statistical methods of model identification that resulted in inconsistencies of findings due to the use of different identifying constraints and assumptions (e.g., Clayton and Schifflers 1987; Kupper et al. 1985; Jemal, Chu, and Tarone 2001; Tarone and Chu 2000; Tarone, Chu, and Gaudette 1997). In general, there is a lack of guidelines for estimating linear APC models of rates.

We integrate descriptive and statistical analyses to more comprehensively assess the distinct impacts of age, period, and cohort on the incidence rates and mortality rates of 19 leading cancers for males and females. We implement *a three-step procedure* that can be readily applied in other studies using the LMs. *Step 1* is to conduct descriptive data analyses using graphics. The objective is to provide qualitative understanding of patterns of A, P, C, and two-way A-by-P and A-by-C variations. *Step 2* is nested model fitting. The objective is to ascertain whether the data are sufficiently well described by any single factor or two-way combination of A, P, and C. If these analyses suggest that only one or two of the three effects are operative, the analysis can proceed with a reduced model that omits one or two groups of variables, and there is no identification problem. If, however, these analyses suggest that all three dimensions are at work, we proceed to *Step 3*—applying the IE to make statistical estimates of the net A, P, and C effects. We now illustrate these steps in analyses of cancer trends.

6.2.1 Step 1: Descriptive Analysis Using Graphics

The graphical presentations of cancer incidence and mortality data described in Chapter 3, Sections 3.3.2 and 3.3.3 are shown in Figure 6.1 for 20 cancer sites. We focus for now on the age-specific rates arrayed by four time periods corresponding to four decades: 1970s, 1980s, 1990s, and after 2000. We also analyzed cohort arrays of the age-specific rates but do not present them in the interest of space. As we noted in Chapter 4, the detection of cohort variation can also be achieved by the comparison of age-specific rates across time periods. The results were consistent with similar analyses shown in the work of Manton, Akushevich, and Kravchenko (2009: Chapter 6), but extended them to include more leading cancers, other races in addition to white and black, and years after 2003.

Total cancer incidence and death rates showed rapid increases after the age of 45 for both males and females of all three race/ethnicities. The incidence rates leveled off for males and decelerated for females in old age for most individual cancer sites, with a few exceptions (leukemia, melanoma of the skin, and uterine cervix). Total cancer mortality rates showed largely continuous increases throughout the life span. The same pattern was observed for half of the sites examined. For the other sites, old age mortality rates leveled off just as the incidence rates.

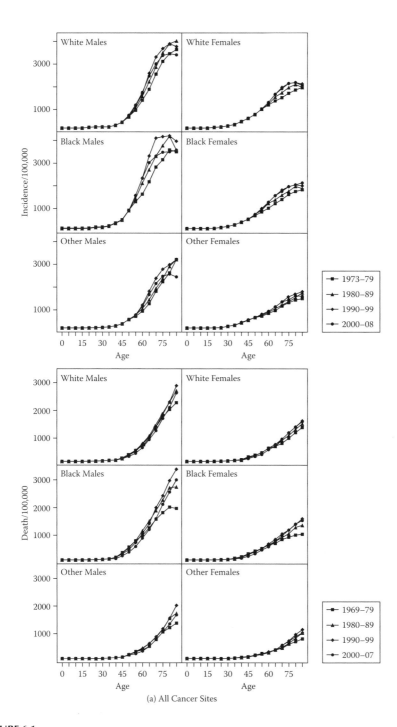

FIGURE 6.1
Age-specific rates of U.S. cancer incidence and mortality by time period.

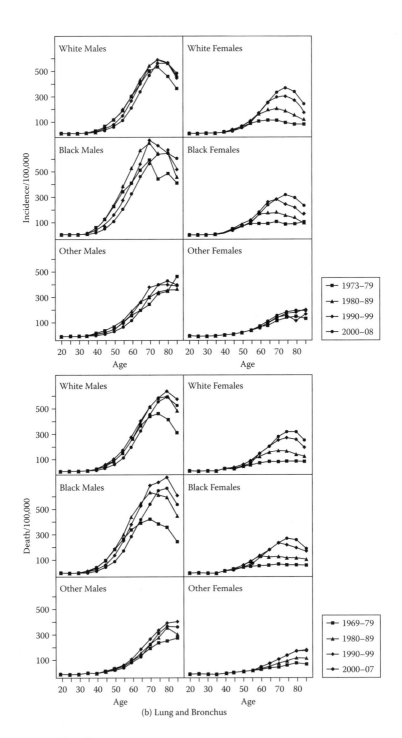

(b) Lung and Bronchus

FIGURE 6.1 (continued)

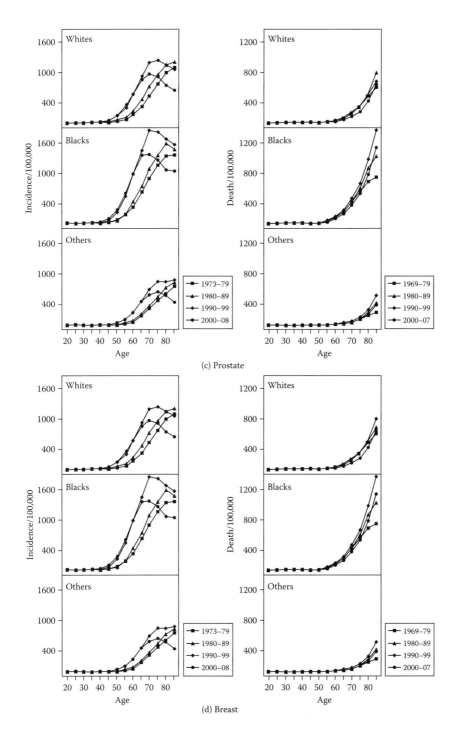

(c) Prostate

(d) Breast

FIGURE 6.1 (continued)

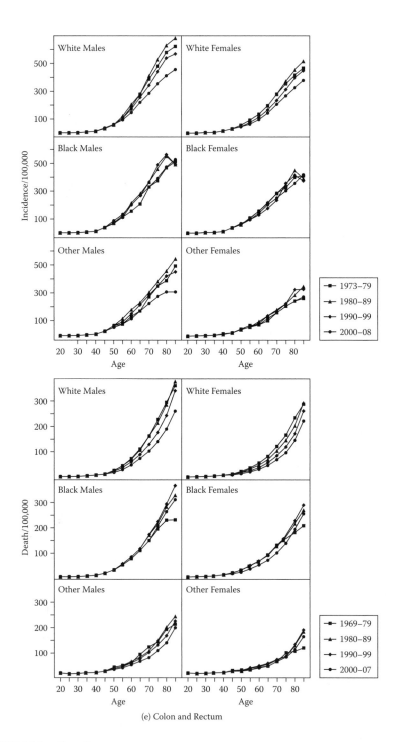

(e) Colon and Rectum

FIGURE 6.1 (continued)

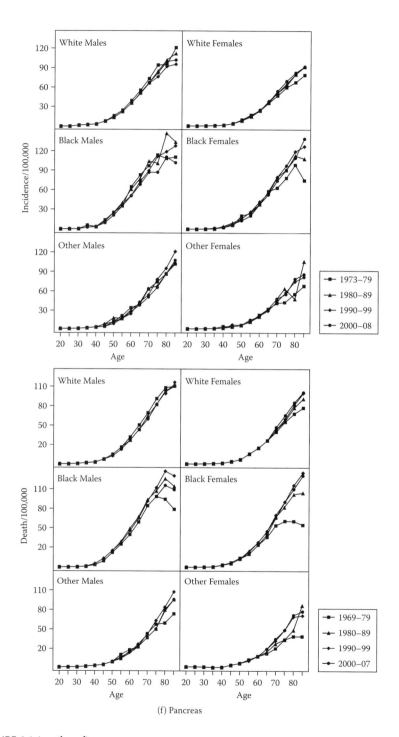

(f) Pancreas

FIGURE 6.1 (continued)

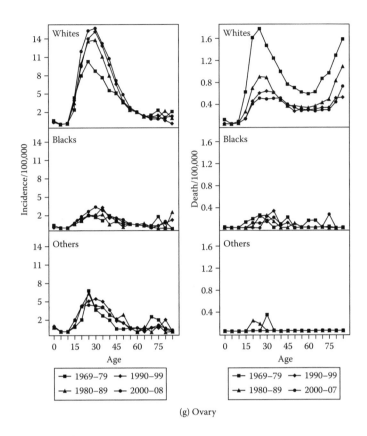

(g) Ovary

FIGURE 6.1 (continued)

Age-specific incidence and mortality rates for all cancers combined increased from the mid-1970s to 1990s and then decreased after 2000. Large increases in incidence rates over this period are evident for cancers of lung and bronchus, breast, prostate, non-Hodgkin lymphoma, esophagus, liver, kidney, brain and other nervous system, melanoma of the skin (for whites), and thyroid (for white males). Decreases in incidence rates over time were prominent for cancers of stomach, uterine cervix, and oral cavity and pharynx. Time trends of age-specific mortality rates for individual sites largely mirrored those of incidence rates. While most cancer sites showed slight fluctuations or continuous changes in age-specific mortality rates throughout the 40 years, increases were more pronounced in the last 10 years for cancers of lung and bronchus, esophagus, and thyroid (for males).

Sex and race differences in age-specific rates of both incidence and mortality were substantial. More males than females were diagnosed with cancers of nonreproductive sites and died from these cancers at most ages. One exception was thyroid cancer, in which case females showed higher incidence and mortality rates. The incidence and death rates were higher for black males

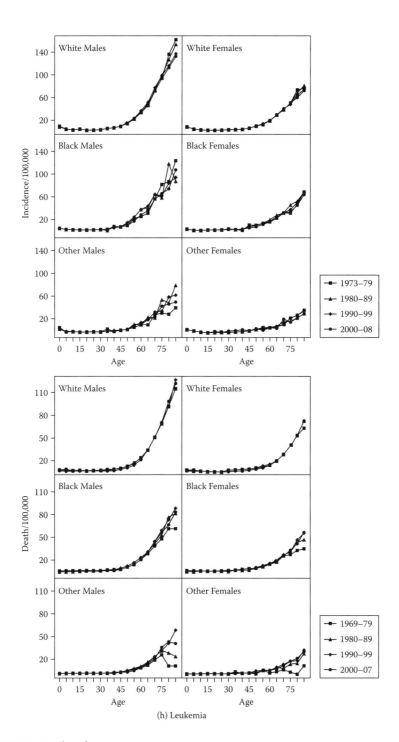

(h) Leukemia

FIGURE 6.1 (continued)

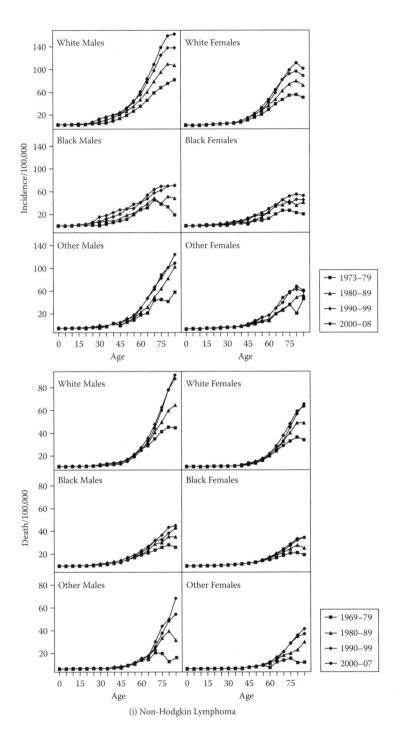

(i) Non-Hodgkin Lymphoma

FIGURE 6.1 (continued)

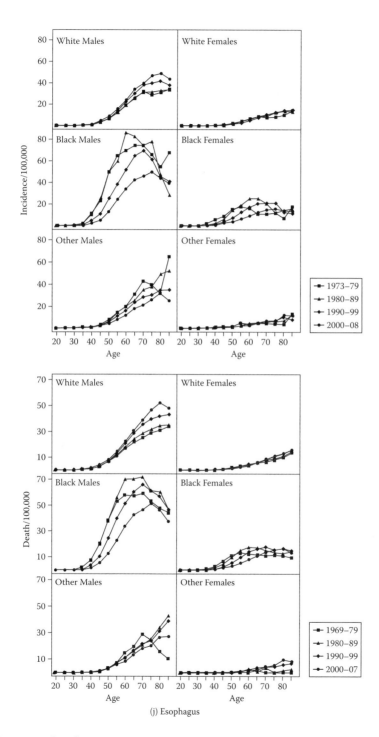

(j) Esophagus

FIGURE 6.1 (continued)

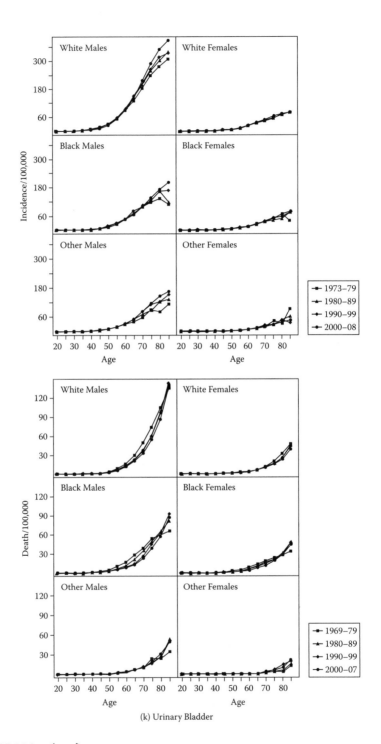

(k) Urinary Bladder

FIGURE 6.1 (continued)

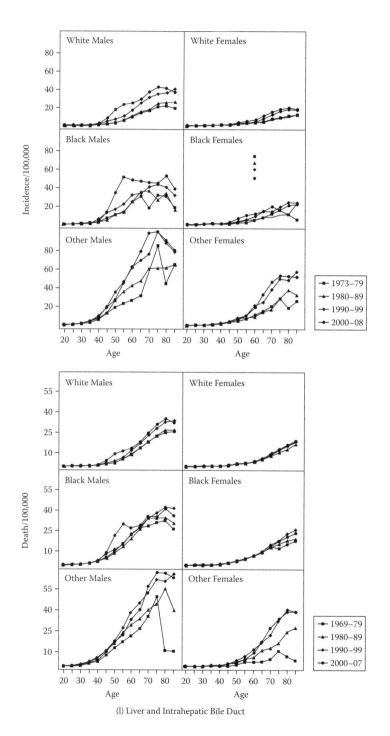

(l) Liver and Intrahepatic Bile Duct

FIGURE 6.1 (continued)

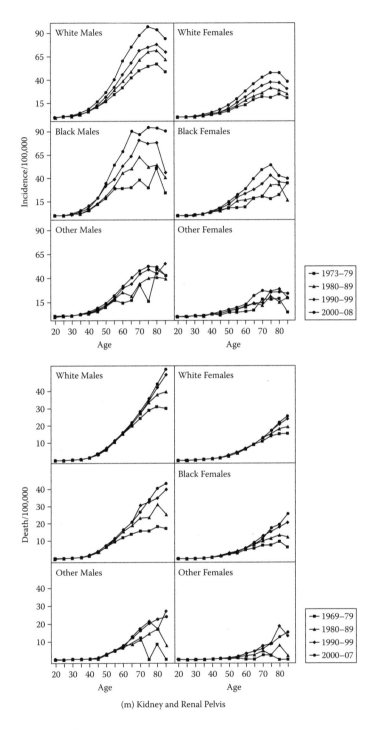

(m) Kidney and Renal Pelvis

FIGURE 6.1 (continued)

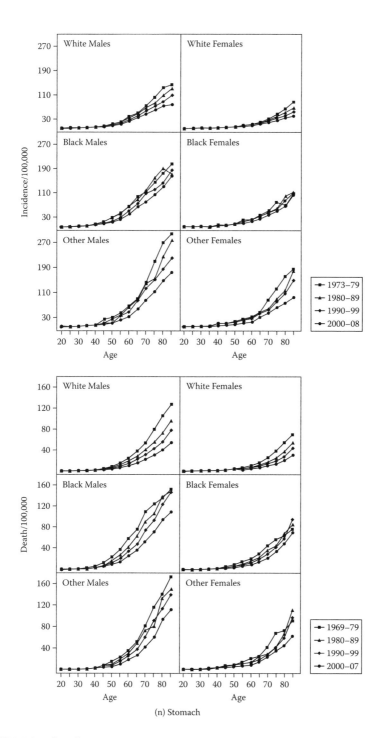

(n) Stomach

FIGURE 6.1 (continued)

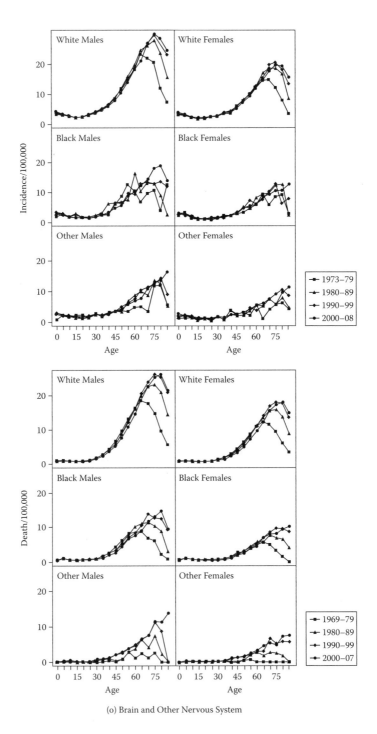

(o) Brain and Other Nervous System

FIGURE 6.1 (continued)

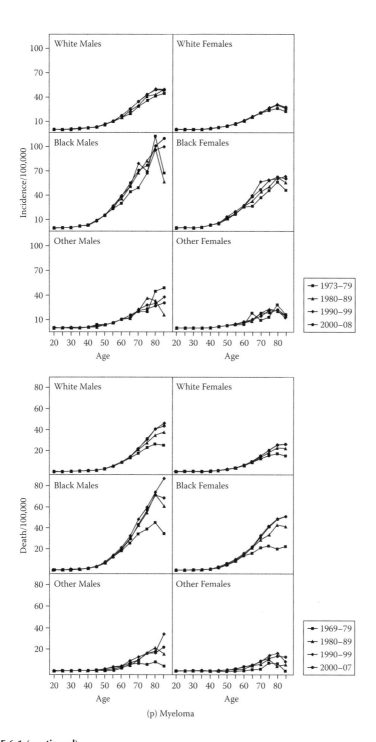

(p) Myeloma

FIGURE 6.1 (continued)

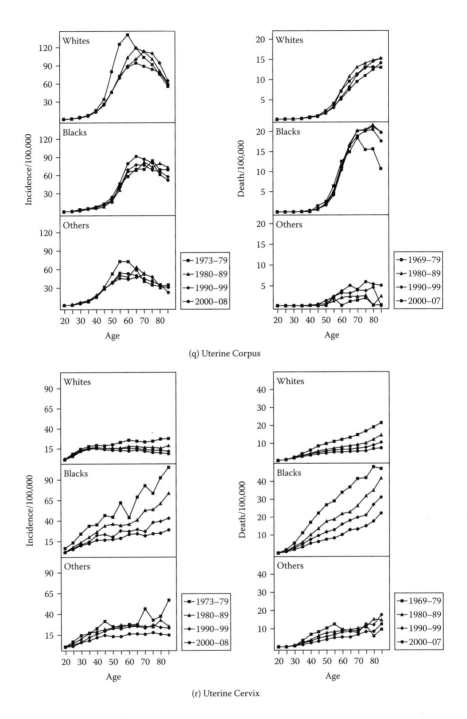

(q) Uterine Corpus

(r) Uterine Cervix

FIGURE 6.1 (continued)

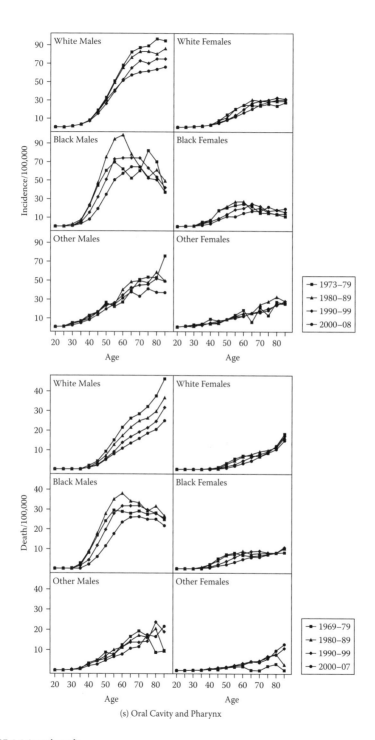

(s) Oral Cavity and Pharynx

FIGURE 6.1 (continued)

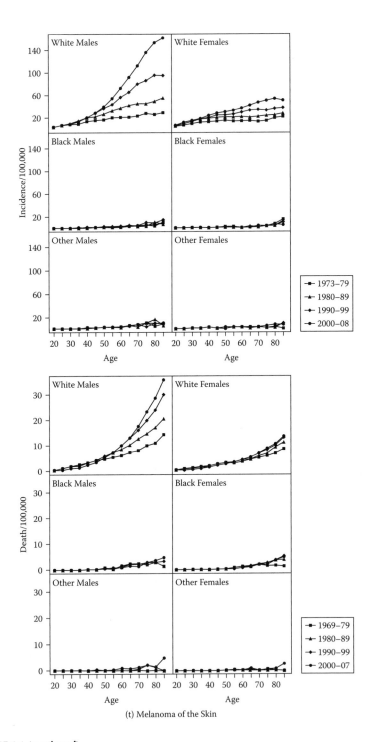

(t) Melanoma of the Skin

FIGURE 6.1 (continued)

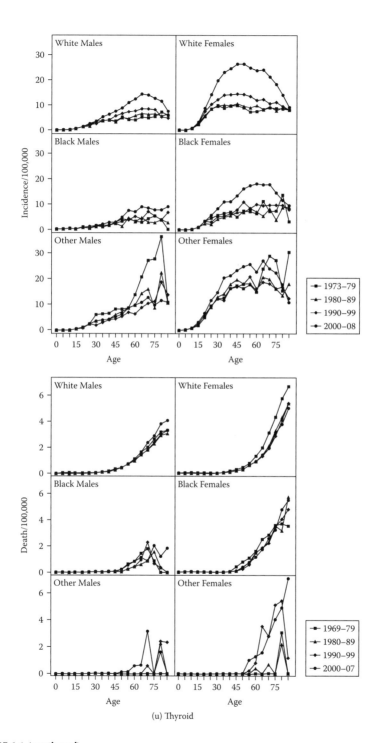

(u) Thyroid

FIGURE 6.1 (continued)

than nonblack males for leading cancer sites at most ages. Males of other races had lower age-specific incidence and mortality rates for all sites except for cancer of liver and intrahepatic bile duct. White females showed higher age-specific incidence rates in cancers of breast, ovary, and uterine corpus than nonwhite females. But, black females had higher age-specific mortality rates of all cancers. There were also sex and race differences in the time trends of cancer incidence and mortality rates. For instance, time period increases in incidence and mortality rates were larger for females than males in lung cancer, but larger for males than females in esophagus and liver cancers. Black and other females experienced increases in mortality from breast and uterine corpus cancers, in contrast to decreases for their white counterparts. In the case of liver cancer, blacks and other races also experienced larger increases in both incidence and mortality rates. On the other hand, blacks and others experienced larger declines in rates of esophagus cancer over this period at most ages and little increase in skin cancer, in contrast to the apparent uptakes of skin cancer incidence and mortality rates over the 40 years for whites.

There is evidence of nonparallelism among age patterns by time period in rates of both incidence and mortality for most cancer sites, indicating cohort variations. And, such evidence is particularly strong for cancers of lung and bronchus, prostate, breast, esophagus, liver, kidney, melanoma of the skin, and thyroid. Therefore, it is likely that cohort effects are more prominent in these cancers than others.

6.2.2 Step 2: Model Fit Comparisons

The foregoing two-dimensional graphical analyses suggested some interesting patterns of the age changes and temporal dynamics of top cancer sites. Although they have implications for the A, P, and C effects that generate the observed data, they are descriptive measures that do not explicitly and effectively assess these effects. We already discussed their limitations in Chapter 4. We next employ statistical models to estimate these effects. The first question that needs to be addressed is which model best summarizes the data. We can compare measures of model fit, such as R squared for linear regression models (LMs) or deviance statistics and penalized functions for GLMs, to determine the relative importance of the A, P, and C dimensions. Table 6.1 exemplifies this model comparison process for fitting log-linear models to total cancer incidence and mortality rates.

We first estimated three reduced log-linear models: a gross age (A) effects model and two two-factor models (i.e., AP and AC effects models). We then estimated the full APC model in which all three factors are simultaneously controlled. We used the IE to estimate the full APC model. We have already noted that the specific identifying constraint chosen does not affect the fit for just-identified models. Goodness-of-fit statistics were calculated and used to select the best-fitting models for male and female mortality data. Because likelihood ratio tests tend to favor models with a larger number of parameters,

TABLE 6.1

Goodness-of-Fit Statistics for Log-Linear Models of All-Site Cancer
Incidence and Mortality Rates

	Incidence							
	Male				Female			
Models	A	AP	AC	APC	A	AP	AC	APC
DF	126	119	102	**96**	126	119	102	96
Deviance	103078	47503	11452	**2643**	75977	64163	6435	3277
AIC	104813	49252	13235	**4438**	77696	65896	8202	5056
BIC	104866	49326	13360	**4581**	77749	65970	8326	5199

	Mortality							
	Male				Female			
Models	A	AP	AC	APC	A	AP	AC	APC
DF	126	119	102	**96**	126	119	102	96
Deviance	21624	8450	17710	**4751**	7720	2954	3467	1881
AIC	23199	10039	19333	**6387**	9304	4552	5099	3525
BIC	23253	10114	19458	**6529**	9358	4626	5224	3668

Note: Best models are indicated in bold.

we calculated two most commonly used penalized-likelihood model selection criteria, namely, Akaike's information criterion (AIC) and Bayesian
information criterion (BIC), which adjust the impact of model dimensions on
model deviances. Because age is a fundamentally important source of variation in morbidity and mortality, models with no A effects (P, C, and PC) have
significantly had lack of fit and are omitted. For both incidence and mortality
data on all cancer sites combined, both AIC and BIC statistics implied that the
full APC models fit the data significantly better than any reduced models.

The comparison of model fit using the BICs for individual cancer sites is
summarized in Tables 6.2a and b Goodness-of-Fit Statistics for Log-Linear
Models by Cancer Site, for incidence and mortality, respectively. We did not
estimate models for thyroid cancer mortality for blacks and others due to
the low death counts and potentially large sampling errors for these groups.
While the full APC models best fit the data for a large portion of cancer sites,
there were variations in best-fitting models by site, outcome, sex, and race.
For incidence rates in males, the AP models are the best models for cancers
of kidney and renal pelvis, pancreas, and myeloma, whereas the AC models
were the best models for cancers of esophagus and brain and nervous system. Whereas the APC models fit the best for incidence of most cancer sites
for white males, A and AP models were the better-fitting models for black
and other males for most sites. Similar patterns held for females, with more
variation by site. In particular, the APC models were the best-fitting models
for only three sites in black females and no site in other females. Results for

TABLE 6.2a

Goodness-of-Fit Statistics for Log-Linear Models by Cancer Site

BICs for Log-Linear Models of Cancer Incidence Rates by Sex and Race

	Male															
	Total				White				Black				Other			
Cancer Site	A	AP	AC	APC	A	AP	AC	APC	A	AP	AC	APC	A	AP	AC	APC
All Cancer Sites	104866.3	49326	13360	**4581**	81068	39757	9694	**4034**	26757	14567	6283	**2655**	2863	1955	1750	**1538**
Prostate	85162	42749	53158	**3665**	73609	37117	45852	**3318**	11021	5304	6883	**1284**	3873	2232	3035	**878**
Lung and bronchus	14135	7541	1684	**1324**	11142	6286	1613	**1281**	3544	2009	1131	**1037**	1522	1306	931	**918**
Colon and rectum	7122	3088	2893	**1520**	6769	2863	2590	**1461**	1062	**995**	1111	1009	1377	1068	1142	**995**
Urinary bladder	2271	2170	1229	**1180**	2206	2084	1201	**1162**	**775**	793	824	831	**747**	765	771	778
Non-Hodgkin lymphoma	5897	2726	3170	**2294**	5620	2610	2949	**2159**	1487	**1142**	1322	1147	1018	**956**	1050	1026
Melanoma of the skin	12528	4494	1192	**1176**	13706	4339	1203	**1156**	**470**	489	545	548	**547**	564	611	621
Kidney and renal pelvis	4283	**1127**	1190	1146	3910	1092	1139	**1088**	1204	**798**	826	832	774	**700**	761	758
Leukemia	1727	1642	1568	**1497**	1626	1584	1522	**1450**	**1026**	1029	1097	1082	**951**	961	1033	1027
Oral cavity and pharynx	2331	1186	1201	**1154**	2071	1288	1154	**1118**	1468	988	917	**863**	830	**795**	852	842
Thyroid	2806	1234	1408	**1156**	2943	1247	1339	**1131**	689	**640**	716	702	792	**740**	847	800
Pancreas	1092	**987**	1126	1063	1047	**958**	1097	1012	758	**746**	797	810	**680**	695	735	747
Stomach	3079	1219	1169	**1134**	2684	1203	1079	**1066**	1029	**800**	842	846	1242	**812**	854	838
Liver and intrahepatic bile duct	7157	1894	1168	**1020**	6104	1871	1072	**966**	1338	914	**742**	748	1000	**815**	834	819
Esophagus	1012	982	**909**	926	1263	**808**	858	833	1654	797	696	**668**	612	**548**	586	593
Brain and other nervous system	1525	1502	**1382**	1387	1481	1457	1356	**1346**	**865**	889	945	951	**813**	834	904	906
Myeloma	950	**851**	925	911	939	**828**	894	861	**683**	703	745	748	**543**	570	599	607

Cancer Site	Total				White (Female)				Black (Female)				Other (Female)			
	A	AP	AC	APC	A	AP	AC	APC	A	AP	AC	APC	A	AP	AC	APC
All Cancer Sites	77749	65970	8326	**5199**	66629	57553	7249	**4802**	13460	11243	2857	**2208**	2326	2132	1671	**1568**
Breast	8650	3379	5682	**2733**	8575	3229	5606	**2640**	1750	1239	1254	**1171**	1525	**1093**	1191	1096
Lung and bronchus	22740	11563	1812	**1475**	22607	10661	1739	**1412**	2404	1863	1001	**959**	913	915	**879**	882
Colon and rectum	6461	2450	1956	**1483**	6360	2324	1845	**1433**	1038	**1004**	1090	1036	937	**898**	953	923
Urinary bladder	1232	1213	1114	**1108**	1202	1174	1094	**1070**	**669**	696	716	727	**631**	646	673	678
Uterine corpus	4660	2948	2141	**1695**	4500	2927	2050	**1659**	888	**839**	869	891	872	**850**	916	897
Non-Hodgkin lymphoma	3559	**1402**	1572	1411	3533	1388	1518	**1368**	1194	**959**	1040	1013	926	**890**	977	967
Melanoma of the skin	5677	1983	1275	**1229**	6769	1961	1258	**1205**	**505**	535	582	590	**546**	558	603	611
Kidney and renal pelvis	3795	**1078**	1116	1110	3552	**1062**	1088	1074	1023	**755**	781	783	717	**638**	684	690
Leukemia	1524	1524	1466	**1457**	1489	1494	1437	**1408**	**975**	988	1045	1047	**882**	915	983	993
Oral cavity and pharynx	1822	1466	**1091**	1094	1633	1388	1077	1066	903	**823**	770	776	**724**	748	782	792
Thyroid	10144	1763	2041	**1324**	9830	1723	1892	**1280**	1245	**823**	938	865	1099	**888**	1019	937
Ovary	2292	1685	1385	**1376**	2141	1682	**1376**	1380	**953**	956	1000	1026	**875**	899	949	977
Pancreas	1086	**1042**	1090	1072	1060	**1008**	1057	1032	**759**	777	806	805	656	**681**	727	729
Stomach	2040	1258	1064	**1059**	2101	1267	989	**988**	783	**764**	799	805	1039	**775**	825	824
Liver and intrahepatic bile duct	2660	**1007**	1096	1013	2017	**937**	1004	953	732	**645**	690	694	805	**702**	730	714
Esophagus	1084	1032	**817**	838	835	846	**737**	742	916	699	597	**592**	**386**	406	448	459
Uterine cervix	3561	1458	1275	**1185**	2666	1534	1229	**1165**	2001	**919**	951	965	1173	**811**	876	877
Brain and other nervous system	1595	1551	**1348**	1354	1578	1513	1338	**1326**	**840**	863	923	929	**774**	794	866	876
Myeloma	856	**826**	891	873	805	**800**	848	828	**646**	656	700	698	**503**	530	564	578

Note: Best models are indicated in bold.

TABLE 6.2b

Goodness-of-Fit Statistics for Log-Linear Models by Cancer Site

BICs for Log-Linear Models of Cancer Mortality Rates by Sex and Race

Male

Cancer Site	Total				White				Black				Other			
	A	AP	AC	APC	A	AP	AC	APC	A	AP	AC	APC	A	AP	AC	APC
All Cancer Sites	23253	10114	19458	6529	21171	9095	16706	5669	4620	2617	4232	2392	2482	1922	2525	1883
Lung and bronchus	88986	57127	7029	2128	71572	47656	5367	1983	19731	12348	3202	1531	1599	1339	1019	962
Prostate	28163	7264	10303	2529	24200	6289	8787	2518	5326	2531	2292	1028	879	628	691	650
Colon and rectum	29299	2636	4169	1887	30510	2411	3774	1758	1827	1408	1498	1258	1190	970	956	948
Pancreas	2855	1602	1659	1260	2654	1470	1745	1235	1202	1097	985	970	636	656	666	675
Leukemia	8621	7309	2291	2284	7765	6735	2204	2180	1826	1732	1360	1339	1246	1224	1234	1218
Non-Hodgkin lymphoma	14831	9161	5013	2746	13850	8603	4476	2503	2096	1657	1705	1461	1265	1054	1133	1039
Esophagus	4299	2449	1195	1205	6616	1415	1274	1160	5866	2128	1389	931	641	639	634	630
Urinary bladder	5972	2317	2045	1147	5000	2046	1934	1118	1583	1157	918	852	510	522	543	540
Liver and intrahepatic bile duct	12893	4803	2066	1501	10336	4282	2105	1509	2576	1882	1130	1087	1189	976	992	1877
Kidney and renal pelvis	3065	2481	1300	1266	2761	2321	1256	1221	1494	1163	1031	1004	797	754	741	740
Stomach	30839	1648	1719	1407	28501	1648	1634	1362	4570	1252	1133	1062	1406	880	873	861
Brain and other nervous system	6284	5916	1733	1682	5648	5327	1707	1641	1522	1508	1189	1192	1248	982	1042	995
Myeloma	3310	1963	1250	1063	2958	1859	1146	1025	1296	998	974	886	664	578	596	552
Oral cavity and pharynx	11496	1330	1432	1290	10569	1309	1254	1203	3502	1724	1361	1043	894	833	846	846
Melanoma of the skin	8712	6208	1406	1271	8839	5778	1408	1243	737	753	725	722	414	296	331	335
Thyroid	983	898	997	879	951	887	961	863	591	576	610	1144	605	606	624	601

Female

Cancer Site	Total				White				Black				Other			
	A	AP	AC	APC	A	AP	AC	APC	A	AP	AC	APC	A	AP	AC	APC
All Cancer Sites	9358	4626	5224	**3668**	10379	4569	5222	**3518**	2186	1858	1805	**1712**	1671	1545	1652	1585
Lung and bronchus	232616	66259	12928	**3313**	218695	58635	11609	**3006**	20072	8104	2286	**1526**	1427	985	950	**893**
Breast	29020	11514	4270	**1624**	29690	12176	3913	**1611**	2666	1913	1494	**1134**	1017	996	899	**895**
Colon and rectum	41094	4358	1907	**1810**	41076	4224	1787	**1726**	2366	1676	1304	**1225**	912	877	875	**863**
Pancreas	2720	2317	1318	**1281**	2175	1909	1288	**1224**	1778	1554	1013	**1004**	693	**664**	675	680
Ovary	10818	9915	2495	**1547**	9447	8879	2329	**1474**	1854	1763	1102	**1012**	742	748	**678**	682
Leukemia	6611	5752	2208	**2137**	5893	5265	2120	**2042**	1753	1701	1327	**1327**	1172	1073	1136	**1064**
Non-Hodgkin lymphoma	12971	7584	3517	**1836**	12589	7409	3233	**1713**	1805	1390	1402	**1287**	1008	810	905	856
Uterine corpus	2275	1519	1250	**1082**	17217	2297	1842	**1347**	8170	1160	1182	**1080**	1016	**747**	767	758
Esophagus	2731	2641	**1161**	1176	1850	1763	1182	**1177**	2763	1992	919	868	639	**496**	526	508
Urinary bladder	2490	1138	1236	**1055**	2128	1042	1168	**1019**	1176	1007	804	766	458	411	450	**408**
Liver and intrahepatic bile duct	2129	1409	1749	**1252**	1837	1312	1659	**1184**	1089	1051	1037	**1006**	1249	823	935	**787**
Kidney and renal pelvis	2801	2168	1235	**1208**	2464	1939	1212	**1183**	1372	1161	932	**924**	718	**520**	612	561
Stomach	19757	1961	1507	**1358**	20080	1810	1392	**1287**	2724	1372	1121	**1068**	1129	**820**	837	828
Brain and other nervous system	7390	6808	1716	**1661**	6839	6180	1695	**1631**	1613	1613	**1163**	1165	1106	848	902	859
Myeloma	3757	2102	1294	**1076**	2976	1866	1151	**1014**	1662	1185	975	**897**	695	540	598	554
Uterine cervix	22790	1966	1984	**1369**	2253	1457	1223	**1043**	916	**924**	**862**	870	635	**535**	540	561
Oral cavity and pharynx	5458	2878	1310	**1251**	4911	2748	1282	**1235**	1707	1255	960	**923**	711	649	702	**671**
Melanoma of the skin	2734	2550	1216	**1203**	2665	2386	1215	**1183**	696	710	**666**	676	320	**224**	263	262
Thyroid	1461	**869**	1125	911	1445	**849**	1049	868	605	606	624	**601**	277	**231**	241	253

Note: Best models are indicated in bold.

mortality rates were largely consistent, with the APC models the best-fitting models for more sites and nonwhite groups.

This was the first systematic analysis of the relative importance of temporal components in cancer incidence to our knowledge. It suggested that cohort effects are not only specific to cancer site and outcome but also strongly dependent on social demographic status. The cohort process of change is more prominent in understanding trends of cancer mortality than cancer incidence, more evident in males than females and for whites than blacks or others. This suggests that the development and diagnosis of cancer were more strongly related to the biological process of aging and period changes in diagnostic techniques, whereas survival from cancer was also strongly influenced by life course accumulation of exposures to risk factors for mortality. Sex and race variations in cohort effects further suggested the importance of the social stratification process. Sociological theories provide support for this finding in that the diffusions of health risk factors such as smoking and obesity and technological innovations are experienced earlier by the socially more advantaged groups (white men with higher socioeconomic status) than the less-advantaged groups (women and racial/ethnic minorities) (Pampel 2005). Differential lengths of exposures at different points in the life course then likely are transformed into cohort differences that are more visible for the former than the latter.

6.2.3 Step 3: IE Analysis

We next examined the A, P, and C effects on cancer incidence and mortality rates based on the results of model comparison analysis using the best-fitting models for each site, outcome, and demographic group. We have already shown the numerical results from estimation of the full APC models using the IE method in Table 5.7 in Chapter 5 for female total cancer mortality rate. To facilitate the comparisons of the age and temporal effects across subpopulations and sites, we plotted the point IE estimates of effect coefficients by sex and race when the full APC models were the best-fitting models for the sex-race groups. The resulting graphs, shown in Figure 6.2, compare the coefficients of successive categories within A, P, and C classifications and show the net A, P, and C effects of incidence and mortality. Within each site and cancer outcome (incidence or mortality), we used the same scale so that the relative magnitude of A, P, and C effects can be easily discerned. The age and cohort ranges of these results varied by site, sex, and race because the counts of incidence and death at younger ages for more recent cohorts were too low to yield reliable estimates for some groups. A caveat for interpreting the sex- and race-specific results is that these are not absolute levels of mortality rates, but model effect coefficients. The former can be compared directly based on the same scale, such as deaths per 100,000 persons, but the latter only indicate relative trends within each population. The model-based results on A, P, and C effects largely accord with the observed patterns from

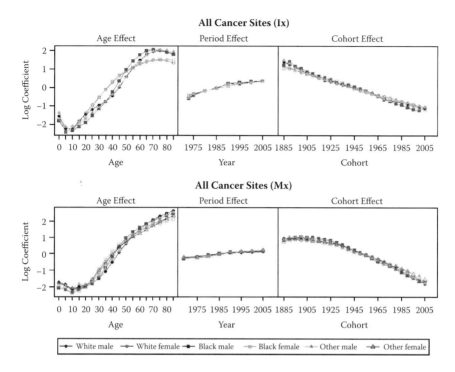

FIGURE 6.2
Intrinsic Estimator coefficient estimates of the age, period, and cohort effects on cancer incidence and mortality for selected cancer sites.

descriptive displays of the data, but they also offer additional insights of and succinct summaries of distinct sources of incidence and mortality changes.

6.2.3.1 All Cancer Sites Combined

The IE analysis of the incidence and mortality changes of all cancer sites showed the dominance of A and C effects over P effects. The age curves of incidence rates showed exponential increases before old ages but decreases during old ages for most groups. Similar exponential increases occurred for cancer mortality rates at all ages. While there appeared some slight decelerations in the rate of increase around the age of 50, there were no signs of declines in mortality rates for any group in old ages. The A effects showed more pronounced sex and race differences than the P or C effects. Specifically, females showed slightly less-steep age accelerations in incidence and mortality rates than males, and black males showed steeper age accelerations after middle ages than nonblack males.

While the A effects were highly consistent with age patterns of rates shown in the graphical analysis in Step 1, the results regarding distinct P and C effects were not immediately obvious in descriptive plots. There is

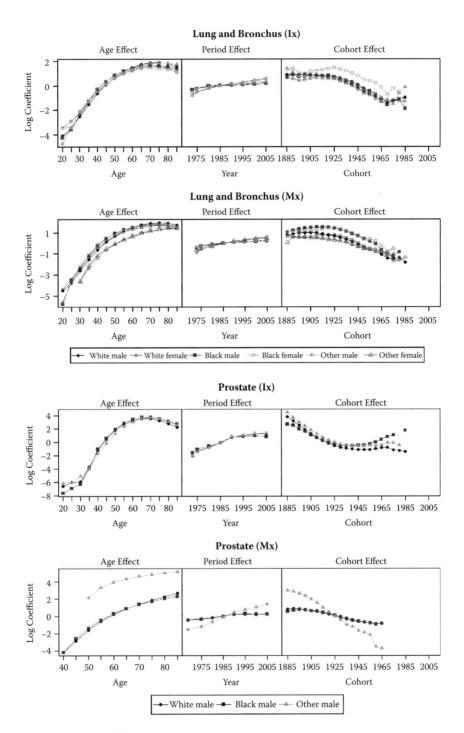

FIGURE 6.2 (continued)

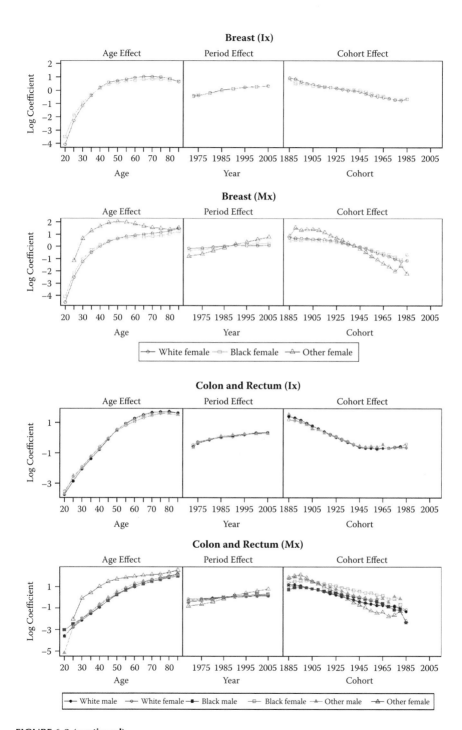

FIGURE 6.2 (continued)

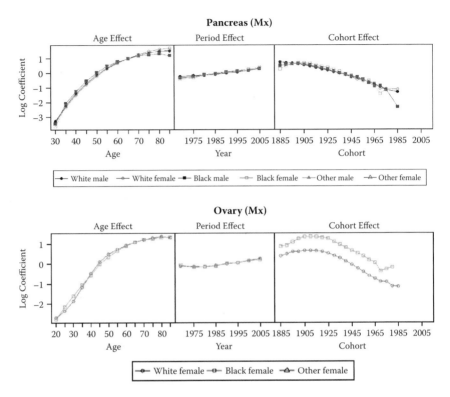

FIGURE 6.2 (continued)

an important distinction of the period changes observed from the Step 1 graphical display of data and those based on APC models. That is, the period changes in the observed rates may be confounded with changes that are age or cohort related, whereas those from the model estimates represent true P effects that are net of A and C effects. When all three factors are considered in the APC models, there seem to be only modest P effects, but substantial C effects. The P effects showed log-linear trends, indicating increases in cancer incidence and mortality over the last 40 years. On the other hand, the cohort declines in cancer incidence were monotonic and continuous, and those in cancer mortality started later, reflecting a lagged effect of incidence on mortality. The successive cohort declines largely support the contention that nutrition, reduced exposure to carcinogens, and medical measures all contributed to increasing health capital in more recent cohorts that reduced the onset of malignancies and improved survival.

6.2.3.2 Age Effects by Site

The A, P, and C effects showed different patterns across sites, but the dominance of A effects was evident for most sites, suggesting the influence of cancer biology on age patterns of both incidence and mortality. The incidence

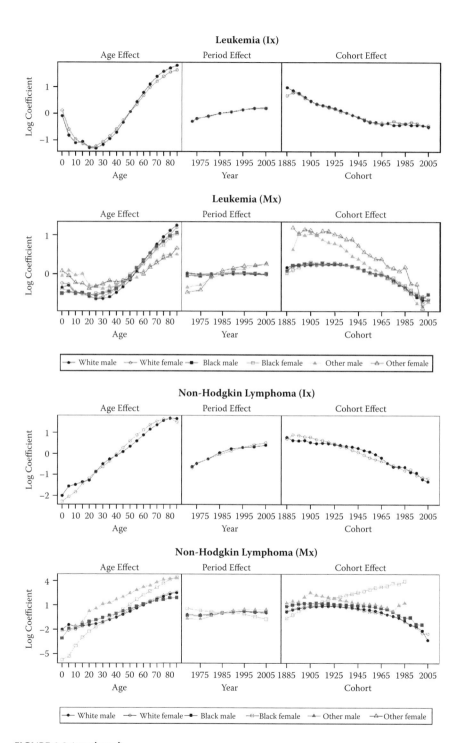

FIGURE 6.2 (continued)

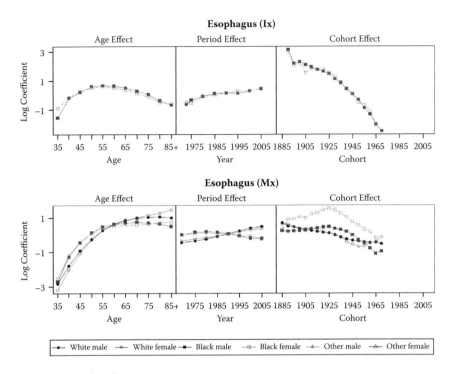

FIGURE 6.2 (continued)

rates of malignancy in most tissues followed the same age pattern of
increases, with four notable deviations. The A effects of female breast cancer
incidence were indicative of the etiology of the disease: The sharp increases
in disease risk in early ages were more strongly genetically determined, and
the increases in risk slowed at the age of menopause and plateaued with a
peak around the ages of 65–70. The results for cervical cancer were similar,
but the plateau appeared much earlier, starting around the age of 40. The
A effects of esophageal cancer incidence showed modest increases between
ages 35 and 55 and gradual declines afterward. The relative magnitude of
the A effects, in this case, was similar to that of the P effects and smaller than
the C effects. The A effects of the thyroid cancer incidence showed a bimodal
pattern for white males, with the first peak in the 20s and the second higher
peak around the age of 65 and a long plateau for white females after the
age of 20. Mortality in nine sites showed continuously increasing A effects
that were similar to those for all sites combined. These sites include prostate,
colorectal, urinary bladder, stomach, leukemia, non-Hodgkin lymphoma,
oral cavity, and melanoma of skin. Mortality in the remaining 11 sites showed
a similar pattern with incidence data, that is, a concave pattern of A effects
that increased from early adulthood to peaks in old ages and then leveled off.

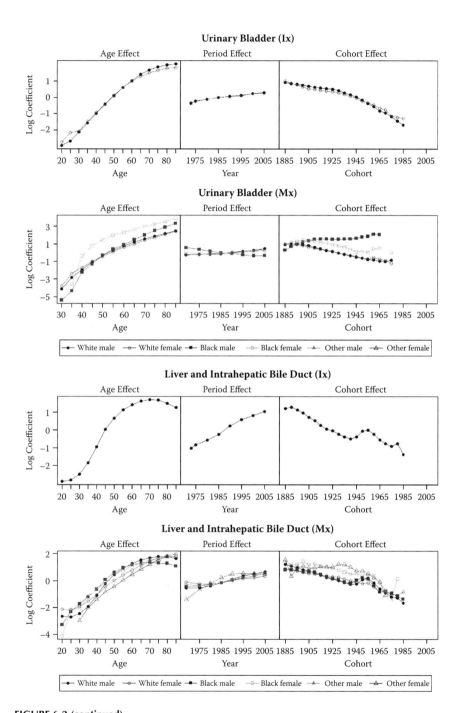

FIGURE 6.2 (continued)

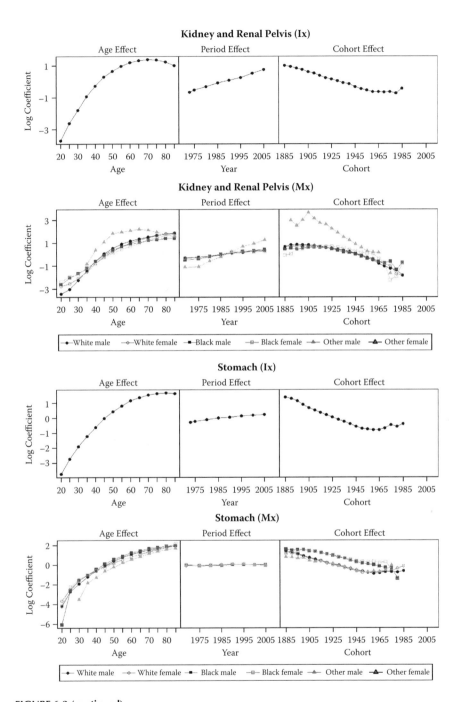

FIGURE 6.2 (continued)

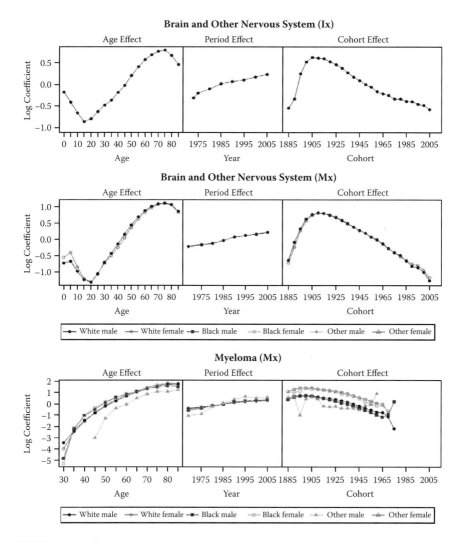

FIGURE 6.2 (continued)

6.2.3.3 Period Effects by Site

The IE estimates of the P effects on cancer incidence showed increasing trends for all but one site. The period trend for cervical cancer incidence was clearly downward, reflecting the success of cancer screening through Pap smear in prevention and control of new cases of malignancies at this site. Period changes in incidence rates at other sites can be related to changes in exposures to carcinogens and behavioral risk factors. For example, the increases in lung cancer incidence can be due to increases in tar and nicotine

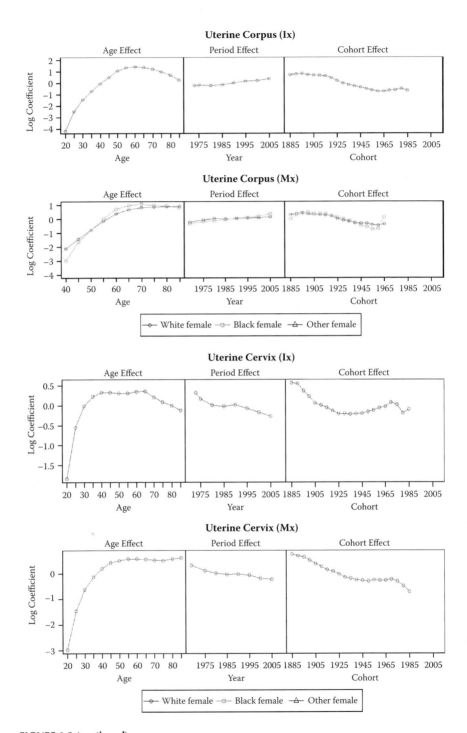

FIGURE 6.2 (continued)

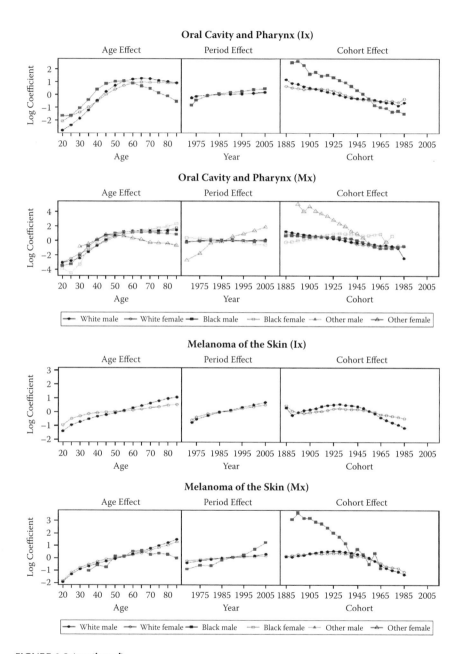

FIGURE 6.2 (continued)

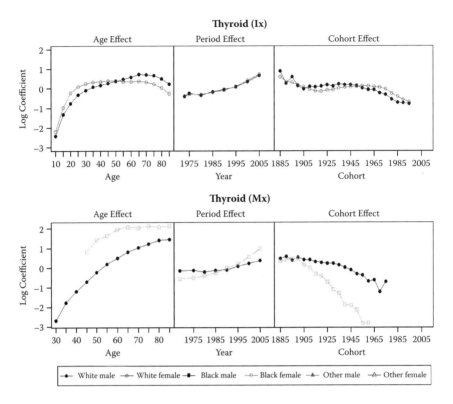

FIGURE 6.2 (continued)

in cigarettes (particularly for females), drug use, and air pollution. The obesity epidemic can have widespread impacts on cancer sites that are linked to fat and glucose metabolism and reproductive hormones. And, period changes can also result from changes in the use of diagnostic and screening procedures. For instance, the increases in breast cancer incidence accompanied increasing use of mammography, which increased detection, whereas the slower increment in breast cancer incidence between 2000 and 2005 may have followed from reduced use of hormonal replacement therapy (HRT). The increases in the colorectal cancer incidence rates could have resulted from increased use of colonoscopy and detection of cancerous adenomatous polyps. Trends in the frequency of prostate-specific antigen (PSA) testing have also influenced prostate cancer incidence trends. The increasing incidence of several other cancers, such as melanoma of the skin, kidney, renal pelvis, and thyroid cancers, could all be related partly to diagnostic and imaging technology (Jemal et al. 2008).

   The IE estimates of P effects on cancer mortality were more varied by site and demographic group. While mortality in most sites showed increasing trends of P effects, mortality due to cancers of stomach and oral cavity (for whites and black males) changed little over time, and leukemia (for whites and blacks) and cervical cancer mortality decreased over time. For lung cancer mortality, females had a steeper increase than males throughout the 40-year time period, which was consistent with the results on lung cancer incidence. For breast, colorectal, liver, and oral cavity cancer mortality, females of other races showed steeper increases than the rest of the groups. For prostate, kidney, and myeloma cancer mortality, males of other races showed steeper increases than the other groups. For leukemia mortality, both males and females of other races showed increases, while whites and blacks showed decreases in period trends. In the cases of urinary bladder and esophageal cancer mortality, blacks had decreases, while other groups had increases in period trends. In the case of non-Hodgkin lymphoma mortality, black females also showed decreases over time. On the other hand, black males and females showed steeper period increases in skin cancer and thyroid cancer mortality than nonblacks, respectively.

As shown in the Step 2 model comparison analysis, the C effects that significantly contributed to the overall temporal variation in incidence and mortality did not occur for all sites and groups. However, they accounted for changes in a substantial proportion of major sites for both sexes. The IE estimates of the C effects revealed myriad patterns of cohort changes that require a more detailed description. We discuss the results of cancer incidence and mortality separately in the following material.

6.2.3.4 Cohort Effects on Cancer Incidence

Pronounced cohort declines in rates of newly diagnosed malignancies were evident for a vast majority of sites. The shapes of the cohort curves of these sites, however, varied. The C effects of lung cancer incidence were consistent with the characteristic inverse-U shape that showed increases in earlier cohorts and decreases for subsequent cohorts (Gardner and Osmond 1984; Jemal, Chu, and Tarone 2001; Lee and Lin 1996; Tarone and Chu 2000). The increases and peaks of the C effects were delayed but more pronounced for females, especially black females. Such effects suggest the influence of gender-specific risk factor exposure and behavioral differences, especially cigarette smoking, in the life experience of specific birth cohorts. Studies have found diverse patterns of smoking across cohorts and the presence of sex differences, with recent increases in smoking in later cohorts of females (Harris 1983; Zang and Wynder 1996). Male cohorts adopted cigarettes in large numbers earlier than females, and the pattern preceded that of females by a decade or two (Lopez 1995). The decreases in incidence for subsequent cohorts may reflect the benefits of smoking cessation. The increases in the most recent three birth cohorts were unexpected and may be attributed to sparse data for these cohorts. In general, the model estimates for the earliest and most recent birth cohorts should be interpreted with caution because they were based on the few observations at the lower left and upper right corners of the age-by-period matrices, respectively, and were accompanied by larger standard errors than estimates for middle cohorts.

Cohort declines in incidence rates leveled off in more recent cohorts for four sites: colorectal, stomach, kidney, and uterine corpus cancers. Increases in more recent cohorts appeared for prostate, liver, and cervical cancers. The slowdowns or reversals of continuous cohort improvement seem to have initiated in the baby boomers born between 1946 and 1964, suggesting the need for better strategies of prevention and control that are specific to these and succeeding cohorts. On the other hand, incidence rates for melanoma of the skin and thyroid cancer showed little changes for earlier cohorts and small declines for the baby boomer cohorts and successors. The cohort-related changes in these two sites, however, were modest compared to age and period-related changes and confined to whites. While the C effects for the majority of sites are relevant only for whites, especially white males, the C effects apply only to blacks in the case of esophageal cancer incidence.

6.2.3.5 Cohort Effects on Cancer Mortality

Cohort-related changes in cancer mortality were prominent and consistent in trends for most sites and demographic groups. Site-specific cohort changes were remarkably similar to the cohort changes in all sites combined and showed largely continuous declines in mortality across successive birth cohorts. There were, however, sex and race differences in these trends within and across sites. The cohort changes in lung cancer mortality mirrored those in lung cancer incidence in terms of sex differences. That is, the lag between the peaking male and female birth cohorts can be attributed to the sex differences in stages of cigarette diffusion (Yang 2008). For cohorts born before 1925, racial groups diverged in trends of change, with blacks showing more increases than whites and others. Compared to the lack of black and white differences in male incidence rates, this finding suggests the black disadvantage in lung cancer prognosis and survival in the earlier cohorts. The black disadvantages were pronounced for four additional sites. Specifically, while other groups showed cohort declines, black females showed continuous cohort increases in non-Hodgkin lymphoma and oral cavity cancer mortality; black males showed cohort increases in urinary bladder mortality; and blacks also showed increases in esophageal cancer mortality for cohorts born before 1930.

In some other cases, however, racial ethnic minorities showed more improvement across cohorts than whites. Black males and females experienced much stronger cohort declines in skin and thyroid cancer mortality, respectively. And the other races showed steeper cohort declines in prostate, breast, colorectal, leukemia, kidney, and oral cavity cancer mortality. The one site in which the other race group showed a clear disadvantage was liver cancer mortality: Females born between 1925 and 1945 experienced an increase. The differences discussed so far pertain to data for which the full APC models are the best fit. It should be noted, then, that the C effects do not add to the model fit for the other race group for most sites or for the blacks for some sites. Therefore, cohort declines in cancer mortality from leading sites were generally more prevalent for whites than blacks or the other races.

6.2.4 Summary and Discussion of Findings

Although U.S. cancer trends for the past few decades have been widely documented, the sources of cancer incidence and mortality experiences attributable to distinct impacts of age, period, and cohort are not clear or understood. Findings from most previous studies using APC analysis were tentative due to limitations in data and analytic methods. And, evidence regarding the relative importance of P and C effects in recent cancer trends was particularly scant. We used a set of comprehensive analyses, including a new method of estimation for the conventional GLM approach to APC analysis, to shed some new light on our progress during the 40 years of the war on cancer. These analyses improved previous studies in several regards. First, the new model estimation approach yielded results that can serve as objective criteria for selecting the best among alternative summaries of data and provided a test of whether the observed pattern revealed in descriptive analysis was real or random. Second, delineation of different sources of variations in incidence and death rates brought clarity to the specific components of cancer dynamics in the recent historical period and highlighted important period- and cohort-specific factors that are operational in the changes of the population risks of malignancies and improvement in survival. Third, the empirical assessment of cancer trends using the most recent data available may help to generate theoretical insights into continuing cancer mortality reductions and reveal previously unknown forces affecting health and mortality conditions in this country that likely will continue into the future.

A major finding is the strong presence of cohort differences in cancer incidence and mortality from major sites. Whenever the C effects emerged, they tended to dominate the P effects in explaining recent cancer trends. This was particularly the case for cancer mortality. This finding largely supports the theory of technophysio evolution, which implies large cohort improvements in health capital and physiological capacities (Fogel 2004), and the "cohort morbidity phenotype" hypothesis, which emphasizes the long-lasting impacts on morbidity and mortality of reduced early life exposures to infection and inflammation (Finch and Crimmins 2004). The effects associated with birth cohorts reflected processes of differential cohort accumulation of lifetime exposures to risk factors for cancer, such as education, diet and nutrition, physical activity, and smoking. If the finding on the C effects is valid, then the standard models of period changes such as those shown in descriptive analyses and Equation (4.8) in Chapter 4 are misspecified. The role of C effects in cancer trends and beyond undeniably has serious implications for measurement and analysis. The lack of cohort declines in incidence and stagnations of declines starting from the baby boomer cohorts for some cancer sites have important implications for future trends and are worthy of additional attention.

It merits discussion that the P effects were generally small or modest when the C and A effects were simultaneously controlled. The model comparison analysis showed numerous cases for which there were virtually no P effects (e.g., ovarian cancer incidence for white females and brain cancer mortality for blacks). Most site-specific results showed monotonically increasing, albeit small, P effects that suggested log-linear trends. The finding of weak P effects relative to the C effects is an insight from the IE analysis that cannot be obtained from descriptive analyses that characterize most cancer trend studies (e.g., Manton, Akushevich, and Kravchenko 2009). This means that the anticipated cancer incidence and mortality declines have been mostly captured by or attributed to the C effects rather than P effects. This result is largely consistent with previous epidemiologic studies using APC models (e.g., Osmond 1985) and comprehensible in light of the different mechanisms underlying independent P and C effects. Because successive cohorts experienced more favorable historical and social conditions, they not only had lower exposure levels to socioeconomic, behavioral, and environmental risk factors but also benefited from reduction in the exposure earlier in life course than previous cohorts. In this sense, mortality reductions as a consequence of advancements in medical measures—measures that, once introduced and found to be effective, tend to persist—manifested as cohort declines because the cumulative effects were more pronounced in successive cohorts compared to periods.

The increasing P effects of cancer mortality are intriguing. They could be related to increases in cancer incidence rates that reflect actual increases in exposures to carcinogens and biobehavioral risk factors or expanded use of effective diagnostic techniques. Future research should further investigate the role of these and related factors in explaining this trend. An alternative explanation to the increasing P effects is a demographic one. That is, the P effects represent a special age-cohort interaction implied by the APC modeling framework. It has been shown that recent cohorts, when exposed to new favorable epidemiologic conditions earlier in life, may experience greater delays of deaths to more advanced ages than cohorts already old at the time of exposure to these conditions (Guillot 2006). The postponement of proportions of deaths in succeeding cohorts to older ages may thus create a seemingly increasing period trend. Such a trend may result from the beneficial effects of early detection and treatment of cancer patients and not represent worsening conditions. There is evidence that the composition of death distributions actually changed by cause. Data from an earlier study by Yang (2008) showed a 50% increase of proportions of lung cancer deaths and a 42% increase of proportions of breast cancer deaths from the elderly population 65 and over in the United States between 1960 and 1999. Combining the estimated P and C effects, it can be inferred that there were not only marked decreases in the level of cancer mortality across cohorts, but also postponement in the timing of cancer-related deaths across cohorts during this period.

We conducted analyses on a comparative basis between different tumor sites and populations, such as two sexes and three race groups. Mortality declines across cohorts occurred for some groups and major sites that are leading causes of cancer deaths. But, the overall improvement was less impressive for blacks and other races. While the latter group had generally lower cancer incidence and mortality rates and less to improve on, blacks suffered substantial disadvantages in cancer survival. The results on the heterogeneity in temporal dynamics can thus be used to identify populations with significantly higher or lower incidence of malignancy in various tissues so that the search for risk factors associated with the disease may be more selective and specific.

In sum, the systematic study of sources of variations in cancer incidence and mortality has established patterns and phenomena not known before that require further explanations. It thus can be used as a triage for refined analysis of explanatory factors underlying period and cohort processes of change. Prominent factors include cigarette smoking, diet, sedentary lifestyles and related obesity, among other debilitative behaviors and unfavorable environmental factors amenable for modifications. Taking these variables into account in APC analysis will further improve our understanding of the prospect of future cancer mortality reductions. How to accomplish this goal using a different family of models is shown in the next chapter.

6.3 APC Model-Based Demographic Projection and Forecasting

The methods and techniques introduced all pertain to the identification of individual coefficients of A, P, and C effects in an APC model. A different, but related, question is how the inclusion or exclusion of A, P, or C effects may change the overall fitted values using the model. We have shown in the Step 2 analysis that the nested model-fitting exercise is most helpful in the choice of best set of variables to describe the data. Building on the results of this analysis, one can further obtain the predicted values from the best-fitting models for the purpose of projection in the outcome of interest. We now extend the study of recent trends in cancer mortality to show how to use APC models for projection and forecasting of future trends. Before we describe the method and present empirical results, we first introduce the theoretical rationale for this approach to forecasting (Reither, Olshansky, and Yang 2011).

6.3.1 Two-Dimensional versus Three-Dimensional View

The fact that both period and cohort effects are important in explaining the trends in cancer mortality motivates the use of APC models in the estimation of future rates of mortality because they take into account both effects and are likely to be more accurate than methods that ignore cohort changes. The advantage of this approach to forecasting can be best understood in the context of health and longevity.

Many researchers have simply extrapolated the historical trend forward in time as the basis for conclusions about population health and prospect of life expectancy. We term this two-dimensional forecasts. They are derived from age-specific health statistics such as death rates from specific causes, which are plotted over a series of periods of observation. The only two variables in use are age and time period. These data are then extrapolated into the future via various methods. Commonly used metrics of population health such as life expectancy based on these mortality rates then provide an estimate of average duration of life under the stationary population assumption that a hypothetical birth cohort experiences the death rates observed at all ages in a given calendar year throughout their lives. Therefore, the implicit assumption underlying two-dimensional forecasting is that younger cohorts, still alive, will be identical to their predecessors as they age, and that future improvements in health and longevity, already experienced by older cohorts, will continue at an identical pace for younger cohorts throughout the remainder of their lives.

By ignoring the third dimension of variation, namely, birth cohort variation, the two-dimensional approach fails to account for the time lag between risk factors or improved conditions now present among younger cohorts and their likely influence on the health status of these cohorts as they age. In other words, although death rates observed today provide an accurate reflection of a lifetime of past behavioral, environmental, and epigenetic factors accumulated by the recently *deceased*, these statistics provide a limited two-dimensional vision of the future because they fail to take into account the health conditions of people still *alive*—particularly children and adolescents. A more accurate gauge of the future of health and longevity requires a three-dimensional view that accounts for this time lag.

If environmental and health conditions were constant or always improving on a steady and predictable course, the underlying assumption in two-dimensional forecasts would be plausible. However, history shows that two-dimensional forecasts have been inaccurate—sometimes with dire policy consequences. Under conditions of declining cohort mortality, they will underestimate actual life expectancy. For example, many pension plans have become strained or bankrupt because they failed to account for declines in smoking and improvements in lifestyles and medical technologies that led to improved survival among employees from these birth cohorts (Olshansky, Carnes, and Mandell 2009). Under conditions of rising mortality, the same approach will overestimate actual life expectancy. For example, the childhood and adult obesity epidemics have indicated rapid and dramatic changes in health conditions among recent birth cohorts that may be eroding hard-won gains in health and longevity, as predicted by the three-dimensional model (Reither, Olshansky, and Yang 2011).

6.3.2 Forecasting of the U.S. Cancer Mortality Trends for Leading Causes of Death

Previous forecasters have generally relied on simple extrapolative methods applied directly to age-specific mortality rates. We show how the extrapolation of coefficients from APC models fit to these rates can produce significant gains in forecasting accuracy. We chose mortality data on several leading cancer sites that showed strong cohort effects, including lung and colorectal cancer mortality for men and women, female breast cancer, and male prostate cancer mortality for whites and blacks. The example of APC models using the IE as the method of identification has provided a summary of the observed rates and attributed changes to both time period and birth cohort in addition to age. The model fit statistics (i.e., the BICs) showed that the full models accounting for the C effects are the models of choice. The strong presence of cohort effects, therefore, is likely to affect the predicted mortality rates and produce different future trends compared to methods that ignore them.

6.3.2.1 Methods of Extrapolation

Two approaches conventionally implemented in demographic projection and forecasting are simple linear extrapolation and time series methods. The simple linear extrapolation employs linear regression models to extrapolate the mortality rates in the base period to the future. It has been used as the benchmark for comparison in the previous study of the APC model-based forecasting by Osmond (1985). An alternative is to use the autoregressive integrated moving average (ARIMA) model for forecasting time series trends (Box, Jenkins, and Reinsel 1994). The simple linear extrapolation focuses on the linear trends and imposes a fixed rate of change based on past data, whereas the ARIMA model may apply to phenomena experiencing variable rates of change by incorporating local stochastic trends or universal fixed trends, thus providing flexibility in accounting for most variations (McNown and Rogers 1989). The ARIMA approach has been the basis for many recent forecasting studies using the time series method (Lee and Carter 1992; McNown and Rogers 1989; Shibuya, Inoue, and Lopez 2005). Instead of directly projecting age-specific mortality rates into the future, we apply these methods to projections of APC model coefficients to obtain fitted values of mortality rates. Because the APC models provide the best summary and characterization of the forces underlying the mortality trend data, coefficient-based extrapolation should be superior to the conventional methods of data extrapolation, which fail to take into account birth cohort variation.

To test this proposition, we compared the performance of two methods: the non-APC model-based, or "simple," method and the APC model-based method. For the simple projection, we fit, for each age group, (1) a linear regression model of mortality rates as a function of period year and (2) an ARIMA model of the time series of mortality rates and obtained predicted mortality rates from these models for the projection period. For the APC model coefficient-based projection, we first fit a linear regression model and an ARIMA model to the set of period effect coefficients and cohort effect coefficients estimated by the IE method, assuming the age effects to be constant over the projection period. We then combined the age effect coefficients and predicted period and cohort effect coefficients from the previous step to obtain the predicted mortality rates. In the application of the linear coefficient extrapolation method, the number of coefficients used or the weight applied to each of these coefficients would affect the prediction outcomes (Osmond 1985). So, we compared different weighted linear extrapolations of period and cohort coefficients. The numbers of period coefficients in use ranged from all to the last two periods. We examined three options of weighting cohort coefficients, including all, the last 10 cohorts, and the last 5 cohorts. For the time series extrapolation, we estimated and identified the ARIMA models that yielded the best fit for both period and cohort coefficients according to the method of Box, Jenkins, and Reinsel (1994).

When using linear extrapolations, the particular choice of constraints to identify the model coefficients does not affect the fitted values because all just-identified APC models have an identical fit to the data. That is, the use of the CGLIM (constrained generalized linear model) or IE affects only the estimates of individual effect coefficients of the A, P, and C variables, but not the predicted mortality rates. When using nonlinear extrapolations such as the ARIMA, the predictions using CGLIM and IE coefficients will be somewhat different, however. We employed the IE coefficients throughout the forecasting analyses given previous theoretical findings on the IE's desirable statistical properties.

6.3.2.2 Prediction Intervals

The uncertainty associated with both the model coefficients and their extrapolation will lead to uncertainty in the point forecasting of future mortality rates. It is, therefore, important to quantify this uncertainty through prediction intervals (PIs) or interval forecasting. We derived the PIs using the bootstrap method. Readers who are not interested in the algebraic details can skip this section and directly apply the codes we provide online. We adopt the scheme of resampling residuals instead of the case resampling scheme usually implemented in the bootstrap method (Wu 1986). Because the age, period, and cohort components of our model have a special dependency structure, the use of a case resampling scheme would miss some specific cases in each run and make it impossible to estimate certain periods or cohorts efficiently. The residual resampling scheme, on the other hand, has the advantage of maintaining the structure of the regressors and thus is more appropriate in the context of APC models. The algorithm of the bootstrap method is included in Appendix 6.1 for reference. Note that the PIs generated by the bootstrap procedure are more conservative (wider) than model-based ones. Therefore, although the IE produces statistically more efficient point estimates with smaller standard errors than other estimators, the PIs for predictions using different APC model estimators derived from the bootstrap method are asymptotically similar.

6.3.2.3 Internal Validation

To examine how well simple and APC model-based methods perform in forecasting, we first conducted an internal validation analysis. That is, we withheld mortality rates of the most recent two periods as the actual or observed rates, applied the simple and APC model-based extrapolations to the mortality rates in the base period and then compared the predictions generated from these two methods with the observed rates in the projection period. In addition, this analysis also allowed us to assess how sensitive the predicted values were to the choice of assumptions in coefficient extrapolation. The results of the comparison are summarized in Table 6.3 for the case of male lung cancer mortality.

For ages 30 to 85+ in the eight periods from 1969–1974 to 2005–2007, the observed death rates are listed in the first line. We fit the log-linear APC models to the observed death rates in the base period of 1969 to 1999. The fitted values are shown in the second line. We then extended period effect coefficients to the next two prediction periods of 2000–2004 and 2005–2007 and cohort effect coefficients to the 1965–1969 and 1970–1974 cohorts. The coefficients used and results of linear and ARIMA extrapolations of these coefficients are shown in the left three panels in Table 6.4. The ARIMA algorithm automatically chooses the best model that fits the data (IE coefficients). For the linear coefficient extrapolation, we found that different numbers of cohort coefficients used only affected the first few age groups (e.g., 30–34 and 35–39) in short-term forecasting and produced similar results even within these age groups, which is consistent with the previous study by Osmond (1985) and the IE coefficient estimates of cohort effects that showed relatively linear trends (Figure 6.3). Thus, we present results that used the last 10 cohort coefficients. Different weighted linear extrapolations of period coefficients made considerable differences in predicting mortality rates, with the best predictions generated from weights applied to the last few periods. In the case of lung cancer mortality internal forecasting, the best predicted values were from extrapolations using the last three period coefficients for males and the last two for females. The analysis of other cancer sites showed

TABLE 6.3

Mortality rates per 1,000,000 for Lung Cancer in Males in the U.S.—Observed, Fitted and Predicted Values

Age Group				Period							
				1969-1974	1975-1979	1980-1984	1985-1989	1990-1994	1995-1999	2000-2004	2005-2007
30–34	observed			19.6	15.8	11.3	11.9	11.2	9.5	6.2	5.6
	fit			18.2	15.8	13.1	12.4	10.7	9.0		
		APC	linear							8.0	6.9
			ARIMA							8.4	7.6
		simple	linear							**6.8**	**5.0**
			ARIMA							7.4	5.4
35–39	observed			80.2	63.2	53.4	43.0	40.8	33.4	26.9	19.8
	fit			71.5	64.7	55.0	45.1	41.7	34.1		
		APC	linear							31.5	26.6
			ARIMA							30.5	27.2
		simple	linear							21.2	12.3
			ARIMA							**24.1**	**14.7**
40–44	observed			222.4	203.8	175.0	147.8	120.8	108.5	98.4	75.1
	fit			213.6	204.6	180.8	151.9	121.5	106.9		
		APC	linear							90.4	82.8
			ARIMA							**87.4**	**78.1**
		simple	linear							78.5	54.3
			ARIMA							85.7	62.9

Age		Model	Method								
45–49	observed			495.0	509.2	469.2	405.5	337.8	261.4	243.2	215.8
	fit	APC	linear	495.0	504.8	472.0	412.4	338.3	257.5	**234.8**	**196.4**
			ARIMA							227.2	185.4
		simple	linear							238.4	188.6
			ARIMA							214.7	168.0
50–54	observed			904.9	999.8	992.3	921.9	787.4	617.8	510.7	471.3
	fit	APC	linear	919.0	997.9	993.5	918.3	783.4	611.6	**480.6**	**435.1**
			ARIMA							464.9	410.6
		simple	linear							656.3	595.1
			ARIMA							424.8	208.5
55–59	observed			1549.6	1641.2	1737.7	1725.8	1560.2	1267.6	1055.5	880.2
	fit	APC	linear	1563.4	1646.3	1745.1	1717.4	1550.1	1258.5	**1015.5**	**791.4**
			ARIMA							982.4	746.8
		simple	linear							1413.8	1366.3
			ARIMA							1345.9	1548.4
60–64	observed			2345.4	2553.3	2624.9	2746.2	2646.1	2239.1	1904.6	1666.4
	fit	APC	linear	2364.1	2543.0	2614.2	2739.2	2632.3	2261.1	**1897.4**	**1518.4**
			ARIMA							1835.6	1432.9
		simple	linear							2512.7	2508.9
			ARIMA							2525.8	2525.8

continued

TABLE 6.3 (continued)

Mortality rates per 1,000,000 for Lung Cancer in Males in the U.S.—Observed, Fitted and Predicted Values

Age Group						Period				
			1969-1974	1975-1979	1980-1984	1985-1989	1990-1994	1995-1999	2000-2004	2005-2007
65–69	observed		3134.2	3473.3	3679.4	3735.3	3813.3	3437.8	3018.1	2709.3
	fit		3122.3	3485.1	3659.8	3718.9	3805.1	3479.9		
	APC	linear							3089.4	2571.1
		ARIMA							2988.8	2426.4
	simple	linear							3804.9	3879.0
		ARIMA							3545.5	3545.5
70–74	observed		3642.6	4224.7	4629.5	4827.7	4750.5	4626.3	4214.2	3814.1
	fit		3632.4	4243.3	4623.9	4799.7	4762.6	4637.5		
	APC	linear							4383.3	3859.5
		ARIMA							4240.5	3642.2
	simple	linear							5119.6	5310.9
		ARIMA							4338.9	3950.6
75–79	observed		3757.0	4490.0	5108.5	5457.5	5497.1	5260.5	5136.3	4852.8
	fit		3719.7	4461.7	5088.4	5480.9	5555.5	5246.2		
	APC	linear							5279.4	4949.2
		ARIMA							5107.5	4670.6
	simple	linear							6017.2	6328.3
		ARIMA							4874.2	4510.6

80–84	observed		3230.5	4247.9	4988.7	5597.2	5935.8	5710.3	5419.8	5376.2
	fit	APC linear	3241.2	4257.8	4985.9	5620.6	5911.9	5702.8	5565.6	5555.0
		ARIMA							**5384.4**	**5242.4**
	simple	linear							6758.8	7275.1
		ARIMA							5484.7	5259.2
85+	observed		2358.1	3211.1	4008.5	4637.5	5208.8	5346.5	4904.5	4619.7
	fit	APC linear	2358.1	3192.4	4094.1	4738.9	5216.7	5221.8	5205.7	5039.0
		ARIMA							**5036.2**	**4755.4**
	simple	linear							6284.8	6900.9
		ARIMA							5944.1	6541.8

TABLE 6.4

APC Model Coefficients and Projections for Lung Cancer in Males in the U.S.

	Internal Validation			Forecasting		
	Coefficient	linear	ARIMA	Coefficient	linear	ARIMA
Period						
1969–1974	−0.34			−0.27		
1975–1979	−0.17			−0.15		
1980–1984	−0.03			−0.06		
1985–1989	0.09			0.03		
1990–1994	0.20			0.08		
1995–1999	0.25			0.09		
2000–2004		0.34	0.31	0.11		
2005–2007		0.42	0.36	0.17		
2010–2014					0.22	0.22
2015–2019					0.28	0.28
2020–2024					0.33	0.33
2025–2029					0.39	0.39
Cohort						
−1885	0.46			0.26		
1890	0.60			0.43		
1895	0.71			0.58		
1900	0.72			0.64		
1905	0.71			0.68		
1910	0.66			0.67		
1915	0.57			0.62		
1920	0.46			0.56		
1925	0.37			0.52		
1930	0.23			0.42		
1935	0.02			0.26		
1940	−0.24			0.07		
1945	−0.54			−0.16		
1950	−0.87			−0.41		
1955	−1.05			−0.55		
1960	−1.31			−0.72		
1965	−1.48			−1.01		
1970		−1.73	−1.65	−1.37		
1975		−1.96	−1.82	−1.50		
1980					−1.70	−1.62
1985					−1.91	−1.75
1990					−2.13	−1.87
1995					−2.35	−1.99

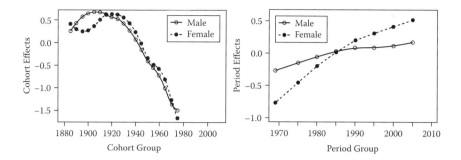

FIGURE 6.3
The IE coefficient estimates of lung cancer mortality by sex: cohort and period effects.

similar findings. The best predictions were from weighted LMs fit to the last three period coefficients for white female breast cancer mortality and the last two coefficients for all the others (male and female colorectal cancer, white and black prostate cancer, and black breast cancer). These findings are consistent with previous cancer-forecasting studies (Bray and Møller 2006; Møller et al. 2003) that suggested that rapid changes in cancer-screening techniques and treatments have occurred in more recent years, reducing the relevance of data from the more distant past for forecasting. In this case, it is more reasonable to allow greater weight to be given to more recent data. In addition, the IE estimates of period effects based on all periods of data illustrated in Figure 6.3 further suggest empirically the advantage of using the last several coefficients. The period coefficients for female lung cancer mortality had one inflection point at year 1990. Therefore, including only points at years 1990 and 1995 yielded a better prediction of the coefficients at years 2000 and 2005. For males, there were multiple inflection points, and including only the last two coefficients would lead to underestimates of the values for the years 2000 and 2005, but adding one previous coefficient at year 1985 would reduce the bias.

The predicted male lung cancer mortality rates using the linear and ARIMA extrapolated coefficients from the APC models are then included in Table 6.3 for comparison with the simple extrapolations of mortality rates. For each age group, the best average predictions in the two periods (2000–2004 and 2005–2007) are highlighted in **bold**. With the exception of the first two age groups, the APC model-based projections performed much better than the simple extrapolations of age-specific mortality rates. This can be more clearly seen in Figure 6.4, which plots the age patterns of observed and predicted mortality rates for the two projection periods together with the 95% PIs. For simple extrapolation, only the linear extrapolation results are shown in the interest of space as the results hold for the ARIMA extrapolations. The figure shows that the predicted

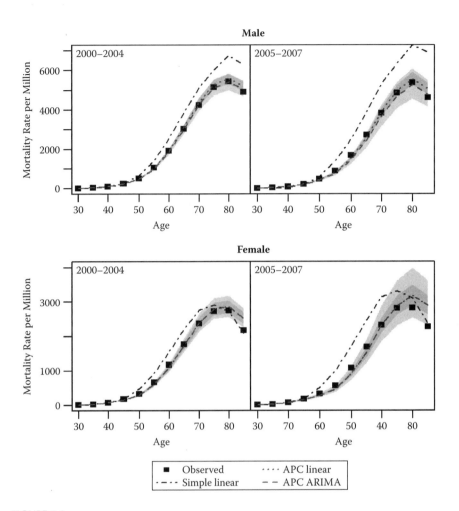

FIGURE 6.4
Age-specific lung cancer mortality rate predicted by different methods with 95% PIs: 2000–2004 and 2005–2007.

mortality rates from the simple linear extrapolation were far higher than the actual mortality rates for both periods, and this difference increased with age. In contrast, the two APC model-based projections yielded much better predictions. The comparison of the interval predictions of the last two methods showed that the linear extrapolation produced narrower PIs than the ARIMA method. This is because the ARIMA models automatically incorporated the stochastic process, allowing for appropriate random structure of the prediction. The ARIMA method thus provided better coverage of the true values. For example, the point predictions from both methods overestimated female death rates at age 85+, but the ARIMA PIs covered the true values, whereas the linear PIs did not. Similar results of internal validation were found for the other cancer sites. While the linear extrapolation method yielded better point forecasts, the ARIMA method produced wider PIs that covered the true values when the point forecasts were biased.

Figure 6.5 further compares the observed and projected lung cancer mortality rates by time period for selected age groups between 40–44 and 70–74. Figure 6.6 shows the corresponding results for other cancer sites. Both suggest that the APC model-based projections were vastly superior to the simple extrapolation projections. This was especially pronounced when the mortality trends changed directions in certain periods due to cohort changes. The simple linear projections extended previous period trends but ignored cohort changes. For example, the lung cancer and white female breast cancer mortality rates were predicted to increase in the projection period (2000–2007) at older ages by the simple linear extrapolation method, whereas the observed rates declined. While the female colorectal cancer mortality rates were predicted to decrease in the projection period at ages 40–44 by the simple linear extrapolation method, the observed rates slighted increased. The simple linear projections deviated from observed mortality rates by large measures for the other cases. In contrast, the APC model-based projections were able to capture actual mortality rates much better by accounting for cohort changes that partly generated the observed trends.

6.3.2.4 Forecasting Results

Based on the results of the foregoing internal validation, we forecast the mortality rates of major cancer sites into the next 20 years from 2010 to 2029. Using the full set of observed mortality data from 1969 to 2007, we first estimated the APC models using the IE to obtain the period and cohort coefficients and fit linear regression models and ARIMA models to these coefficients. The results of the lung cancer mortality forecasting by sex are shown in the right three panels of Table 6.4. For the linear coefficient extrapolations, we used the last two period coefficients and the last 10 cohort coefficients as the base for projection based on conclusions from previous research and the internal validation findings. The results obtained from this approach that emphasized most recent period influences can then be compared with those from the ARIMA approach that utilized information from all previous periods and cohorts.

Table 6.5 presents the point forecasting results for the lung cancer mortality rates using the two APC model-based methods. The forecasted age-specific rates with 95% PIs are then displayed in Figure 6.7a. The male lung cancer mortality rates were predicted to continue to decline for all ages, with the rates of decline slowing down in 2025–2029 for ages 50–54 and 70–74. The female lung cancer death rates were predicted to continue to increase for the next 10 years for ages 80–84 and decline afterward. The lags between

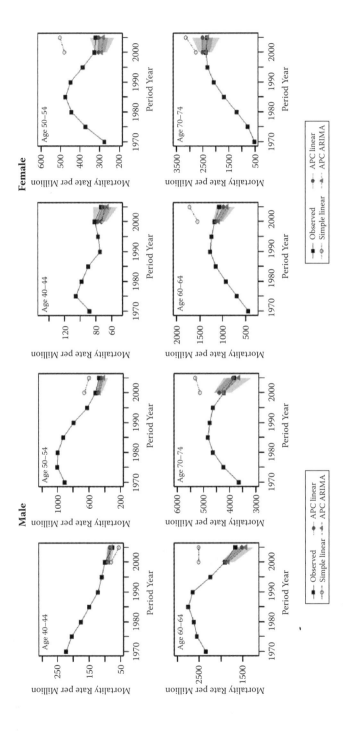

FIGURE 6.5
Lung cancer mortality rate predicted by different methods with 95% PIs by period for selected age groups.

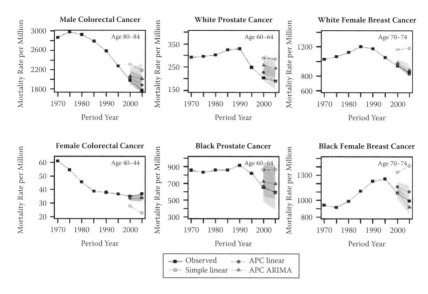

FIGURE 6.6
Colorectal, prostate, and breast cancer mortality rate predicted by different methods with 95% PIs by period for selected age groups.

the male and female declines reflect the sex difference in cigarette smoking for cohorts born in the early twentieth century widely observed in previous studies and suggested by the IE estimates of cohort trends in lung cancer mortality (Figure 6.3). Figures 6.7b–6.7d display the forecasted age-specific rates with the 95% PIs for colorectal, prostate, and breast cancer mortality, respectively. Overall, point forecasts showed mortality declines for all cancer sites across most ages. Since the period effects were mostly projected to increase or remain constant, this was largely due to the cohort effects that showed decreases in cohort mortality. The point forecasts from the two projection methods were similar for lung mortality. For the other three cancer sites, the two methods demonstrated larger differences. The differences were particularly large for blacks, with the ARIMA forecasts predicting less declines in prostate cancer mortality and evident increases in breast cancer mortality. With the exception of male colorectal cancer mortality, the predicted values from the ARIMA models were consistently higher than those from the LMs. The ARIMA models and the LMs yielded much similar results of lung cancer mortality rates because the period effect coefficients had a stronger linear trend (Figure 6.3). In contrast, the period effect coefficients showed considerable slowdowns in mortality increases for prostate and breast cancer mortality in more recent periods. While the LMs fit to the last two periods extended such slowdowns into the future, the ARIMA models incorporated earlier periods of mortality increases and resulted in extensions of higher period coefficients, which in turn translated into less-optimistic prospects of predicted rates.

TABLE 6.5

Mortality rates per 1,000,000 for Lung Cancer in the U.S.—Predicted Values (2010–2029)

Age	APC	Male				Female			
		2010–2014	2015–2019	2020–2024	2025–2029	2010–2014	2015–2019	2020–2024	2025–2029
30–34	linear	4.9	4.2	3.5	3.0	4.6	4.0	3.5	3.1
	ARIMA	5.3	4.9	4.6	4.3	3.4	2.7	2.1	1.7
35–39	linear	18.8	16.3	13.9	11.8	14.1	15.2	13.3	11.6
	ARIMA	18.8	17.6	16.4	15.4	14.1	11.2	8.9	7.0
40–44	linear	55.7	52.0	45.1	38.4	49.4	36.9	39.7	34.7
	ARIMA	55.7	52.0	48.6	45.4	49.3	36.7	29.1	23.1
45–49	linear	170.3	125.1	116.9	101.4	152.2	106.6	79.6	85.6
	ARIMA	170.3	125.1	116.9	109.3	151.9	106.1	78.8	62.5
50–54	linear	414.7	327.6	240.7	224.9	317.0	273.7	191.7	143.1
	ARIMA	414.7	327.6	240.7	224.9	316.4	272.3	189.9	141.1

55–59	linear	789.5	704.2	556.3	408.7	504.0	515.7	445.3	311.9
	ARIMA	789.5	704.2	556.3	408.7	503.0	513.1	441.2	307.6
60–64	linear	1311.4	1210.2	1079.5	852.8	792.5	759.1	776.7	670.6
	ARIMA	1311.4	1210.2	1079.5	852.8	791.0	755.1	769.5	661.4
65–69	linear	2220.5	1814.6	1674.6	1493.6	1382.5	1100.1	1053.6	1078.0
	ARIMA	2220.5	1814.6	1674.6	1493.6	1379.8	1094.4	1043.9	1063.2
70–74	linear	3346.7	2820.6	2305.0	2127.1	2099.5	1805.2	1436.4	1375.7
	ARIMA	3346.7	2820.6	2305.0	2127.1	2095.4	1795.9	1423.2	1356.8
75–79	linear	4437.8	3864.1	3256.7	2661.4	2733.2	2492.9	2143.5	1705.6
	ARIMA	4437.8	3864.1	3256.7	2661.4	2727.9	2480.0	2123.7	1682.0
80–84	linear	5263.2	4759.5	4144.2	3492.8	3084.6	3004.8	2740.6	2356.4
	ARIMA	5263.2	4759.5	4144.2	3492.8	3078.6	2989.2	2715.3	2323.9
85+	linear	5007.4	4788.2	4330.0	3770.3	2804.4	2897.8	2822.9	2574.7
	ARIMA	5007.4	4788.2	4330.0	3770.3	2799.0	2882.9	2796.9	2539.2

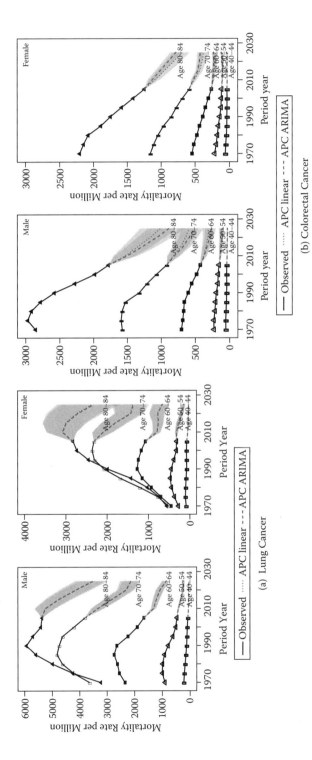

FIGURE 6.7
Forecasting of age-specific cancer mortality rates for 2010–2029 with 95% PIs.

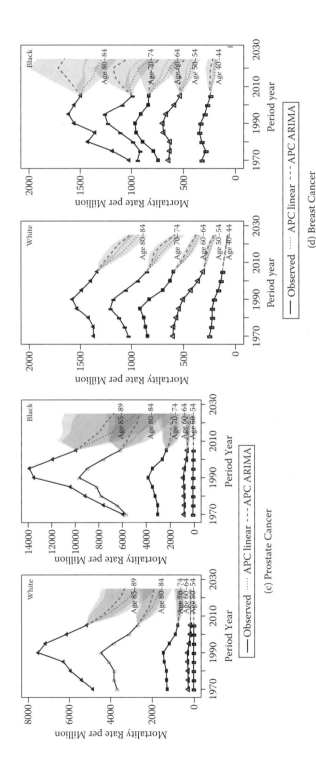

FIGURE 6.7 (continued)

The interval forecasting results showed increases in widths of the PIs with age and time period. The PIs of the forecasts for lung cancer mortality were considerably tighter for males than females, whereas the PIs of the forecasts for colorectal cancer mortality were narrower for females than males. The PIs for the forecasts of prostate and breast cancer mortality were substantially larger for blacks than whites, particularly at older ages, indicating that the mortality data were less robust for blacks than for whites at these ages.

In sum, we have demonstrated the utility of the APC models for forecasting using cancer mortality as examples. The APC model-based forecasting methods showed substantial advantages over the simple data extrapolation methods in accuracy of point forecasts and assessments of uncertainty in these forecasts through PIs. Like any forecasting studies, the APC model-based forecasting should rest on reasonable assumptions of how future rates relate to the past. Extrapolations of model coefficients yielded results highly sensitive to the choice of assumptions and must be evaluated empirically. We have shown how to achieve a range of predictions based on alternative assumptions. The ARIMA approach produced wider PIs and thus may be more appropriate for conservative forecasting if no other information exists that indicates stronger influences of more recent periods on the future. In the case of cancer, changes in period conditions can occur frequently. So, weighted linear regression models based on the most recent periods can be highly effective. The APC model-based forecasting methods introduced here concern the extrapolations of model coefficients of period and cohort effects. Recent studies of cancer incidence also suggested the use of health-based forecasting that incorporates health-related risk factors such as smoking rates and treatment effects (Manton, Akushevich, and Kravchenko 2009). How to conjoin these two modeling approaches to further improve the accuracy of forecasting of cancer incidence and mortality rates remains a challenge for future development of forecasting methodology.

Appendix 6.1: The Bootstrap Method Using a Residual Resampling Scheme for Prediction Intervals

1. Fit the APC model and retain the fitted link values $\hat{\mu}_i$ and the residual $\hat{\varepsilon}_i = \mu_i - \hat{\mu}_i = \log(y_i) - \hat{\mu}_i$, $i = 1, 2, ..., n$, where n indicates the number of observations or the number of age groups a times the number of period groups p.

2. For each pair x_i, μ_i in which x_i is the explanatory variable, add a randomly resampled residual $\hat{\varepsilon}_j$ to the link variable μ_i. In other words, create the synthetic link and response variables $\mu_i^* = \hat{\mu}_i + \hat{\varepsilon}_j$ and $y_i^* = \exp(\mu_i^*)$ where j is selected randomly from the list $(1, ..., n)$ for every i.

3. Refit the model using the fictitious response variables y_i^*, make forecasting based on the synthetic $\hat{y}_i^*$ as from the prediction procedure, and retain the predicted $\hat{y}_i^*$.

4. Repeat Steps 2 and 3 a large number of times (10,000 times here) to obtain (10,000) samples of $\hat{y}_i^*$ and derive the mean, standard errors, and PIs.

References

Box, G. E. P., G. M. Jenkins, and G. C. Reinsel. 1994. *Time series analysis: Forecasting and control*. Vol. 3. Upper Saddle River, NJ: Prentice Hall.

Bray, F., and B. Møller. 2006. Predicting the future burden of cancer. *Nature Reviews Cancer* 6:63–74.

Clayton, D., and E. Schifflers. 1987. Models for temporal variation in cancer rates. II: Age-period-cohort models. *Statistics in Medicine* 6:469–481.

Finch, C. E., and E. M. Crimmins. 2004. Inflammatory exposure and historical change in human life-spans. *Science* 305:1736–1739.

Fogel, R. W. 2004. Health, nutrition, and economic growth. *Economic Development and Cultural Change* 52:643–658.

Gardner, M. J., and C. Osmond. 1984. Interpretation of time trends in disease rates in the presence of generation effects. *Statistics in Medicine* 3:113–130.

Guillot, M. 2006. Tempo effects in mortality: An appraisal. *Demographic Research* 14:1–26.

Harris, J. E. 1983. Cigarette smoking among successive birth cohorts of men and women in the United States during 1900–80. *Journal of the National Cancer Institute* 71:473–479

Jemal, A., K. C. Chu, and R. E. Tarone. 2001. Recent trends in lung cancer mortality in the United States. *Journal of the National Cancer Institute* 93:277–283.

Jemal, A., M. J. Thun, L. A. G. Ries, et al. 2008 Annual report to the nation on the status of cancer, 1975–2005, featuring trends in lung cancer, tobacco use, and tobacco control. *Journal of the National Cancer Institute* 100:1672–1694.

Kupper, L. L., J. M. Janis, A. Karmous, and B. G. Greenberg. 1985. Statistical age-period-cohort analysis: A review and critique. *Journal of Chronic Diseases* 38:811–830.

Lee, R. D., and L. R. Carter. 1992. Modeling and forecasting U. S. mortality. *Journal of the American Statistical Association* 87:659–671.

Lee, W. C., and R. S. Lin. 1996. Autoregressive age-period-cohort models. *Statistics in Medicine* 15:273–281.

Lopez, A. D. 1995. The lung cancer epidemic in developed countries. In *Adult mortality in developed countries: From description to explanation*, ed. A. D. Lopez, G. Caselli, and T. Valkonen, 111–143. New York: Oxford.

Manton, K. G. 2000. Gender differences in the cross-sectional and cohort age dependence of cause-specific mortality: The United States, 1962 to 1995. *Journal of Gender-Specific Medicine* 3:47–54.

Manton, K. G., I. Akushevich, and J. Kravchenko. 2009. *Cancer mortality and morbidity patterns in the U.S. population: An interdisciplinary approach*. New York: Springer-Verlag.

McNown, R., and A. Rogers. 1989. Forecasting mortality: A parameterized time series approach. *Demography* 26:645–660.

Møller, B., H. Fekjær, T. Hakulinen, et al. 2003. Prediction of cancer incidence in the Nordic countries: Empirical comparison of different approaches. *Statistics in Medicine* 22:2751–2766.

Olshansky, S. J., B. A. Carnes, and M. S. Mandell. 2009. Future trends in human longevity: Implications for investments, pensions and the global economy. *Pensions: An International Journal* 14:149–163.

Osmond, C. 1985. Using age, period and cohort models to estimate future mortality rates. *International Journal of Epidemiology* 14:124–129.

Pampel, F. C. 2005. Diffusion, cohort change, and social patterns of smoking. *Social Science Research* 34:117–139.

Reither, E. N., S. J. Olshansky, and Y. Yang. 2011. New forecasting methodology indicates more disease and earlier mortality ahead for today's younger Americans. *Health Affairs* 30:1562–1568.

Shibuya, K., M. Inoue, and A. D. Lopez. 2005. Statistical modeling and projections of lung cancer mortality in 4 industrialized countries. *International Journal of Cancer* 117:476–485.

Tarone, R., and K. Chu. 2000. Age-period-cohort analyses of breast-, ovarian-, endometrial- and cervical-cancer mortality rates for Caucasian women in the USA. *Journal of Epidemiology and Biostatistics* 5:221–231.

Tarone, R. E., K. C. Chu, and L. A. Gaudette. 1997. Birth cohort and calendar period trends in breast cancer mortality in the United States and Canada. *Journal of the National Cancer Institute* 89:251–256.

Wu, C. F. J. 1986. Jackknife, bootstrap and other resampling methods in regression analysis. *The Annals of Statistics* 14:1261–1295.

Yang, Y. 2008. Trends in U.S. adult chronic disease mortality, 1960–1999: Age, period, and cohort variations. *Demography* 45:387–416.

Zang, E. A., and E. L. Wynder. 1996. Differences in lung cancer risk between men and women: Examination of the evidence. *Journal of the National Cancer Institute* 88:183–192.

7

Mixed Effects Models: Hierarchical APC-Cross-Classified Random Effects Models (HAPC-CCREM), Part I: The Basics

7.1 Introduction

In Chapter 5, we discussed the advantages and disadvantages of various approaches to identification and estimation of the A, P, and C effect coefficients for the conventional APC accounting/multiple classification model. In this context, the Intrinsic Estimator method has a number of desirable statistical properties and often yields good estimates of the effect coefficients. It is important to point out, however, that the classical APC identification problem is not inevitable. It is largely a function of the application of conventional linear or generalized linear models to APC data, often but not exclusively in the form of the first research design (tabulated tables of population rates or proportions). These models not only are certain to produce the identification problem but also may be a poor approximation to the process of social change that a researcher seeks to model. We now discuss major limitations of linear models and introduce alternative model specifications through the mixed effects models increasingly employed in demography, epidemiology, and the social sciences to address these limitations.

We then define and discuss the properties of a hierarchical age-period-cohort (HAPC) modeling framework. An HAPC framework does not incur the identification problem because the three effects are not assumed to be linear and additive at the same level of analysis. In addition, HAPC models can capture the contextual effects of periods and cohorts and thus stimulate new conceptualizations of processes of change. This chapter focuses on the exposition of the basic cross-classified random effects modeling (CCREM) specifications of HAPC models through algebraic analysis and empirical analyses of examples of the second research design (repeated cross-sectional sample surveys), including the General Social Survey (GSS) and National Health and Nutrition Examination Survey (NHANES) described in Chapter 3, to show

how HAPC models for such designs avoid the identification problem and offer more opportunities for testing explanatory hypotheses.

7.2 Beyond the Identification Problem

In many previous applications of linear models to age-period-cohort (APC) analysis, the identification problem could not be avoided. First, because most of these applications were focused on aggregate population data using the first research design, the three variables relate to one another in an exact linear fashion. For example, for each age and calendar year, there is a unique birth cohort. This linear dependency, however, does not constitute a problem until certain assumption is made about how the data simultaneously affect an outcome of interest. The linear models fit to the APC data in this form, namely, APC accounting models, induce this problem by imposing the assumption of additivity and treating the three temporal variables as fixed effects that are independent of each other. Since the formalization of accounting/multiple classification models for APC analysis in the early 1970s (Mason et al. 1973), the voluminous literature in social sciences and biostatistics has focused on the resulting underidentification "conundrum" as a methodological problem. However, the problem is also a theoretical one. That is, a linear APC model may not accurately describe age-, period-, and cohort-related phenomena. It should be noted that additivity of the effects of these three temporal dimensions is only one simple approximation to the process generating time-related changes. Some cohort or period phenomena involve problems that can hardly be handled by any version of APC accounting models (Hobcraft, Menken, and Preston 1982; Smith 2004). So, there is a need to develop new models that accommodate both additive and nonadditive processes.

Our point of departure then is to develop such models that present a more thorough solution to the identification problem (by not incurring it); better characterize the cohort and period effects; stimulate new conceptualizations of processes of social, demographic, and epidemiologic changes; and are capable of addressing new theoretical questions about these changes. We begin with the observation in Chapter 3 that the repeated cross-sectional design is in fact a multilevel design in which individual-level observations are nested in, and cross-classified simultaneously by, the two higher-level social contexts defined by time period and birth cohort.[*] Applications of linear models to such a design based on the assumption of fixed period and

---

[*] Parts of Sections 7.2–7.5 are adapted and updated from Yang, Y., and K. C. Land. 2006. *Sociological Methodology* 36:75–97 and Yang, Y., and K. C. Land. 2008. *Sociological Methods & Research* 36:297–326.

cohort effects ignore the multilevel structure of the data design and may not be adequate substantively or statistically.

We have noted before that when data become available to allow age intervals to differ from period intervals, such as the case of sample surveys in the second research design, one can utilize the unequal intervals to group one or more of the three variables to break their exact linear dependency. As explained in Chapter 4, this solution is not capable of resolving the identification problem completely when used in the conventional linear models in that the results may be sensitive to specific interval lengths and hence degrees of overidentification. We further note here that such differential groupings are also problematic for statistical inference. To illustrate this, consider the application of the fixed effects APC regression model of Equation (4.5) to, say, the following five sample members, ages 31, 32, 33, 34, and 35 at last birthday, each of whom is a member of the same 5-year birth cohort, the 1955–59 birth cohort, and each of whom is a survey respondent in the 1990 GSS:

$$Y_{1,1990,1955-59} = \mu + \alpha(31) + \varepsilon_{1,1990,1955-59} \tag{7.1}$$

$$Y_{2,1990,1955-59} = \mu + \alpha(32) + \varepsilon_{2,1990,1955-59} \tag{7.2}$$

$$Y_{3,1990,1955-59} = \mu + \alpha(33) + \varepsilon_{3,1990,1955-59} \tag{7.3}$$

$$Y_{4,1990,1955-59} = \mu + \alpha(34) + \varepsilon_{4,1990,1955-59} \tag{7.4}$$

$$Y_{5,1990,1955-59} = \mu + \alpha(35) + \varepsilon_{5,1990,1955-59} \tag{7.5}$$

where the five individual sample respondents are numbered from 1 through 5, respectively, their respective ages (31 through 35) have been entered into the age term of the model whose coefficient is α. To complete the APC model specification of Equation (4.5) in Chapter 4, we have the following specification on the error terms:

$$\varepsilon_{i,1990,1960-64} = \beta_{1990} + \gamma_{1955-59} + e_{i,1990,1955-59}, \text{ for } i = 1, 2, ..., 5 \tag{7.6}$$

The error terms specifically include fixed effect coefficients to measure the impact of the time period (β_{1990}) and the birth cohort ($\gamma_{1956-59}$) to which these sample respondents belong.

The fixed effects model of Equations (7.1)–(7.6) could be estimated straightforwardly by using dummy variables to control for the period and cohort effects in a conventional multiple linear regression model using the differential grouping as the identifying constraint (single-year-of-age and 5-year cohorts). The key assumption used here is that impacts of cohort and period (survey year) on the responses of sample members are adequately modeled as fixed. This ignores the possibility that the effects of cohort membership and survey year may have shared random as well as, or instead of, fixed effects on the responses; that is, that the observations are correlated. That is, sample respondents in the same cohort group or survey year may be similar in their responses due to the fact that they share unmeasured contextual random error components (i.e., through random cohort or period components of $e_{i,1990,1955-59}$) unique to their cohorts or periods of the survey. Note that the sharing of common elements in the error terms may result in such weak covariation among responses that there are no serious complications for estimates of APC coefficients for standard regression models estimated by ordinary least squares (OLS). But, a failure to assess this potentially more complicated error structure adequately in APC analysis may have serious consequences for statistical inferences. The standard errors of estimated coefficients of fixed effects regression models like Equations (7.1)–(7.6) may be underestimated, leading to inflated t ratios and actual alpha levels that are larger than the nominal .05 or .01 levels (Hox and Kreft 1994).

This heterogeneity problem can be addressed by modifying the fixed effects specification toward a random effects or random coefficients regression model. That is, to take into account the possibility that the common period and cohort elements of the error terms of Equations (7.1)–(7.6) are statistically significant, we should allow for the possibility that at least some of the effect coefficients μ, α, β_{1990}, and $\gamma_{1955-59}$ in those equations are not fixed but instead vary randomly by cohort or time period. This implies that we should modify the fixed effects linear APC model (4.5) in Chapter 4 to a mixed effects model.

The *mixed (fixed and random) effects model* is thus a reasonable alternative to the linear model because it does not assume fixed age, period, and cohort effects that are additive and therefore avoids the identification problem. More important, it can statistically characterize contextual effects of historical time and cohort membership and reveal the process by which individuals' lives are shaped by their environment. Prior studies utilizing multilevel mixed effects models in the analysis of temporal change are few but exist in demographic research on health (Lynch 2003) and developmental psychology and aging research on cognitive skills (e.g., Alwin 2009). These examples did not explicitly embody a full-blown APC analysis and focused instead on age patterns in the context of specific cohorts only. The breakthrough of the approach to cohort analysis developed initially by Yang and Land (2006) and described in the following is the simultaneous modeling of all three factors using multilevel data and mixed effects or hierarchical models.

In addition to its ability to account for the multilevel heterogeneity in the data, the mixed effects modeling approach has a second important advantage of being able to explain random variability by the incorporation of additional covariates at both the individual level (especially for microsurvey data design) and the higher level. That is, through the specification and modeling of individual attributes and contextual characteristics of time period and birth cohort, this approach allows for the test of explanatory hypotheses of

A, P, and C effects not possible in the APC accounting model framework. It also improves on the proxy variable approach mentioned in Chapter 4. That is, whereas the proxy variable approach equalizes period or cohort effects with period- or cohort-specific variables and thus is vulnerable to problems of misspecification, the mixed effects model approach accounts for the uncertainty of corresponding effects through explicit error terms and provides means to ascertain how much of the period or cohort variance can actually be explained by measured period- or cohort-level characteristics. All these features of the mixed effects modeling framework facilitate mechanism-driven analyses that move far beyond the conventional APC analysis of trend identification.

7.3 Basic Model Specification

The simplest form of model specification of the HAPC model as introduced by Yang and Land (2006) is a member of the class of *linear mixed models* (LMMs).* This model specification is the most widely used form of hierarchical or multilevel linear models as presented in such standard expositions as those of Raudenbush and Bryk (2002) and Snijders and Bosker (1999). It consists of a two-component model: The level 1 component is a regression of an individual-level outcome variable on a set of individual-level explanatory variables (regressors, covariates) with an intercept term, fixed regression slope coefficients, and an individual-level random error term. Level 2 models use level 1 regression coefficients as outcomes and contain intercepts and specification of random effect coefficients for the effects of each cohort and time period distinguished in the model. The level 2 model may also contain cohort or time period explanatory variables with fixed effect coefficients that are hypothesized to explain, at least in part, the cohort or period effects (Yang 2006), which are illustrated in the next chapter.

To describe the HAPC model, we focus for now on the example of trends of verbal ability in the GSS data. The previous findings on trends in verbal scores are interesting and suggestive. But until age, period, and cohort effects are simultaneously estimated, the question of whether the trends are due to any of these components remains unresolved. In recognition of the multilevel characteristics of the GSS data structure shown in Chapter 3 (Table 3.5), we formulate a CCREM specification of the HAPC model to assess the

---

* General matrix algebraic representations of LMMs and generalized linear mixed models (GLMMs) are given in Appendix 7.1. The exposition in the chapter text focuses on scalar representations accessible to readers who do not have knowledge of matrix algebra.

relative importance of the two contexts, cohort and period, in understanding individual differences in verbal test outcome.

Because the WORDSUM outcome variable has a relatively bell-shaped sample frequency distribution, it is reasonable to use an HAPC mixed model specification that has a conventional normal errors level 1 fixed effects regression model. In the absence of evidence to the contrary, this level 1 model can be combined with a conventional normal period and cohort residual random effects specification at level 2. In such an LMM model applied to the verbal test data, variability in WORDSUM associated with individuals, cohorts, and periods is specified as follows:

Level 1 or "within-cell" model:

$$WORDSUM_{ijk} = \beta_{0jk} + \beta_1 AGE_{ijk} + \beta_2 AGE_{ijk}^2 + \beta_3 EDUCATION_{ijk}$$

$$+ \beta_4 SEX_{ijk} + \beta_5 RACE_{ijk} + e_{ijk} \tag{7.7}$$

$$\text{with } e_{ijk} \sim N(0, \sigma^2)$$

Level 2 or "between-cell" model:

$$\beta_{0jk} = \gamma_0 + u_{0j} + v_{0k} \text{, with } u_{0j} \sim N(0, \tau_u) \text{, } v_{0k} \sim N(0, \tau_v) \tag{7.8}$$

Combined model:

$$WORDSUM_{ijk} = \gamma_0 + \beta_1 AGE_{ijk} + \beta_2 AGE_{ijk}^2 + \beta_3 EDUCATION_{ijk}$$

$$+ \beta_4 SEX_{ijk} + \beta_5 RACE_{ijk} + u_{0j} + v_{0k} + e_{ijk} \tag{7.9}$$

for
$i = 1, 2, \ldots, n_{jk}$ individuals within cohort j and period k;
$j = 1, \ldots, 20$ birth cohorts;
$k = 1, \ldots, 17$ survey years;

where, within each birth cohort j and survey year k, respondent i's verbal score is modeled as a function of his or her age, age-squared, educational attainment, and two covariates, gender and race, that have been found in previous research to be related to verbal ability (see, e.g., Hedges and Nowell 1995; Campbell, Hombo, and Mazzeo 2000). This *random intercepts model* specification allows only the level 1 intercept to vary randomly from cohort to cohort and period to period, but not the level 1 slopes. One can also specify the cross-classified *random coefficients model* specification of the HAPC model (see Section 7.6) wherein level 1 slope coefficients also have random varia-tions across cohorts and periods. Because supplemental analyses of the ver-bal test score data did not show significant random cohort or period effects of any level 1 slope coefficients, we estimated only the random intercepts model. We illustrate this more sophisticated specification in other examples.

In model (7.7)–(7.9), β_{0jk} is the intercept or "cell mean," that is, the mean verbal test score of individuals who belong to birth cohort j and surveyed in year k; $\beta_1, \ldots, \beta_5$ are the level 1 fixed effects; e_{ijk} is the random individual effect, that is, the deviation of individual ijk's score from the cell mean, which is assumed normally distributed with mean 0 and a within-cell variance σ^2; γ_0 is the model intercept or grand-mean verbal test score of all individuals; u_{0j} is the cohort effect or residual random effect of cohort j, that is, the contri-bution of cohort j averaged over all periods, on β_{0jk}, assumed normally dis-tributed with mean 0 and variance τ_u; and v_{0k} is the period effect or residual random effect of period k, that is, the contribution of period k averaged over all cohorts, assumed normally distributed with mean 0 and variance τ_v. In addition, $\beta_{0j} = \gamma_0 + u_{0j}$ is the cohort verbal test score random effect averaged over all periods; and $\beta_{0k} = \gamma_0 + v_{0k}$ is the period verbal test score random effect averaged over all cohorts.

The HAPC-CCREM of Equations (7.7)–(7.9) is defined by its statistical parameters: the regression parameters $\gamma_0, \beta_1, \beta_2, \beta_3, \beta_4, \beta_5$ and the variance components, $\sigma^2, \tau_u,$ and τ_v. Two major approaches to the estimation of these parameters (Longford 1993; Jiang 2007) are *maximum likelihood* (ML) and *restricted maximum likelihood* (REML). The two methods differ little with respect to estimation of the regression coefficients. For estimation of the vari-ance components, however, the REML method takes into account the loss of degrees of freedom resulting from the estimation of the regression param-eters, whereas the ML method does not. The consequence is that the ML esti-mators for the variance components have a downward bias, and the REML estimators do not. For mixed models with relatively small numbers, say less than 30, of groups, this can be important. Since the numbers of time periods and cohorts in HAPC models almost always will be less than 30, REML is the preferred estimator for HAPC models.

This model specification can be used to highlight an important differ-ence between the fixed effects coefficients of the level 1 model (7.7) and the random effects of the level 2 model (7.8). In brief, the individual-level fixed effects regression coefficients are *parameters* to be estimated in a conventional statistical sense. As Snijders and Bosker (1999: 58) emphasized, by contrast the random group (time period and cohort) effects are *latent variables* (i.e., not directly observable) rather than statistical parameters and accordingly are not an integral part of the statistical parameter estimation (see Appendix 7.1). In statistical terminology, the random effects are "predicted" rather than esti-mated, as the term estimation is reserved for finding likely values of statistical parameters. The method used to predict the random effects is *empirical Bayes*

(EB), which produces *posterior means* (Efron and Morris 1975; Jiang 2007: 86). For a specific group in model (7.7)–(7.9), say a specific cohort, the EB estimator predicts its random effect as a *weighted average* of the adjusted grand mean or population average of the WORDSUM outcome variable across the pooled sample of all observations and the estimated average impact of that cohort on the expected value of the outcome variable across all time periods. The weighted average takes the form of a convex combination of these two estimates, where the weights are a function of the estimates of the variances of the cohort effects and the errors of the level 1 model. The resulting predicted random effect *shrinks the cohort-specific estimated effect toward the adjusted grand mean*, which is considered to be more reliably estimated (because it is based on a larger sample of observations) than the average impact of a specific cohort. EB estimators thus are biased toward the population averages but have a smaller mean-squared error for a randomly drawn group or cohort. When used in combination with the REML method for estimation of the HAPC model parameters, the result is a REML-EB algorithm, which is the method programmed into most conventional software programs for mixed model estimation.

An important decision in mixed effects/hierarchical linear model analyses pertains to "centering" or choosing the location of the individual-level explanatory variables (Raudenbush and Bryk 2002). The main choices are (a) using the *natural metric* (NM) of the variables, (b) *grand mean centering* (GMC) by subtracting the complete sample or grand mean from the observed values; and (c) *centering within subgroups or contexts* (CWC) studied by subtracting subgroup means from observed values. For hierarchical models in which only the intercept but not the slopes is random at level 1, as is the case for the models mentioned, Snijders and Bosker (1999: 81) showed that all three of the NM, GMC, and CWC approaches led to models that are statistically equivalent in terms of the parameterizations of the combined models. In fact, we found empirically in our analyses of the WORDSUM data that there is not a great deal of difference among estimated coefficients under the three different approaches (although there are some variations in terms of variance decompositions and fit statistics). Thus, in the absence of methodological guidelines that privilege one of the three alternatives, substantive-theoretical reasoning guided the choice of centering. Because the minimum value of the key explanatory variable of age does not include zero (since the GSS sample frame is for ages 18 and over) in the model of Equation (7.9), one of the other options should be used. Furthermore, the literature on the effects of age on vocabulary knowledge (Wilson and Gove 1999b: 257–258) cites a pure physiological age effect that does not vary by cohort context. Therefore, we applied centering on the grand mean to the individual-level age variable in Equation (7.9). In the case of education, by contrast, Wilson and Gove (1999b: 255–256) argued that changing average levels of school years completed varies very substantially across the cohorts surveyed in the GSS. To take this changing cohort context of education into account, we therefore centered education on the cohort means.

7.4 Fixed versus Random Effects HAPC Specifications

An alternative model specification for HAPC analysis of repeated cross-sectional survey data would specify the time period and cohort effects as fixed rather than random. Under what conditions should an analyst use a fixed rather than a random effects specification? The literature on hierarchical/multilevel regression models contains some general guidelines (not specific to HAPC models) on when certain effect coefficients should be treated as fixed or random. To articulate these guidelines, consider a simple *hierarchical/multilevel linear model* (HLM) that has a level 1 or individual-level model:

$$Y_{ij} = \beta_0 + \beta_1 x_{ij} + e_{ij} \tag{7.10}$$

where Y_{ij} is a response variable for individual i in group j, x_{ij} is an explanatory variable or regressor for individual i in group j, β_0 is the intercept parameter for the regression model, β_1 is the slope parameter of the regression, and e_{ij} is a random error term. Suppose that the intercept β_0 is group dependent, and that it varies randomly among J observed groups. To model this random variation, we specify the level 2 or group-level model:

$$\beta_0 = \gamma_{00} + r_{0j} \tag{7.11}$$

This level 2 model separates the group-dependent intercept into an average intercept among the groups γ_{00} plus a group-level deviation or error r_{0j}. Substitution of Equation (7.11) into Equation (7.10) then yields the combined model:

$$Y_{ij} = \gamma_{00} + \beta_1 x_{ij} + r_{0j} + e_{ij} \tag{7.12}$$

The values of the r_{0j} are the main effects of the groups: Conditional on having a specific X-value and being in group j, the expected Y-value for individual i deviates by r_{0j} from the average expected value for all individuals over all groups. Note again that this is the simplest possible formulation of a hierarchical model; a more general formulation would allow for the possibility that the slope coefficient in Equation (7.10) could vary among the groups, and there could be more than one explanatory variable in the level 1 model.

As a statistical model, Equations (7.10)–(7.12) can be treated in two ways:

1. As a *fixed effects model*, wherein the r_{0j} are treated as *fixed* parameters, J in number.[*] This approach leads to a specific instance of a fixed effects regression model, namely, the conventional *analysis-of-covariance model*, in which the grouping variable is a covariate.

2. As a *random intercepts model*, wherein the r_{0j} are assumed to be independent identically distributed random variables (more generally, if the slope coefficient β_1 is specified as interacting with the level 2 random effects, then the model is a *random coefficients model*). These errors now are assumed to be randomly drawn from a population with zero mean and an *a priori* unknown variance. This assumption is equivalent to the specification that the group effects are governed by mechanisms or processes that are roughly similar from group to group and operate independently among the groups. This is termed the *exchangeability* assumption. The random coefficients model also requires the assumption that the random level 2 or contextual effects (i.e., the r_{0j} coefficients) are distributed *independently* of the level 1 regressors.

These two approaches to the model of Equations (7.10)–(7.12) imply that hierarchical data generally can be analyzed in two different ways using models with fixed or random group-level coefficients. Which of these two specifications is the most appropriate in a given situation depends on a number of considerations.

Goldstein (2003: 3–4) and Snijders and Bosker (1999: 43–44) provided summaries of conventional statistical wisdom and methodological guidelines for choosing between the fixed or random specifications. They pointed out that

1. If the groups are regarded as *unique entities* and the objective of the analysis is primarily to draw conclusions pertaining to each of the J groups, then it is appropriate to use the conventional analysis-of-covariance model.

2. If the groups are regarded as a *sample* from a (real or hypothetical) population and the objective of the analysis is to make inferences about this population, then the random coefficients model is appropriate.

3. The *fixed effects model explains all differences among the groups* by the fixed effect adjustments (through the use of indicator or dummy variables to represent the group-level adjustments) to the intercept coefficient of Equation (7.10). This implies that there is no between-group variability left that could be explained by group-level variables.

---

[*] In applications to cross-classified data from a repeated cross-section survey design, the fixed effects specification would lead, in parallel to the cross-classified random effects and random coefficients model forms of HAPC models, to the class of *cross-classified fixed effects model* (CCFEM) form of HAPC models.

Therefore, if the objective of the analysis is to test effects of group-level variables, the random coefficient model should be used. The exception to this guideline pertains to the case wherein the analyst introduces explicitly measured group-level variables that are hypothesized to account for the group-level effects. In this case, however, the model cannot at the same time incorporate indicator variables for the group-level fixed effect adjustments. Rather, the analyst must assume that the group-level fixed effect adjustments are completely accounted for by the explicitly measured group-level variables.[*]

4. The random coefficients model typically is used with some additional assumptions. Most important, as noted, it requires that the *level 2 random residuals be distributed independently of the level 1 regressors/explanatory variables*, which implies that $corr(r_{0j}, X) = 0$. In addition, in conventional normal errors HLM models, it is assumed that the *random coefficients r_{0j} and e_{ij} are normally distributed*. If these assumptions are poor approximations to the characteristics of a specific set of empirical data (e.g., the regressors are not independent of the random coefficients, or there is high density in the tails of the distributions of the errors), then these assumptions should be modified.

These, then, are several of the main considerations that conventional statistical wisdom indicates should be taken into account in deciding on fixed versus random effects formulations of hierarchical statistical models.

Applied to HAPC models, this reasoning leads to the following guidelines:

1. The range of the age categories for contemporary human populations is essentially fixed at 0 to 125, and most empirical studies utilize only a part of this fixed range (e.g., the adult ages 18 to 89 in the GSS WORDSUM data). Therefore, the individual ages or age categories may be regarded as unique entities, and it is reasonable to specify the age effects as fixed.

2. On the other hand, the time period and cohort categories available for any specific empirical analysis typically are only a sample of periods and cohorts for any human population. Therefore, it also is reasonable to specify the period and cohort effects as random.

3. In addition, a key problem with the fixed effects specification is the assumption that the indicator/dummy variables representing the fixed cohort and period effects fully explain all of the cohort and period effects. That is, the fixed effects model does not allow for the possibility of any additional random variance associated with the individual cohort and period effects. This implies that there is

---

[*] This assumption is the same as that of the proxy variables approach to identification of the conventional APC accounting model; see Section 4.4.4 of Chapter 4.

no unexplained between-cohort or between-period variability left beyond that captured by the fixed cohort and period effects. In the context of HAPC models, this appears not to be the best assumption with respect to statistical estimation of the parameters of the models. First, the fixed effects model tends to produce substantially larger standard errors for the intercept, indicating much more uncertainty in the mean estimate (for an empirical example, see Yang and Land 2008: 314). The CCREM can explain the variance for the intercept better than the indicator variables representing the fixed period and cohort effects. Second, fixed effects models require estimating unique effect coefficients for each period and cohort: $(J-1) + (K-1)$ parameters in all. Random effects models instead estimate one variance parameter that represents the distribution of the random effects for the periods and one variance parameter that represents the distribution of random effects for the cohorts. The latter usually yields a better model fit. Third, repeated cross-sectional survey data tend to be highly unbalanced, where *unbalanced* means that, when the sample members in a repeated survey design are cross classified by cohort (arrayed, say, in rows) and time period of observation (arrayed in columns), the numbers of observations in cells above the diagonal are not symmetric with those below the diagonal. And, under such data designs, mixed effects models use the available information in the data more efficiently and show better statistical efficiency relative to fixed effects models (e.g., Duchateau and Janssen 1999).

All of these considerations point toward advantages of the specification of HAPC models as mixed effects models rather than as pure fixed effects models. There are exceptions, however. One exception is the situation in which a very small number of time periods of repeated cross-sectional surveys, say only two or three, is available.* In this case, it may be quite reasonable to specify the effects of time periods as fixed with the cohort effects specified as either fixed or random.

A second exception pertains to situations in which the assumption that the level 2 random effects are distributed independently of the level 1 regressors is not valid. Note that most conventional empirical applications of hierarchical linear models proceed without a careful examination of the empirical veracity of this assumption. By contrast, the comparative performance of the fixed and random effects model specifications is a standard part of model criticism and assessment in longitudinal panel models (often referred to as pooled time series cross-sectional models) in econometrics (see, e.g., Greene 2003: 301–303). This is due to the general results in

---

* Using a thinned sample of five time periods of GSS verbal test score data and five cohorts, Yang and Land (2008) found that the CCREM specification produced results similar to those with 19 cohorts and 15 time periods of data.

statistical theory for mixed fixed-random effects models that, under the null hypothesis of zero correlation between the individual-level regressors and the contextual effects coefficients, both the OLS estimator of the individual-level coefficients in the fixed effects model and the REML estimator of those coefficients in the random effects model are consistent, but the OLS estimator is inefficient. Therefore, under the null hypothesis, the two estimators should produce estimates of the individual-level coefficients that do not differ systematically.

To describe this and a corresponding statistical test for the tenability of the independence assumption in HAPC models in more detail, consider the *cross-classified fixed effects model* (CCFEM) corresponding to the CCREM of Equations (7.7)–(7.9) where the effects of the cohorts u_{0j}, $j = 1, \ldots, J$ and the effects of the time periods (years) of the surveys v_{0k}, $k = 1, \ldots, K$ are assumed fixed and unique to each of the respective cohorts and period rather than variable and random. In practice, the fixed effects of the cohorts and periods are estimated by the incorporation of two sets of indicator/dummy variables for $J - 1$ cohorts and $K - 1$ periods. Therefore, Equation (7.8) changes to

$$\beta_{0jk} = \gamma_0 + \gamma_{1j} \sum_{j=2}^{19} Cohort_j + \gamma_{2k} \sum_{k=2}^{15} Period_k \qquad (7.13)$$

where the variance in the intercept β_{0jk} is assumed to be completely captured by the indicator variables for cohorts and periods. Substituting this expression into Equation (7.7) yields the combined CCFEM:

$$WORDSUM_{ijk} = \gamma_0 \beta_1 AGE_{ijk} + \beta_2 AGE_{ijk}^2 + \beta_3 EDUCATION_{ijk}$$
$$+ \beta_4 FEMALE_{ijk} + \beta_5 BLACK_{ijk} + \gamma_{1j} \sum_{j=2}^{19} Cohort_j + \gamma_{2k} \sum_{k=2}^{15} Period_k + e_{ijk} \qquad (7.14)$$

This fixed effects HAPC specification is in the form of a cross-classified analysis of covariance model. Because of the nesting of the individual-level observations within the time period and cohort groups and the flexibility of differential temporal intervals facilitated by the presence of individual-level observations in repeated cross-sectional survey designs as described in Section 7.2, however, the CCFEM does not have the underidentification conundrum of the classical APC accounting model described in Chapter 5. Nonetheless, because the individual-level observations corresponding to a specific cohort or time period may have correlated errors or variances that differ from those of observations from other periods or cohorts, a heteroscedasticity-corrected estimator (White 1980; Greene 2003: 198–199) of the variance-covariance matrix should be estimated. A comparison of estimates of the resulting standard errors of the regression coefficients with the noncorrected estimates permits an analysis of the extent to which the clustering of the individual-level observations within periods and cohorts affects the variances.

Following along the lines of comparisons of random and fixed effects models in longitudinal panel studies, we now can make a similar comparison of the CCREM and CCFEM models for the repeated cross-sectional data on verbal ability in the GSS. Note that the present comparison differs from standard longitudinal panel designs in that the same individuals are not repeatedly surveyed in consecutive waves of the GSS. However, given the temporal dimensions embedded in the cohort and time period contextual variables as we have defined them, it is important to address the independence assumption explicitly.

Yang and Land (2008) recommended a two-step procedure for assessing the assumption of the independence of the random effects and the individual-level regressors. First, estimate both the CCREM and the CCFEM models and qualitatively assess the resulting model fits and the parameter estimates and performance of each with respect to the data. Second, calculate a statistical test by applying a form of what is known in the econometric analysis of pooled time series cross-sectional regression models as a Hausman specification test (see Hausman and Taylor 1981; Baltagi 1995). The Hausman test is a Wald chi-squared test of the form

$$W = \chi^2 [K] = \left[b - \hat{\beta} \right]^T \hat{\Sigma}^{-1} \left[b - \hat{\beta} \right] \tag{7.15}$$

where, in the present case, b denotes the vector of individual-level regression coefficients estimated from the CCFEM model, $\hat{\beta}$ denotes the corresponding vector of regression coefficients estimates from the CCREM model, and $\hat{\Sigma} = Var[\hat{b}] - Var[\hat{\beta}]$ is the difference of the variance-covariance matrices of the two estimators (the constant term is excluded from all vectors and matrices). Under the null hypothesis that the cohort and period random effects in the CCREM model are independent of the individual-level regressors, W is distributed as chi squared with K degrees of freedom, where K is the dimension of the b and β vectors. Applied to the GSS verbal test score data for the years 1974–2000, Yang and Land (2008: 317–319) found that the null hypothesis was not rejected; thus, the level 1 explanatory variables in the CCREM model of Equations (7.7)–(7.9) can be assumed to be distributed independently of the random effects for time periods and cohorts. In addition, Yang and Land (2008: 314–418) showed that the weighted averages of the estimated partial regression coefficients for time periods and cohorts in the CCFEM of Equation (7.14) displayed patterns of changes across the respective periods and cohorts that were very similar to those from the CCREM, which is consistent with the stochastic independence assumption.

In sum, the various substantive and statistical advantages of mixed effects model specifications of HAPC models described imply that these specifications should be used under most circumstances. The exceptions, as indicated here, pertain to empirical applications in which there are very small numbers of repeated cross-sectional surveys available for analysis or in which the independence assumption of mixed models is not tenable.

7.5 Interpretation of Model Estimates

For comparative purposes, Table 7.1 reports baseline OLS estimates of pooled repeated cross-sectional regression models without controls for period and cohort effects. It shows significant curvilinear age effects as hypothesized by Wilson and Gove (1999b). And, consistent with prior research on verbal ability, being female is positively associated with one's expected score on WORDSUM, whereas being black is negatively associated with the response variable. More years of education is associated with greater verbal ability. These covariates together explain 31% of variance in WORDSUM.

Table 7.2 reports the parameter estimates and model fit statistics for the CCREM [Equation (7.9)] estimated on the 17 GSS repeated cross-sectional surveys. These results were obtained using the REML-EB estimation method via the application of the SAS PROC MIXED (see the sample codes). Examining first the model fit statistics reported at the bottom of the table, it can be seen that the model deviance is very large compared to the degrees of freedom of the model, thus indicating a highly significant association of the explanatory variables with the WORDSUM response variable. The variance components show that most of the variance in WORDSUM is accounted

TABLE 7.1

Fixed-Effects Regression Model for Pooled GSS WORDSUM Data, 1974–2006, without Controls for Period and Cohort Effects ($N = 22,042$)

Fixed Effects	Coefficient	se	t Ratio	p Value
Intercept	6.22	0.02	281.63	<0.001
AGE[a]	0.18	0.01	22.99	<0.002
AGE^2	−0.05	0.00	−12.10	<0.003
EDUCATION[a]	0.36	0.00	86.30	<0.006
SEX (1 = male)	0.22	0.02	9.22	<0.004
RACE (1 = black)	−1.06	0.03	−31.04	<0.005
Adjusted R^2	0.31			

[a] Centered around grand means.

TABLE 7.2

HAPC-CCREM of the GSS WORDSUM Data: 1974–2006

Fixed Effects	Parameter	Coefficient	se	*t* Ratio	*p* Value
INTERCEPT	γ_0	6.18	0.06	112.50	<0.001
AGE	β_1	0.03	0.02	1.71	0.087
AGE^2	β_2	−0.06	0.01	−11.87	<0.001
EDUCATION	β_3	0.37	0.00	86.57	<0.001
SEX	β_4	0.23	0.02	9.49	<0.001
RACE	β_5	−1.03	0.03	−30.07	<0.001
Random Effects		**Coefficient**	**se**	**t Ratio**	**p Value**
Cohort					
1894	u_1	−0.21	0.14	−1.48	0.140
1895	u_2	−0.11	0.12	−0.93	0.353
1900	u_3	−0.05	0.10	−0.49	0.625
1905	u_4	−0.29	0.09	−3.27	0.001
1910	u_5	0.02	0.08	0.26	0.797
1915	u_6	0.16	0.07	2.22	0.027
1920	u_7	−0.08	0.07	−1.15	0.249
1925	u_8	0.08	0.07	1.23	0.220
1930	u_9	0.00	0.07	0.01	0.990
1935	u_{10}	0.07	0.06	1.06	0.289
1940	u_{11}	0.24	0.06	3.91	<.001
1945	u_{12}	0.45	0.06	7.50	<.001
1950	u_{13}	0.18	0.06	3.10	0.002
1955	u_{14}	−0.04	0.06	−0.57	0.568
1960	u_{15}	0.00	0.07	0.04	0.970
1965	u_{16}	−0.16	0.07	−2.20	0.028
1970	u_{17}	−0.14	0.08	−1.70	0.090
1975	u_{18}	0.00	0.09	−0.01	0.990
1980	u_{19}	0.06	0.11	0.55	0.583
1985	u_{20}	−0.20	0.15	−1.34	0.180
Period					
1974	v_1	0.03	0.04	0.77	0.442
1976	v_2	0.06	0.04	1.41	0.158
1978	v_3	0.00	0.04	−0.04	0.967
1982	v_4	−0.01	0.04	−0.36	0.718
1984	v_5	0.02	0.04	0.37	0.709
1987	v_6	−0.06	0.04	−1.52	0.129
1988	v_7	−0.13	0.05	−2.76	0.006
1989	v_8	−0.06	0.05	−1.34	0.182
1990	v_9	0.02	0.05	0.43	0.670
1991	v_{10}	0.04	0.05	0.92	0.358
1993	v_{11}	0.00	0.05	−0.09	0.926

TABLE 7.2 (continued)

HAPC-CCREM of the GSS WORDSUM Data: 1974–2006

Random Effects		Coefficient	se	t Ratio	p Value
1994	v_{12}	0.02	0.04	0.49	0.623
1996	v_{13}	−0.06	0.04	−1.52	0.128
1998	v_{14}	0.04	0.04	1.02	0.306
2000	v_{15}	0.01	0.04	0.11	0.915
2004	v_{16}	0.04	0.04	0.88	0.381
2006	v_{17}	0.05	0.05	1.16	0.247
Variance Components		**Variance**	**se**	**z Statistic**	**p Value**
COHORT	τ_{u0}	0.03	0.01	2.56	0.010
PERIOD	τ_{v0}	0.01	0.00	1.49	0.135
Individual	σ^2	3.12	0.03	104.87	<0.001
Model Fit					
Deviance		87707.2	df = 21,999		

for by the individual-level regressors. Level 2 variance components results indicate that variation by cohorts is 0.03 and statistically significant ($p < .01$), whereas there is little variation by time periods after controlling for age and other individual covariates. Examining further the estimated average effect coefficients for cohorts (see also Figure 5.3), it can be seen that the estimated effects are particularly negative for the 1905–09 cohort and particularly positive for the 1940–44, 1945–49, and 1950–54 cohorts. There also is a negative trend from the 1960–64 to the 1980–84 cohort.

Examining next the estimated individual-level coefficients in Table 7.2, it can be seen that the qualitative results are similar to those reported in Table 7.1—a quadratic age effect, a positive effect for females, a negative effect for blacks, and a highly significant positive effect for education. Taken together, these regressors account for about 30% of the unconditional level 1 (individual-level) variance. The estimated regression coefficients and their standard errors are numerically quite similar between the two tables for the sex, race, and education variables. Estimates for the linear component of the quadratic age curve are quite another story, however. The estimated coefficient for this term is reduced from a highly statistically significant .18 in the pooled regression model without controls for periods or cohorts of Table 7.1 to a marginally significant .03 in Table 7.2, after cohort and time period effects are taken into account. This implies that a failure to control for the effects of cohort and period variation in vocabulary knowledge could lead to large overestimates of the increases in verbal ability that are related to aging from young adulthood into the middle-age years.

In sum, our results lend support for some aspects of both sides of the debate on the intercohort decline in vocabulary knowledge in the United States. First, the HAPC analyses found evidence in support of the quadratic age effect on vocabulary knowledge hypothesized by Wilson and Gove (1999a, 1999b). However, the linear effect (which indicates the extent to which the quadratic age curve of vocabulary knowledge increases with age) was reduced to statistical insignificance when controls were introduced for the random effects of time periods and cohorts. Furthermore, controlling for the effects of key individual characteristics in the HAPC analyses (namely sex, race, and education) did not explain away all the age effects. We found that about 1% of variation in verbal scores at the individual level was due to the quadratic effect of aging after controlling for the random effects of cohorts and periods as well as the individual-level covariates of sex and race. This was about three times the "one-third of 1%" found by Alwin and McCammon (2001) in regressions that controlled for cohort effects but not for period effects or individual-level covariates.

Second, we found only evidence of modest time period effects. This supports the contentions of Alwin and McCammon (1999) that period effects in the GSS vocabulary knowledge data are relatively minor. The presence of these effects, however, affects the estimates of age and cohort effects.

Third, the HAPC analyses found evidence in support of the contentions of Alwin (1991) and Glenn (1994, 1999) that there has been an intercohort decline in vocabulary knowledge. In fact, we found evidence of a bimodal curve of cohort effects. There was evidence of a peak in vocabulary knowledge for cohorts born in the 1940s and perhaps the early 1950s. But, our analyses also suggested a deficit for birth cohorts from the first decade of the twentieth century. Relative to this early century decline, vocabulary knowledge showed a secondary peak in the immediately following cohorts, thus yielding a bimodal cohort curve not found in previous studies.

7.6 Assessing the Significance of Random Period and Cohort Effects

The development of HAPC models may provide a useful apparatus for modeling and estimating distinct age, period, and cohort effects in repeated cross-sectional survey designs. In this context, however, the question arises: How can one assess or judge the significance of estimates of cohort and period effects in such models?* This question may be addressed by examining the

---

* Parts of Section 7.6 are adapted from Frenk, S. M., Y. Yang, and K. C. Land. *Assessing the significance of cohort and period effects in hierarchical age-period-cohort models with applications to verbal test scores and voter turnout in U.S. Presidential elections.* Under review.

statistical significance of the estimated effect coefficients for each individual cohort and time period in a study. But, it may be the case that, say, some cohorts have statistically significant effect coefficients and some do not, and the same may be true for the estimated period coefficients. In such a case, how does one assess the overall statistical significance of the cohort or the period effects? Beyond statistical significance is the question of substantive significance. It could be the case, for example, that most of the individual estimates of cohort or period effects are not statistically significant at a conventional level of significance, but they exhibit an interesting trend or pattern that merits substantive interpretation.

In the context of empirical applications of LM and GLMM specifications of HAPC models, we describe a two-step approach and set of guidelines to address these questions that build on a large body of literature on methods for hypothesis testing in mixed (fixed and random effects) models in statistics. We claim no originality for these general statistical methods. Rather, the object of this section is to organize them into a set of methods specifically adapted to the features of HAPC models and to illustrate their application in the context of empirical analyses of two specific datasets: GSS data on trends in verbal ability and NHANES data on trends in obesity. In addition, substantive findings from the empirical applications clearly demonstrate the dominance of cohort effects in the former case and period effects in the latter and thus help to resolve long-standing empirical questions and disputes in each case. The procedure and guidelines that are articulated and illustrated in these two empirical analyses can be readily adapted and applied more generally to other empirical HAPC analyses.

7.6.1 HAPC Linear Mixed Models

To assess and test the statistical significance of the cohort and time period effects estimated from the HAPC-CCREM analysis of the GSS data shown in Table 7.2, we suggest a two-step approach.

7.6.1.1 Step 1: Study the Patterns and Statistical Significance of the Individual Estimated Coefficients for Time Periods and Birth Cohorts

As an initial step, the individual estimated period and cohort effects should be studied both for substantively meaningful patterns and statistical significance. This can be done in two parts.

Step 1.1: Graphically Plot the Temporal Sequences of Estimated Cohort and Period Effect Coefficients

While the numerical values of the estimated cohort and period effects in Table 7.2 contain the same information, as a first step in the analysis of

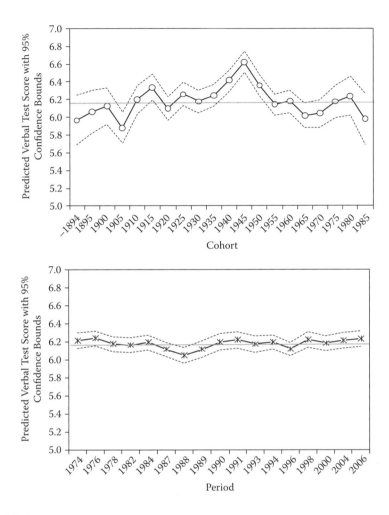

FIGURE 7.1
Estimated cohort and period effects and 95% confidence bounds for GSS verbal ability model.

their substantive and statistical significance, we also graphically plotted the estimates. This facilitates a quick visual check of the extent to which the estimated effects exhibit temporal trends or patterns that are of substantive significance. Particular periods or birth cohorts that stand out also may be identified.

Figure 7.1 contains graphs of the temporal sequences of the estimated cohort effects (i.e., the estimated $\beta_{0j} = \gamma_0 + u_{0j}$ cohort verbal test score effects averaged over all time periods for each cohort j) and the time period effects (i.e., the estimated $\beta_{0k} = \gamma_0 + v_{0k}$ period verbal test score effects averaged over all cohorts for each time period k) with their 95% confidence bounds. Each graph also has a horizontal line at 6.175, the numerical value of the estimated γ_0 or intercept coefficient reported in Table 7.2. This line facilitates a visual inspection of those cohorts and time periods, if any, that have substantial deviations from the overall intercept.

With 20 and 17 observations, respectively, the 95% confidence bounds for the birth cohort and time period effects are relatively broad. The pattern of the estimated time period effects does not show substantial variations or systematic temporal patterns, with only the 1988 value deviating substantially from the overall average. By comparison, it can be seen that the estimated temporal pattern of birth cohort effects contains some gyrations that are quite pronounced and relevant to substantive debates concerning historical trends in verbal ability, which we have already discussed.

Step 1.2: Examine the Statistical Significance of Individual Cohort and Period Effect Coefficients

Turning from visual and substantive assessments of estimated cohort and period effects, the next step is to examine the statistical significance of the individual effect coefficients for the birth cohorts and time periods—the estimates of the u_{0j} and the v_{0k} random effects with the null hypothesis in each case being that the respective coefficient is zero, that is:

$$H_0 : u_{0j} = 0 \text{ versus } H_a : u_{0j} \neq 0 \text{, and } H_0 : v_{0k} = 0 \text{ versus } H_a : v_{0k} \neq 0.$$

If these null hypotheses are not rejected, this implies that the mean of the WORDSUM outcome variable for the jth time period or the kth cohort is no different from the overall average. The coefficients and their standard errors for the birth cohorts and time periods in Table 7.2 were estimated by the REML-EB method, and their ratios can be interpreted as asymptotic/large-sample t ratios in the conventional way (Raudenbush and Bryk 2002: 57–58).

As will often be the case with HAPC models, some cohorts and time periods may have statistically significant effects, as measured by t ratios, while others do not. For instance, in Table 7.2, the effect coefficients for the 1905–9, 1915–19, 1940–45, 1945–49, 1950–54, 1965–69, and 1970–74 cohorts are statistically significant at conventional levels, whereas those for the other cohorts are not. For time periods, only the effect coefficient for 1988 attained conventional levels of statistical significance.

These assessments of statistical significance of the individual cohort and period estimated coefficients are consistent with the graphical representations in Figure 7.1. Given that the estimated cohort and period coefficients are displayed in the figure with their 95% confidence bounds, there is a correspondence between those cohort and period coefficients with asymptotic t ratios that are statistically significant at the .05 level in Table 6.2 and those for which the 95% confidence bounds do not cross the 6.175 horizontal line in Figure 7.1—that is, the 1905–9, 1915–19, 1940–45, 1945–49, 1950–54, 1965–69, and 1970–74 cohorts and the 1988 period. In general, however, even when none of the individual birth cohort and time period coefficients are statistically significant, it often is useful to examine graphically the patterns of each set of coefficients for trends that could be of substantive interest.

7.6.1.2 Step 2: Test for the Statistical Significance of the Period and Cohort Effects Taken as a Group

When some cohort and period effect coefficients attain statistical significance but some or most do not, we next may address the question of whether the birth cohort or time period effects, taken as a set, are statistically significant. This question is one of whether these effects, taken together, contribute to explained variance in the model.

Step 2.1: Deviance and Variance Components Analysis

A first approach to answering this question is to study the model deviance statistic and variance components. The deviance is defined as minus twice the natural logarithm of the likelihood of an estimated model and can be regarded as a measure of lack of fit between model and data. In Table 7.2, the bottom rows show that the deviance statistic is very large compared to the degrees of freedom of the model, thus indicating a highly significant association of the explanatory variables with the WORDSUM response variable. In addition, the variance components show that most of the variance in WORDSUM is the individual-level regressors at level 1. Level 2 variance components results indicate that variation by cohorts was statistically significant, whereas that for time periods was not. This variance component analysis based on z scores is consistent with the results noted for the individual cohort and period coefficients. That is, a sufficient number of estimated cohort effects were statistically significant for the cohort variance component to attain statistical significance. But, since only one of the time period effects was statistically significant, the overall contribution of the random effects for time periods was not sufficiently large for its variance component to attain statistical significance.

Step 2.2: An F Test for the Presence of Random Effects

The previous results were obtained using REML, which rests on the assumption that the error terms are asymptotic normally distributed and yield random effect estimators with good large-sample properties. When the number of level 2 units, in this case cohorts and periods, is not large, this assumption may not be appropriate. And, the z scores for the REML estimates of the variance components are only proximate. To further test whether the birth cohort and time period effects make statistically significant contributions to explained variance in an outcome variable, a general linear hypothesis may be applied. Specifically, one can use an F test to test the hypothesis of the presence of random effects. The sampling distribution of the F statistic is exact when the random effects are independently distributed as normal random variables. This test statistic is preferred over the z score when the sample sizes for random effects are small (Littell et al. 2006). The statistical theory for such tests has been developed in a very general LMM context (Demidenko 2004).

In the present case, for the CCREM model of Equation (7.9), there are only two sets of random effect coefficients that are estimated, namely, the set of residual random effects of cohort j, u_{0j}, and the set of residual random effects of period k, v_{0k}. Each of these sets of random coefficients is assumed to be independently, normally distributed with mean 0 and variances τ_u and τ_v, respectively. Thus, for a CCREM model with random intercepts, the exact F test amounts to testing null hypotheses for the relevance of either the birth cohort random effects:

$$H_0: \tau_u = 0 \text{ versus } H_a: \tau_u > 0 \tag{7.16}$$

or the time period effects:

$$H_0: \tau_v = 0 \text{ versus } H_a: \tau_v > 0. \tag{7.17}$$

Alternatively, one can test for the joint relevance of both the cohort and the period effects:

$$H_0: \tau_u = \tau_v = 0 \text{ versus } H_a: \tau_u > 0 \text{ or } \tau_v > 0 \tag{7.18}$$

These null hypotheses correspond to situations, respectively, in which the levels of variation in the cohort effects, the period effects, and the cohort and period effects taken together do not differ significantly from zero.

The results of the F tests for hypotheses (7.16)–(7.18) for the GSS data are summarized in Table 7.3. Consider the case of the null hypothesis (7.16) in which the birth cohort effects are not relevant to explaining variation in the verbal test score outcome variable of Equation (7.9). The idea of the F test is that, when the variance of the random birth cohort effects $\tau_u = 0$, the difference between the minimum sum of squares (SS) of the model (7.9) with random cohort and period effects S_{min} and the minimum SS without the

TABLE 7.3

F-tests for the Presence of Random Effects, GSS WORDSUM Data (1974–2006)

	Cohort Effects $\tau_u = 0$ vs. $\tau_u > 0$	Period Effects $\tau_v = 0$ vs. $\tau_v > 0$	Cohort and Period Effects $\tau_u = \tau_v = 0$ vs. τ_u or $\tau_v > 0$
S_{OLS}	69,377	69,377	69,377
S_{min}	68,696	69,268	68,558
R	25	22	42
M	5	5	5
N_T	22,042	22,042	22,042
$(S_{OLS} - S_{min})/(r - m)$	34.05	6.41	22.14
$S_{min}/(N_T - r)$	3.12	3.15	3.12
F	10.9	2.03	7.10
$f_{0.95}(r - m, N_T - r)$	1.57	1.62	1.41

random effects, as estimated by an OLS regression on the level 1 explanatory variables S_{OLS}, should be close. Accordingly, we compute the residual SS:

$$S_{OLS} = \sum_{i=1}^{N} \left\| y_i - X_i \hat{\beta}_{OLS} \right\|^2 \tag{7.19}$$

for an ordinary fixed effects regression model that assumes no random effects of cohorts or time periods, as shown in the pooled regression model without controls for periods or cohorts of Table 7.1. Next, the minimum SS in the presence of the random effects, that is, the minimum

$$S_{\min_\delta} = \min_\delta \left\| y - W\delta \right\|^2 \tag{7.20}$$

where the matrix W consists of the matrix X of observations on the individual-level explanatory variables adjoined with a design matrix Z for the random cohort effects, that is, $W = [X, Z]$, and $\delta = (\beta', u_0')'$. In this example, $S_{OLS} = 69{,}377$ and $S_{\min} = 68{,}696$. Under the null hypothesis (7.10), it can be shown that the ratio of two quadratic forms has an F distribution, or more precisely:

$$\frac{(S_{OLS} - S_{\min})/(r - m)}{S_{\min}/(N_T - r)} \sim F(r - m, N_T - r) \tag{7.21}$$

where N_T denotes total sample size, r is the rank of the matrix W, and m is the number of explanatory variables in the OLS regression (Demidenko 2004: 137). When random cohort effects are present in a LMM model, such as Equation (7.9), that is, when τ_u is nonzero, $S_{\min}$ should be relatively small so that the ratio (7.15) becomes large. Thus, we reject the null hypothesis (7.10) if the left-hand side of (7.15) is large. More precisely, let $1 - \alpha$ be a chosen significance level, for example, $\alpha = 0.05$, and $f_{0.95}$ be the quantile of the F distribution with $r - m$ and $N_T - r$ degrees of freedom. Then, the H_0 is rejected when the ratio in (7.15) exceeds $f_{0.95}$.

To apply the F test (7.15) to model (7.9), note first that, under the assumption that the explanatory variables in the X matrix are linearly independent, the rank of X is m, and the number of explanatory variables is five in this case. In addition, since individuals in the pooled GSS data may be members of different birth cohorts, the columns of the design matrix Z for the random effects will be linearly independent and thus have rank 20. Therefore, in the numerator of (7.15), $r = 25$ and $m = 5$, which gives $r - m = 20$ degrees of freedom. In the denominator, $N_T - r = 22{,}042 - 25 = 22{,}017$. Under the null hypothesis that the cohort effects have zero variance in the GSS verbal test score analysis, the F ratio is 10.9. With 20 and 22,017 degrees of freedom, this far exceeds the critical value $f_{0.95} = 1.571$. The corresponding F ratio for hypothesis (7.11) of period effects is 2.03, which also exceeds the critical value of 1.623. Thus, we reject the null hypotheses that the variance of either the birth cohort random effects or the period random effect is zero and conclude that inclusion of these sets of random effects is relevant to the explanation of variation of the GSS verbal test score data. In addition, the F test can be applied to the sets of random effect coefficients taken as a whole, that is, to test the null hypothesis (7.12). In this case, the Z matrix is expanded to include both u_0' and v_0', which changes the rank of W to 42 (= 5 + 20 + 17). The F ratio is 7.096, which is significant at the .05 level.

The foregoing analyses indicate that there is evidence that the two sets of random effects taken together contribute significantly to the explained variance. Note that the z score reported in Table 7.2 for the variance component for period effects is 1.49 ($p = .135$), indicating failure to reject the null hypothesis of zero variance for the period coefficients. The F tests described here indicate the opposite, that is, that the period effects contribute significant variability that should not be ignored in the model. Because there are only a few levels of the period random effects, the F SS method is a more statistically sensitive method for testing hypotheses about the variance components than the z-score method. In particular, the F tests typically will indicate statistical significance of either the cohort or the period effects, taken as a whole, when at least one of the members of these sets of effects is statistically significant, as is the case for the estimated period effects reported in Table 7.2.

Statistical significance does not, of course, equate to substantive importance, but it does indicate a contribution to explained variance. Substantively, taking into account all of the foregoing assessments of individual coefficients and sets of coefficients, it must be concluded that, while there is evidence of statistical significance of one time period effect, and while this is sufficient to conclude that the period effects make a statistically significant contribution to explained variance, the dominant explanation on trends in verbal ability, as measured by the GSS WORDSUM data, is a cohort as opposed to a period one. That is, net of the effects of individual-level covariates in accounting for temporal trends in individuals' verbal test scores, cohort effects are much more prominent than period effects, and researchers should indeed study cohort-based explanations for these trends.

7.6.2 HAPC Generalized Linear Mixed Models

The HAPC approach to modeling age, birth cohort, and time period effects developed by Yang and Land (2006) is not restricted to applications to normally distributed outcome variables that can be modeled by a Gaussian LMM model specification. Rather, the HAPC approach can be applied to dichotomous and multiple categorical outcome variables. For such outcome variables, the HAPC framework takes the form of GLMM specifications. Our suggested approach to testing for the statistical significance of the random effects again has two steps.

To illustrate the significance testing guidelines for the class of GLMM formulations of the HAPC-CCREM mode, we study an application of one of the most important classes of GLMM models, namely, that of a logistic response function for a dichotomous outcome variable. Specifically, consider the example of obesity epidemic in the United States and the contributions of time period and birth cohort effects to the epidemic. For this analysis, we pooled data from the NHANES 1971–2008 values on obesity status and a number of individual-level covariates that may affect the probability of being obese. We focus for now on key social demographic and SES variables as other risk factors, such as health behaviors and diet, are not available for all survey years, and their inclusion would substantially reduce the sample size. To model the likelihood of obesity, we specified the following HAPC mixed model:

Level 1 or "within-cell" model:

$$Logit\,Pr(OBESE_{ijk} = 1) = \beta_{0jk} + \beta_1 AGE_{ijk} + \beta_2 AGE_{ijk}^2 + \beta_3 SEX_{ijk} + \beta_4 RACE_{ijk}$$
$$+ \beta_5 EDUCATION_{ijk} + \beta_6 INCOME_{ijk} \tag{7.22}$$

Level 2 or "between-cell" model:

$$\beta_{0jk} = \gamma_0 + u_{0j} + v_{0k}, \quad u_{0j} \sim N(0, \tau_u), \quad v_{0k} \sim N(0, \tau_v) \tag{7.23}$$

Combined model:

$$Logit\,Pr(OBESE_{ijk} = 1) = \gamma_0 + \beta_1 AGE_{ijk} + \beta_2 AGE_{ijk}^2 + \beta_3 SEX_{ijk} + \beta_4 RACE_{ijk}$$
$$+ \beta_5 EDUCATION_{ijk} + \beta_6 INCOME_{ijk} \tag{7.24}$$
$$+ u_{0j} + v_{0k}$$

for
  $i = 1, 2, \dots, n_{jk}$ individuals within cohort j and period k;
  $j = 1, \dots, 17$ birth cohorts;
  $k = 1, \dots, 9$ survey years.

This model, similar to that for the GSS verbal test score, specifies that the intercept or the expected mean has random period and birth cohort effects. The level 1 model, however, is specified in terms of the logit of the probability of being obese (p) modeled as the log-odds of obesity, logit(p) = log[$p/(1 - p)$], that is, as a logistic response function. This moves the HAPC-CCREM model from the LMM family of statistical models into the GLMM family. As indicated by previous studies of GLMMs, we typically assume multivariate normality for random effects (Raudenbush and Bryk 2002). As described previously, we centered the age variable around the grand mean.

Table 7.4 reports parameter estimates and model fit statistics for models (7.22)–(7.24) obtained from the SAS PROC GLIMMIX (Littell et al. 2006) (see the sample codes). All individual-level covariates had statistically significant coefficients that were consistent with prior research. Specifically, the effect of age was curvilinear, indicating increases in the risk of obesity with age that decelerated at older ages. Males, whites, those with some college education or college degrees, and those in the highest income quartile were less likely to be obese than females, blacks, those with 12 years of education or less, and those in the middle income quartiles. People in the lowest income quartile are more likely to be obese than others. Beyond the effects of the individual-level covariates of this model, what is interesting about the results in Table 7.4, however, is that, as contrasted to those for the GSS verbal test score data described previously, of the two sets of random effects, period effects are more relevant to the explanation of obesity risk than are birth cohort effects, as measured by contributions to explained variance.

To reach this conclusion, we applied the same sequence of graphical displays and statistical tests of significance as identified for the LMM form of the HAPC-CCREM model.

As in Step 1.1, graphically plot the temporal sequences of estimated cohort and period effect coefficients. As was the case for the estimated random effects of time periods and cohorts in LMM HAPC models, a first step is to examine graphical displays of the temporal patterns of the effects.

Symmetrically with Figure 7.1, Figure 7.2 contains graphs of the temporal sequences of the estimated cohort effects and the time period effects with their 95% confidence bounds. The cohort effect is calculated as $\hat{\beta}_{0j} = \hat{\gamma}_0 + u_{0j}$, where $\hat{\gamma}_0 = -0.587$ is the intercept or estimated overall mean and u_{0j} are the cohort-specific random effects coefficients, and converted to probabilities of obesity $= \exp(\hat{\beta}_{0j})/(1 + \exp(\hat{\beta}_{0j}))$. The period effect is calculated as $\hat{\beta}_{0k} = \hat{\gamma}_0 + v_{0k}$, where v_{0k} are the period-specific random effects coefficients, and converted to probabilities of obesity. The graphs also have a horizontal line at 0.363, the transformed probability of obesity of the intercept estimate of $\hat{\gamma}_0$, of sample respondents at mean age and in the reference group. In contrast to what was observed for the GSS verbal test score cohort estimates in Figure 7.1, the pattern for estimated cohort effects in Figure 7.2 is relatively constant, with cohorts born after 1935 showing more higher-than-average probabilities of obesity and a slightly more apparent upward trend for those born between 1960 and 1974. The graph of the estimated obesity time period effects, on the other hand, shows quite pronounced variations. The average probabilities of obesity increased rapidly from the early 1970s to 2000 and gradually but continuously increased until 2008.

Following the procedure of Step 1.2, examine the statistical significance of individual cohort and period effect coefficients. Of the individual estimated random effect coefficients for birth cohorts and survey periods given

TABLE 7.4

HAPC-CCREM of Obesity Trends, NHANES 1971–2008

Fixed Effects	Parameter	Coefficient	se	*t* Ratio	*p* Value
Intercept	γ_0	−0.56	0.16	−3.56	0.007
AGE	β_1	0.01	0.00	8.11	<0.001
AGE2	β_2	−0.00	0.00	−11.06	<0.001
SEX (male = 1)	β_3	−0.42	0.03	−17.31	<0.001
RACE (black = 1)	β_4	0.46	0.03	16.49	<0.001
EDUCATION	β_5				
SOME COLLEGE (13–15 years)		−0.08	0.03	−2.49	0.013
COLLEGE (≥ 16 years)		−0.40	0.04	−9.74	<0.001
FAMILY INCOME	β_6				
LOWEST QUARTILE		0.13	0.03	4.51	<0.001
HIGHEST QUARTILE		−0.15	0.03	−4.37	<0.001
Random Effects		**Coefficient**	**se**	***t* Ratio**	***p* Value**
Cohort					
1899–1904	u_1	0.02	0.04	0.60	0.550
1905–1909	u_2	−0.03	0.04	−0.79	0.432
1910–1914	u_3	0.03	0.04	0.88	0.378
1915–1919	u_4	0.00	0.03	0.08	0.936
1920–1924	u_5	−0.03	0.03	−0.82	0.411
1925–1929	u_6	0.00	0.03	0.10	0.917
1930–1934	u_7	−0.02	0.03	−0.58	0.561
1935–1939	u_8	0.03	0.03	0.86	0.390
1940–1944	u_9	0.02	0.03	0.70	0.483
1945–1949	u_{10}	0.01	0.03	0.17	0.065
1950–1954	u_{11}	−0.04	0.03	−1.14	0.253
1955–1959	u_{12}	0.01	0.03	0.15	0.879
1960–1964	u_{13}	−0.06	0.03	−1.64	0.102
1965–1969	u_{14}	−0.01	0.04	−0.13	0.895
1970–1974	u_{15}	0.04	0.04	1.12	0.264
1975–1979	u_{16}	0.02	0.04	0.46	0.642
1980–1982	u_{17}	−0.01	0.04	−0.17	0.864
Period					
1971–1975	v_1	−0.73	0.16	−4.66	<0.001
1976–1980	v_2	−0.70	0.16	−4.45	<0.001
1989–1991	v_3	−0.26	0.16	−1.61	0.107
1991–1994	v_4	−0.04	0.16	−0.26	0.794
1999–2000	v_5	0.34	0.16	2.08	0.038
2001–2002	v_6	0.26	0.16	1.60	0.109
2003–2004	v_7	0.36	0.16	2.23	0.026
2005–2006	v_8	0.41	0.16	2.58	0.010
2007–2008	v_9	0.37	0.16	2.33	0.020

TABLE 7.4 (continued)

HAPC-CCREM of Obesity Trends, NHANES 1971–2008

Variance Components		Variance	se	z Statistic	p Value
COHORT	τ_{u0}	0.00	0.00	1.05	0.147
PERIOD	τ_{v0}	0.22	0.11	1.99	0.024

Model Fit

–2 Res Log Pseudo-Likelihood		186586.7	df=40228

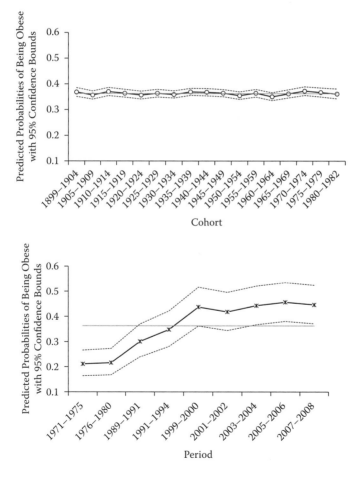

FIGURE 7.2

Estimated cohort and period effects and 95% confidence bounds for NHANES obesity model.

in Table 7.4, it can be seen that only one cohort, the 1960–64 birth cohort, had a *t* ratio that is statistically significant at the α = 0.10 level. By comparison, the time period coefficients for the majority of survey periods were significant at the α = 0.05 level.

Perform deviance and variance components analysis as in Step 2.1. As was the case for the GSS verbal test score data, the deviance statistic reported at the bottom of Table 7.4 shows that the full CCREM model explains much variance in the obesity outcome variable. In contrast to the verbal test score example, however, the variance component analysis indicated that the set of period random effect coefficients had a more statistically significant contribution to explaining obesity than the set of birth cohort coefficients.

Per Step 2.2, perform an *F* test for the presence of random effects. The model of Equations (7.22)–(7.24) is that of a logistic regression model with a random intercept. Demidenko (2004: 374–375, 408–409) showed that the *F* test (7.21) can be generalized to develop an asymptotic *F* test of the null hypothesis that the intercepts are constant or homogeneous in a logistic regression model with random intercepts. We next describe and apply this test.

To generalize the *F* test (7.21) and to specialize the homogeneity test of Demidenko (2004: 374–375, 408) to GLMM formulations of HAPC-CCREM models, recall that the deviance statistic (twice the negative log-likelihood function *l*) asymptotically behaves as the SS (McCullagh and Nelder 1989). Given this, *S* in the test statistic (7.21) can be replaced by $-2l$ to obtain

$$\frac{(l_0 - l_{\max})/(r - m)}{l_{\max}/(N_T - r)} \cong F(r - m, N_T - r) \tag{7.25}$$

where l_0 is the maximum of the log-likelihood of the standard level 1 logistic regression model with no controls for cohort and period effects, and $l_{\max}$ is the maximum of the log likelihood treating the cohort and period effects as fixed parameters. Demidenko (2004: 54–55) showed that a fixed effects model that treats the random effects as fixed corresponds to a random effects model with infinite covariance matrix. Thus, $l_{\max}$ is an upper-bound estimate for the SS of the mixed model. And, as in (7.21), N_T denotes total sample size, *r* is the rank of the matrix *W*, and *m* is the number of explanatory variables in the level 1 model.

To apply the asymptotic *F* test (7.25) to the NHANES obesity model (7.24), note that *m* = 8, the number of explanatory variables in the level 1 logistic regression model, and *r* = 34 = *m* + the number of birth cohorts + the number of election time periods = 8 + 17 + 9. Then, the value of the *F* ratio for the statistical significance of the birth cohort effects alone is

$$F = 47.8, p < .01$$

that for the statistical significance of the period effects alone is

$$F = 96.8, p < .01$$

and that for the statistical significance of cohorts and periods together is

$$F = 34.8, p < .01$$

Numerical details for these F tests for hypotheses (7.22)–(7.24) for the NHANES data are given in Table 7.5.

In agreement with the variance components analysis using the z statistics reported in Step 2, these F-test results show that the incorporation of time period random effects into the model produced a statistically significant variation in the level 1 model intercept. Different from the z-test result, however, incorporation of birth cohort random effects also produced significant variation in the intercept by the standard of the F test. In addition, the incorporation of both sets of effects produced a statistically significant F ratio. These Step 4 results corroborate the inferences regarding the period effects on obesity from Steps 1, 2, and 3 and suggest an additional source of variation contributed by birth cohort. We note that to estimate the mixed logistic models, existing software programs like SAS use a pseudo-ML estimation algorithm in which a consistent and asymptotically normal estimate of certain model parameters is computed rather than a ML estimate. In such contexts, the log-likelihood functions necessary to calculate the asymptotic F test may be only approximate. Substantively, taking into account all of the foregoing assessments of individual coefficients and sets of coefficients, it must be concluded that, while there is some evidence of statistical significance of birth cohort effect, the main story on trends in obesity epidemic in

TABLE 7.5

F-tests for the Presence of Random Effects, NHANES Data (1971–2008)

	Cohort Effects $\tau_u = 0$ vs. $\tau_u > 0$	Period Effects $\tau_v = 0$ vs. $\tau_v > 0$	Cohort and Period Effects $\tau_u = \tau_v = 0$ vs. τ_u or $\tau_v > 0$
l_0	43313	43313	43313
l_{max}	42456	42395	42359
r	25	17	34
m	8	8	8
N_T	40,261	40,261	40,261
$(l_0 - l_{max})/(r - m)$	50.41	102.00	36.69
$l_{max}/(N_T - r)$	1.06	1.05	1.05
F	47.78	96.82	34.85
$f_{0.95}(r - m, N_T - r)$	1.62	1.88	1.50

the U.S. population, as measured by the objective body mass index (BMI) data, is a period story rather than a cohort story—the opposite of what we concluded for trends in verbal ability.

Future applications of the HAPC model in the similar forms as specified for the GSS and NHANES data are likely to find some instances in which either time period or cohort effects are effective, and there may also be instances for which both types of contextual effects may be operative. Use of the HAPC modeling apparatus and decision guidelines to assess statistical and substantive significance described in this section should be quite fruitful in disentangling the age, period, and cohort sources of temporal change in such studies.

It merits noting that the second part of Step 2 in our strategy, that of an application of a formal statistical test of homogeneity of the estimated random effect coefficients of cohort and period effects, may become more challenging or even intractable when more complicated versions of GLMMs are used for the HAPC analysis. The comparisons of the z- and F-test results based on the two models suggest that the former tends to yield results more conservative than the latter, while the benefit added by the application of the F test is the detection of the combined random components of period and cohort. As these tests have only been developed for the random intercept models, we recommend the analyst make a cautious assessment of whether the F test can be reasonably applied and interpreted in other kinds of mixed models, such as the random coefficients APC models we introduce next.

7.7 Random Coefficients HAPC-CCREM

The mixed regression models approach not only is methodologically relevant for APC analyses but also enhances our ability in addressing additional questions that cannot be addressed in linear fixed effects models but bear theoretical importance to studies of social change and heterogeneity. First, the basic random intercept models (7.9) and (7.24) introduced specify fixed effects of all individual-level covariates. The results of the analyses of the GSS verbal test score data and the NHANES obesity data have shown that the fixed effect coefficients of these covariates are statistically significant, and their inclusion also accounts for some of the age effects and variance components. Could these level 1 coefficients also have random variation, just as the intercept, such that the associations between individual-level covariates and the outcome may vary depending on period and birth cohort? While classical fixed effects linear regression/accounting APC models are often confined to the specification of age effects, the HAPC-CCREM can more easily specify temporal variations in other individual-level covariates that may substantially broaden the scope of investigations. In the example that follows, we extend the NHANES obesity analysis in the previous section to show how the specification of random coefficients models enables us to detect increasing social demographic differentials in obesity rates across birth cohorts.

The results of the previous analysis suggested weak cohort variations in the overall intercept or mean probability of obesity. Given findings from an earlier study that suggested population heterogeneity in the period and cohort effects from the National Health Interview Survey (NHIS) data (Reither, Hauser, and Yang 2009), we further modeled random effects of level 1 effect coefficients. That is, we tested the hypothesis about whether sex, race, or socioeconomic status (SES) disparities in obesity varied significantly by time period or birth cohort.

The level 1 model specification is similar to Equation (7.22) but allows the level 1 coefficients to have random effects:

$$Logit \, Pr(OBESE_{ijk} = 1) = \beta_{0jk} + \beta_{1jk} AGE_{ijk} + \beta_{2jk} AGE_{ijk}^2 + \beta_{3jk} SEX_{ijk} \quad (7.26)$$

$$+ \beta_{4jk} RACE_{ijk} + \beta_{5jk} EDUCATION_{ijk} + \beta_{6jk} INCOME_{ijk}$$

The level 2 model is then revised as follows:

Sex effect:

$$\beta_{3jk} = \gamma_3 + u_{3j} + v_{3k} \quad (7.27)$$

Race effect:

$$\beta_{4jk} = \gamma_4 + u_{4j} + v_{4k} \quad (7.28)$$

Education effect:

$$\beta_{5jk} = \gamma_5 + u_{5j} + v_{5k} \quad (7.29)$$

Income effect:

$$\beta_{6jk} = \gamma_6 + u_{6j} + v_{6k} \quad (7.30)$$

These models tested the hypotheses about period and cohort effects through the specifications of random variance components for the random coefficients for sex, race, education, and income effects. The age effects could potentially also vary from cohort to cohort or from period to period. In this analysis, however, such random age effects were not statistically significant or of theoretical interest. We therefore could constrain the age effects to be fixed just as in the analysis of the GSS verbal test score data in Section 7.5. Therefore, we have omitted the level 2 random effects models for these two coefficients, which yielded $\beta_{1jk} = \gamma_1$ and $\beta_{2jk} = \gamma_2$. In this set of equations, γ_3 to γ_6 are the level 2 fixed effects coefficients that represent the fixed effects of sex, race, education, and income, respectively. In the level 1 model, we can test the life course hypotheses about age variations in social gaps in obesity by including interaction terms of age with each of the other level 1 covariates. In fact, the specification of level 1 model that accounts for the functional form of level 1 covariates' effects across age preceded that of the level 2 model. This process is explicated in the analysis of happiness in the next chapter. To test whether these differentials varied by time or birth cohort, the equations specified that their coefficients have cohort effects $u_{3j} - u_{6j}$ and period effects $v_{3k} - v_{6k}$, whose corresponding random variance components are $\tau_{u3} - \tau_{u6}$ and $\tau_{v3} - \tau_{v6}$. Based on the combined models of (7.26)–(7.30), we next estimated logit CCREMs of obesity using SAS PROC GLIMMIX (Littell et al. 2006) (see the sample codes).

Table 7.6 summarizes the results of the random coefficient model estimates. The fixed effects estimates were largely similar to those reported in the previous analysis using the random intercept models (Table 7.4). Also similar to the previous analysis, the estimates of random effects in terms of residual variance components at level 2 indicated significant period but not cohort effects on the overall intercept. Level 1 coefficients of two covariates—sex and race—showed significant random effects, whereas those of SES variables did not. There seem to be cohort changes in sex gaps in obesity and period changes in sex and race gaps in obesity that are significant at $\alpha = .10$ level. The race gaps showed larger variations across cohorts than time periods that are significant at $\alpha = .05$ level. To see more clearly how specific sex and race groups changed in the risk of obesity across cohorts, we display the estimated random cohort effects in Figure 7.3 in terms of the predicted probabilities of being obese for each cohort at the mean age and averaged over all years by sex and race.

For the reference group of white females with 12 years or less of education and in middle income quartiles, the predicted probabilities were calculated the same way as those for Figure 7.2. For black males, we first calculated the random effect coefficients as $\hat{\beta}_{0j} + \hat{\beta}_{3j} + \hat{\beta}_{4j} = \hat{\gamma}_0 + \hat{\gamma}_3 + \hat{\gamma}_4 + u_{0j} + u_{3j} + u_{4j}$ (where $\hat{\gamma}_0 = -0.563$ is the intercept or estimated overall mean; $\hat{\gamma}_3 = -0.424$ is the estimated fixed sex effect coefficient; $\hat{\gamma}_4 = 0.458$ is the estimated fixed race effect coefficient; u_{0j} are the intercept cohort random effects; u_{3j} and u_{4j} are the sex and race-specific random effects, respectively) and converted them to predicted probabilities. Similar calculations yielded results for white males and black females. The figure shows that black cohorts experienced more pronounced increases in risks of obesity than white cohorts, controlling for the

TABLE 7.6

The Random Coefficient CCREM Estimates of Obesity: NHANES 1971–2008

Fixed Effects	Parameter	Coefficient	se	t Ratio	p Value
Intercept	γ_0	−0.57	0.15	−3.79	0.004
AGE	γ_1	0.01	0.00	8.46	<0.001
AGE^2	γ_2	−0.00	0.00	−10.81	<0.001
SEX (male = 1)	γ_3	−0.43	0.06	−7.44	<0.001
RACE (black = 1)	γ_4	0.45	0.07	6.55	0.001
EDUCATION	γ_5				
SOME COLLEGE (13–15 years)		−0.08	0.03	−2.38	0.017
COLLEGE (≥ 16 years)		−0.40	0.04	−9.80	<0.001
FAMILY INCOME	γ_6				
LOWEST QUARTILE		0.13	0.03	4.44	<0.001
HIGHEST QUARTILE		−0.14	0.03	−4.26	<0.001
Variance Components		**Variance**	**se**	**z Statistic**	**p Value**
COHORT					
Intercept	τ_0	0.00	0.00	0.78	0.217
SEX	τ_3	0.01	0.01	1.41	0.079
RACE	τ_4	0.03	0.02	1.68	0.047
PERIOD					
Intercept	τ_{v0}	0.19	0.10	1.97	0.024
SEX	τ_{v3}	0.02	0.01	1.39	0.082
RACE	τ_{v4}	0.02	0.01	1.44	0.075
Model Fit					
−2 Res Log Pseudo-Likelihood		188268.0	df=40180		

age and period effects and other factors. Consistent with the earlier study by Reither, Hauser, and Yang (2009), both black females and males showed particularly large increases for cohorts born after 1955–60. And black females were the most likely to be obese on average among all groups examined, with the peak cohort of 1970–74 having a 52% predicted probability of obesity. The black and white gaps increased for both sexes across cohorts, contributing to the significant random variance of the race coefficient in Table 7.6.

In sum, the findings corroborated those from prior studies using other data sources and analytic methods. We have shown that the obesity epidemic in the overall population is a period-driven phenomenon. There are, however, substantial social demographic variations in cohort trends of obesity. Increasing racial differentials in cohort trends of obesity are particularly worthy of attention as the surge in the obesity risks in most recent cohorts was restricted to blacks.

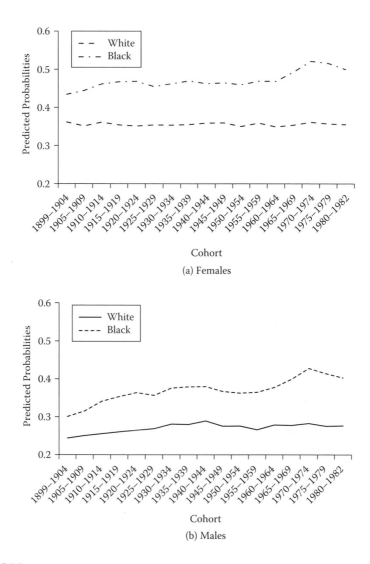

(a) Females

(b) Males

FIGURE 7.3
Cohort effects on the sex and race disparities in obesity: NHANES 1971–2008.

Appendix 7.1: Matrix Algebra Representations of Linear Mixed Models and Generalized Linear Mixed Models

A general linear mixed (fixed and random) effects statistical model (LMM) may be written as

$$y = X\beta + Z\alpha + \varepsilon \tag{7.31}$$

where y is a vector of observations, X is a matrix of observations on known covariates (regressors), β is a vector of unknown regression coefficients to be estimated, Z is a known design matrix, α is a vector of random effects, and ε is a vector of errors. Both α and ε are unobservable random variables. The assumptions of this model are that the random effects and errors have mean zero and finite variances with covariance matrices $G = Var(\alpha)$ and $R = Var(\varepsilon)$ that usually involve some unknown dispersion parameters or variance components to be estimated from the data. Note that this is a vector-matrix version of the "combined model" form of the HAPC models described in this chapter.

In most applications of LMMs, the α's and ε's are assumed to be independent. Equivalently, net of the effects of the observed covariates and the random effects, the observations in the y vector are assumed to be independent. Then, the covariance matrices reduce to scalars. Specifically, under this specification, the random effects $\alpha_1, \ldots, \alpha_m$ are assumed to be independently and identically distributed (*i.i.d.*) with mean 0 and variance σ^2, and the errors are i.i.d. with mean 0 and variance τ^2. And, the usual assumption of LMMs is that the random effects and errors also are Gaussian/normally distributed, that is, $\alpha \sim N(0, \sigma^2)$ and $\varepsilon \sim N(0, \tau^2)$. This specification defines the class of *Gaussian* LMMs, and, while other distributional specifications on the random effects (such as t distributions) have been studied in statistics, it is the form of LMMs typically programmed in LMM software and therefore most widely applied empirically.

To give a corresponding general representation of the class of Gaussian LMMs, we first give an alternative expression of the Gaussian LMMs just defined (Jiang 2007: 121). Given a column vector of random effects α, the Gaussian LMM assumes the observations $y_1, \ldots, y_n$ are (conditionally) independent such that $y_i \sim N(x_i'\beta + z_i'\alpha, \tau^2)$, where the primes denote the matrix/vector transposition operator, x_i and z_i are known column vectors for the ith observation, β is an unknown vector of regression coefficients, and τ^2 is an unknown variance. Suppose also that α is multivariate normal with mean 0 and covariance matrix G, which depends on a vector θ of unknown variance components. Let X and Z be the matrices whose ith rows are x_i' and z_i', respectively. It is easy to see that this leads to the LMM of (7.31) with normality and $R = \tau^2 I$.

In the specification of the Gaussian LMM, the two key assumptions are (1) conditional independence of the observations (given the random effects) and (2) the distribution of the random effects. These two elements can be used to define the class of GLMMs. Given a column vector of random effects α, suppose the responses $y_1, \ldots, y_n$ are (conditionally) independent with conditional distribution a member of the exponential family of distributions with probability distribution function:

$$f_i\left(y_i|\,\alpha\right) = exp\left\{\frac{y_i\varepsilon_i - b(\varepsilon_i)}{\alpha_i(\,\,)} + c_i(y_i,\,\,)\right\} \qquad (7.32)$$

where $b(\cdot)$, $\alpha_i(\cdot)$, $c_i(\cdot,\cdot)$ are known functions, and φ is a dispersion parameter that may or may not be known. The quantity ε_i is associated with the conditional mean $\mu_i = E(y_i|\alpha)$, where E is the expectation operator. This conditional mean in turn is associated with the linear predictor:

$$\eta_i = x_i'\beta + z_i'\alpha \qquad (7.33)$$

where x_i and z_i are known column vectors, and β is a vector of unknown regression coefficients (i.e., the fixed effects), through a known *link function* $g(\cdot)$ such that

$$g\left(u_i\right) = \eta_i \qquad (7.34)$$

As in the LMM class of models, it is assumed that $\alpha \sim N(0, G)$, where the covariance matrix may depend on a vector θ of unknown variance components. In brief, the GLMMs retain the linear predictor function of the LMMs but generalize the family of frequency distributions for the outcome observations in such a way they accommodate nonnormal and discrete variables. The importance of the GLMM class of models is that members of the exponential family of distributions it incorporates include the Gaussian, binomial, Poisson, gamma, and inverse-Gaussian distributions. This facilitates application, for example, of the logistic and Poisson link functions, as illustrated in Sections 7.5.2 and 7.6 and Sections 8.2, 8.3, and 8.5 in Chapter 8.

In the empirical applications of LMMs and GLMMs to the HAPC models and analyses described in this chapter, the elements of the y vectors are the outcome observations for which an analysis of age, period, and cohort effects is sought (e.g., the GSS WORDSUM scores), the elements of the X matrices and corresponding x vectors are the individual-level explanatory variables for which observations are available in the repeated cross-sectional sample surveys and for which corresponding fixed effects coefficients are specified in the β vectors, and the elements of the Z matrices and corresponding z vectors are the specifications of the time periods and cohorts to which specific observations belong and for which corresponding random effects are specified in the α vectors.

As noted, the conventional assumption in specifications of LMMs and GLMMs is that the random effects and errors are normally distributed, and as noted in the text of this chapter, the estimation algorithm typically used in software programs for these models is an ML or REML-EB. Research in statistics has shown that, even if the normality assumptions do not hold, the ML and REML estimators are consistent and asymptotically normal for point estimation of the fixed effects coefficients (in non-Gaussian LMMs, the use of normality-based estimators is termed *quasi-likelihood*; Jiang 2007: 16–17). In addition, however, when normality does not hold, the asymptotic covariance matrix involves parameters other than the variance components, namely, the third and fourth moments of the random effects and errors. This implies that standard errors may be incorrectly estimated. Accordingly, in empirical applications of LMMs and GLMMs, analysts should assess the plausibility of the normal distributional assumptions. If the data are substantially nonnormal and the number of random effects is relatively small, caution is advised regarding inferences about statistical significance as conventional software programs often do not include functions of the third and fourth moments in their algorithms, and standard errors may be underestimated.

References

Alwin, D. F. 1991. Family of origin and cohort differences in verbal ability. *American Sociological Review* 56:625–638.

Alwin, D. F. 2009. History, cohorts, and cognitive aging. In *Aging and cognition: Research methodologies and empirical advances*, ed. H. B. Bosworth and C. Hertzog, 9–38. Washington, DC: American Psychological Association.

Alwin, D. F., and R. J. McCammon. 1999. Aging versus cohort interpretations of intercohort differences in GSS vocabulary scores. *American Sociological Review* 64:272–286.

Alwin, D. F., and R. J. McCammon. 2001. Aging, cohorts, and verbal ability. *The Journals of Gerontology Series B: Psychological Sciences and Social Sciences* 56:S151–S161.

Baltagi, B. H. 1995. *Econometric analysis of panel data*. New York: Wiley.

Campbell, J. R., C. M. Hombo, and J. Mazzeo. 2000. NAEP 1999 trends in academic progress: Three decades of student performance. *Education Statistics Quarterly* 2:31–36.

Demidenko, E. 2004. *Mixed models: Theory and applications*. Hoboken, NJ: Wiley.

Duchateau, L., and P. Janssen. 1999. Small vaccination experiments with binary outcome: The paradox of increasing power with decreasing sample size and/or increasing imbalance. *Biometrical Journal* 41:583–600.

Efron, B., and C. N. Morris. 1975. Data analysis using Stein's estimator and its generalizations. *Journal of the American Statistical Association* 70:311–319.

Frenk, S. M., Y. Yang, and K. C. Land. 2012. *Assessing the significance of cohort and period effects in hierarchical age-period-cohort models with applications to verbal test scores and voter turnout in U.S. presidential elections*. Under review.

Glenn, N. D. 1994. Television watching, newspaper reading, and cohort differences in verbal ability. *Sociology of Education* 67:216–230.

Glenn, N. D. 1999. Further discussion of the evidence for an intercohort decline in education-adjusted vocabulary. *American Sociological Review* 64:267–271.

Goldstein, H. 2003. *Multilevel statistical models*. London: Arnold.

Greene, W. H. 2003. *Econometric analysis*. 5th ed. Upper Saddle River, NJ: Prentice Hall.

Hausman, J. A., and W. E. Taylor. 1981. Panel data and unobservable individual effects. *Econometrica* 49:1377–1398.

Hedges, L. V., and A. Nowell. 1995. Sex differences in mental test scores, variability, and numbers of high-scoring individuals. *Science* 269:41–45.

Hobcraft, J., J. Menken, and S. Preston. 1982. Age, period, and cohort effects in demography: A review. *Population Index* 48:4–43.

Hox, J. J., and I. G. G. Kreft. 1994. Multilevel analysis methods. *Sociological Methods & Research* 22:283–299.

Jiang, J. 2007. *Linear and generalized linear mixed models and their applications*. New York: Springer-Verlag.

McCullagh, P., and J. A. Nelder. 1989. *Generalized linear models*. New York: Chapman and Hall.

Littell, R. C., G. A. Milliken, W. W. Stroup, R. D. Wolfinger, and O. Schabenberrger. 2006. *SAS for mixed models*. Cary, NC: SAS Institute.

Longford, N. T. 1993. *Random coefficient models*. New York: Oxford University Press.

Lynch, S. M. 2003. Cohort and life-course patterns in the relationship between education and health: A hierarchical approach. *Demography* 40:309–331.

Mason, K. O., W. M. Mason, H. H. Winsborough, and W. K. Poole. 1973. Some methodological issues in cohort analysis of archival data. *American Sociological Review* 38:242–258.

Raudenbush, S. W., and A. S. Bryk. 2002. *Hierarchical linear models: Applications and data analysis methods*. Thousand Oaks, CA: Sage.

Reither, E. N., R. M. Hauser, and Y. Yang. 2009. Do birth cohorts matter? Age-period-cohort analyses of the obesity epidemic in the United States. *Social Science & Medicine* 69:1439–1448.

Smith, H. L. 2004. Response: Cohort analysis redux. *Sociological Methodology* 34:111–119.

Snijders, T., and R. Bosker. 1999. *Multilevel analysis: An introduction to basic and advanced multilevel modeling*. Thousand Oaks, CA: Sage.

White, H. 1980. A heteroscedasticity-consistent covariance matrix estimator and a direct test for heteroscedasticity. *Econometrica* 48:817–838.

Wilson, J. A., and W. R. Gove. 1999a. The age-period-cohort conundrum and verbal ability: Empirical relationships and their interpretation: Reply to Glenn and to Alwin and McCammon. *American Sociological Review* 64:287–302.

Wilson, J. A., and W. R. Gove. 1999b. The intercohort decline in verbal ability: Does it exist? *American Sociological Review* 64:253–266.

Yang, Y. 2006. Bayesian inference for hierarchical age-period-cohort models of repeated cross-section survey data. *Sociological Methodology* 36:39–74.

Yang, Y., and K. C. Land. 2006. A mixed models approach to the age-period-cohort analysis of repeated cross-section surveys, with an application to data on trends in verbal test scores. *Sociological Methodology* 36:75–97.

Yang, Y., and K. C. Land. 2008. Age-period-cohort analysis of repeated cross-section surveys: Fixed or random effects? *Sociological Methods & Research* 36:297–326.

8

Mixed Effects Models: Hierarchical APC-Cross-Classified Random Effects Models (HAPC-CCREM), Part II: Advanced Analyses

8.1 Introduction

In this chapter, we illustrate applications of the family of hierarchical age-period-cohort (HAPC) models for more advanced analysis of repeated cross-sectional data designs in the forms of both sample surveys and aggregate rates. In addition to a new example using repeated cross-sectional surveys from the General Social Survey (GSS) on happiness introduced in Chapter 3, we continue to analyze the cancer mortality rate data to relate their temporal trends revealed in Chapter 6 to potential risk factor mechanisms. We also describe extensions to HAPC models useful for solving substantive problems in empirical research. These extensions involve more advanced statistical methods, such as Bayesian methods and heteroscedastic regressions (HRs), and can be skipped by readers with less-advanced statistical methodological knowledge.

8.2 Level 2 Covariates: Age and Temporal Changes in Social Inequalities in Happiness

If there is evidence from the basic HAPC-CCREM (cross-classified random effects model) analysis for clustering effects of random errors or significant random variability across birth cohorts, how can it be explained or what may account for the variance? The same questions may apply to the investigation of period effects. Such problems suggest the importance of explanatory factors related to birth cohort and period effects and cannot be fully accommodated in the classical APC (age-period-cohort) accounting framework introduced in Chapters 5 and 6 (Smith 2004). The HAPC modeling approach, therefore, is an improvement that offers an option for researchers interested in identifying key explanatory factors in addition to age, period, and cohort indicators. The specification of random coefficient HAPC-CCREMs with period-level and cohort-level variables to explain the period and cohort effects we have estimated in Chapter 7 is a next step in model building and assessment.

We now illustrate the use of the more elaborate models that combine the specifications of random coefficients and level 2 covariates through an analysis of trends in social inequalities of happiness.[*] Based on the time series data on happiness in the United States for people 18 years of age and older from the GSS 1972–2004, we conducted an HAPC analysis to address the following questions about changes in the social inequalities of quality of life: Do people get happier with age and over time? Are there any birth cohort differences in happiness? How do social inequalities—sex, race, and education differentials—in happiness vary over the life course and by time? And, what could have contributed to these differences and changes?

The HAPC-CCREM applied to the happiness data was specified as the following:

Level 1 or "within-cell" model:

$$\Pr(HAPPY_{mijk} = 1) = \beta_{0jk} + \beta_1 AGE + \beta_2 AGE^2 + \beta_{3jk} SEX + \beta_{4jk} RACE$$

$$+ \beta_5 EDUC + \sum_{p=6}^{P} \beta_p X_p \tag{8.1}$$

It states that the cumulative logit of happiness score that has $m = 3$ ordinal response categories (1 = very happy; 2 = pretty happy; 3 = not too happy) for respondent i (for $i = 1, 2, \ldots, n_{ij}$) within period j (for $j = 1, 2, \ldots, 22$) and cohort k (for $k = 1, 2, \ldots, 18$ five-year cohorts) is a function of age, age squared, sex, race, education, and other correlates X, where X denotes the vector of other individual-level variables, including age by sex, age by race, age by education interaction terms for the test of hypothesis of differential social inequalities over the life course, and control variables (income, marital status, health, employment status, family structure, and religious attendance) that are known to be strongly associated with happiness (Blanchflower and Oswald 2004; Davis 1984; Easterlin 2001; Hughes and Thomas 1998). In this model, β_{0jk} is the intercept indicating the cell mean for the reference group at mean age surveyed in year j and belonging to cohort k; $\beta_{1jk} - \beta_{5jk}$, and β_p denote the level 1 coefficients; and P is the maximum number of covariates included. Age divided by 10 and all other continuous covariates were grand mean centered for ease of interpretation of the intercept values.

---

[*] Parts of Section 8.2 were adapted and updated from Yang, Y. 2008. *American Sociological Review* 73:204–226.

Level 2 or "between-cell" model:

Overall mean:

$$\beta_{0jk} = \gamma_0 + u_{0j} + v_{0k} \tag{8.2}$$

Sex effect:

$$\beta_{3jk} = \gamma_3 + v_{3k} \tag{8.3}$$

Race effect:

$$\beta_{4jk} = \gamma_4 + v_{4k} \tag{8.4}$$

Equation (8.2) is the model for the random intercept that specifies that the overall mean varies from period to period and from cohort to cohort, where γ_0 is the expected logit of being very happy at the zero values of all level 1 variables averaged over all periods and cohorts; u_{0j} is the cohort effect in terms of residual random coefficients of cohort j averaged over all time periods with variance τ_{u0}; v_{0k} is the period effect in terms of residual random coefficients of period k averaged over all birth cohorts with variance τ_{v0}. We can constrain the age effects to be fixed just as in the previous analysis of verbal test score. The next two equations specify the sex and race coefficients to have random period effects represented by residuals v_{3k} and v_{4k}, whose random variance components are denoted as τ_{v3} and τ_{v4}, respectively. We also tested for random cohort effects for the sex and race coefficients but did not find any. Other level 1 covariates are modeled as fixed across level 2 units because supplementary analysis did not suggest they have significant random effects. Based on the combined models of Equations (8.1)–(8.4), we next estimated ordinal logit CCREMs of happiness using SAS PROC GLIMMIX (Littell et al. 2006) (see the sample codes). Probabilities modeled are cumulated over the lower ordered values based on the assumption of proportional odds, and the coefficients indicate the effects for the probability of being very happy (1) versus others (2 and 3) and the probability of being very happy or pretty happy (1 or 2) versus not too happy (3).

Table 8.1 presents estimates of fixed effects coefficients in the form of odds ratios of being happy and random effects variance components from the ordinal logit CCREMs. Model A includes the key social status variables, and model B adjusts for control variables. Model A shows a significant age effect net of the random period and cohort effects. It suggests that, adjusting for time period and birth cohort variations, the odds of being happy increased 9.8% with every 10 years of age. Adjusting for other covariates in model B, the age effect became significantly quadratic, indicating an acceleration of the age increase at the rate of 2.7% every decade over the life course. Thus, the first important finding is that with age comes happiness. It is important to note

TABLE 8.1

The Ordinal Logit CCREM Estimates of Happiness: GSS 1972–2004

Fixed Effects	Parameter	Model A				Model B			
		Coefficient	Odds Ratio	t Ratio	p Value	Coefficient	Odds Ratio	t Ratio	p Value
Intercept	γ_0	−0.65	0.52		<0.001	−0.98	0.38		<0.001
AGE	γ_1	0.09	1.10	6.23	<0.001	0.08	1.08	4.46	<0.001
AGE2	γ_2	0.00	1.00	−0.68	0.495	0.03	1.03	4.94	<0.001
SEX (female=1)	γ_3	0.07	1.07	1.93	0.067	0.17	1.19	5.72	<0.001
RACE (black=1)	γ_4	−0.64	0.53	−12.25	<0.001	−0.41	0.67	−7.35	<0.001
EDUCATION	γ_5								
EDUC 1 (< 12 yrs)		−0.36	0.70	−11.84	<0.001	−0.04	0.96	−1.39	0.165
EDUC 2 (>= 16 yrs)		0.26	1.30	8.50	<0.001	0.01	1.01	0.17	0.862
AGE x FEMALE	γ_6	−0.11	0.90	−7.86	<0.001	−0.03	0.97	−2.17	0.029
AGE x BLACK	γ_7	0.11	1.12	5.51	<0.001	0.14	1.15	6.71	<0.001
AGE x EDUC 1	γ_8	0.03	1.03	1.96	0.051	0.04	1.04	2.32	0.020
AGE x EDUC 2	γ_9	−0.03	0.97	−1.33	0.184	−0.03	0.97	−1.60	0.109
INCOME	γ_{10}								
MIDDLE QUARTILES							1.00		
LOWER QUARTILE						−0.95	0.87	−26.03	<0.001
UPPER QUARTILE						−1.10	1.26	−21.67	<0.001
MARITAL STATUS	γ_{11}								
MARRIED							1.00		
DIVORCED						−0.75	0.39	−17.52	<0.001
WIDOWED						0.67	0.33	23.77	<0.001
NEVERMAR						−0.60	0.47	−17.37	<0.001

	Parameter	Variance	s.e.	z Statistic	p Value	Variance	s.e.	z Statistic	p Value
HEALTH STATUS	γ_{12}								
GOOD							1.00		
EXCELLENT						−1.21	1.95	−21.00	<0.001
FAIR						−0.11	0.55	−2.83	<0.001
POOR						−0.66	0.30	−9.12	<0.001
WORK STATUS	γ_{13}								
FULLTIME							1.00		
PARTTIME						0.13	0.89	2.67	0.005
UNEMPLOYED						−0.14	0.52	−4.58	<0.001
RETIRED						0.23	1.14	7.52	0.007
CHILDS_0	γ_{14}					0.11	1.11	3.03	0.002
ATTEND	γ_{15}					0.08	1.09	18.11	<0.001
Variance Components	**Parameter**	**Variance**	**s.e.**	**z Statistic**	**p Value**	**Variance**	**s.e.**	**z Statistic**	**p Value**
COHORT	τ_{u0}	0.01	0.00	1.93	0.03	0.01	0.00	1.94	0.026
PERIOD									
Intercept	τ_{r0}	0.00	0.00	1.22	0.11	0.00	0.00	1.63	0.051
SEX	τ_{r3}	0.01	0.01	1.86	0.03	0.01	0.00	1.31	0.095
RACE	τ_{r4}	0.03	0.02	1.63	0.05	0.04	0.02	1.78	0.037
Model Fit									
−2 Res Log-Pseudo-Likelihood		219758.2	(df=28853)			231656.2	(df=28840)		

Source: Adapted from Yang (2008: Table 2).

that in the HAPC models of happiness where all three temporal factors are considered, the age effects dominate, and the period and cohort variations are small. This suggests that studies of temporal changes in subjective well-being that ignore life course changes may be misleading in giving the impression that time trends and cohort differences of happiness are more substantial than they actually are.

Results of other fixed effects in model A show that women, whites, and college graduates on average had higher odds of being happy relative to men, blacks, and people with less education, respectively. Whereas the sex effect was small and only marginally significant, the race and education effects were substantial and highly significant. Being black was associated with 47% lower odds of being happy. Having college education increased the odds of being happy by about 30%, whereas having less than a high school degree decreased the odds by 30%. Comparing models A and B showed that the salient sex and race differentials in happiness persisted even when major correlates of happiness were held constant. But, the race effect decreased in size, and the effects of educational attainment were no longer significant. This means that some of the effect of being black and all of the effect of education were mediated by other covariates. These results were largely consistent with previous cross-sectional studies of social correlates of happiness at a point in time. But, they bear different interpretations in the HAPC analysis. That is, they represent individual-level fixed effects when level 2 heterogeneity by period and cohort are taken into account.

The interaction terms at level 1 in model A suggest that the sex, race, and educational gaps in happiness varied significantly with age. All the interaction terms remained significant when all things were considered in model B, so the life course trajectories of happiness still showed appreciably different trends. When potential explanatory factors were controlled, however, the interaction effects of sex, race, and education with age shrank in size. Figure 8.1a shows the predicted probabilities of being very happy by age from model B for various social groups at the zero values of all other covariates (for the reference group). The age curves of happiness are convex and J shaped, suggesting that the average probability of being happy bottomed out in early adulthood and increased at an increasing rate as one moved through the life course. The life course patterns of happiness were different, however, for men and women, blacks and whites, and the more educated and the less educated. For both blacks and whites, women were happier than men, but the gaps showed convergences with age. The adjustment for marital status, health, and work status decreased the age-by-sex interaction effect. This supports the hypothesis that the changes of the sex effect on happiness with age are partly due to increases in widowhood and worsening health among women, which decrease happiness, and retirement and better health among men, which increase happiness. The racial gap in happiness was larger than the sex gap for a large segment of the life course. Whites' advantage over blacks starts large and gradually decreases with age.

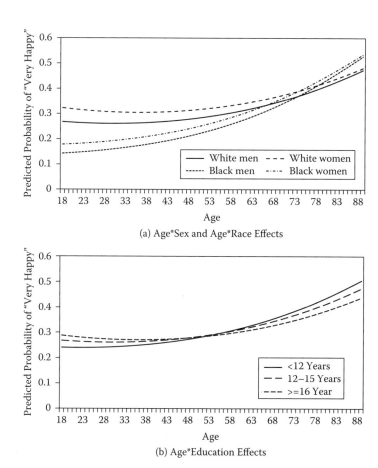

FIGURE 8.1
Predicted age variations in sex, race, and education effects on happiness from model B: (a) Age*Sex and Age*Race effects; (b) Age*Education effects.

When other factors were held constant, the race differentials showed reversals occurring at around the age of 70, after which blacks became slightly happier than whites. The age and education interaction effects are graphically presented in Figure 8.1b. There appear to be crossovers in age effects of happiness among the three educational groups at the age of 50, after which the highly educated were not particularly better off than the less educated. These convergences indicating loss of advantages for women, whites, and the highly educated with age do not seem to be completely explained by marital status, health, or other factors. The persistent social disparities in life course patterns of happiness invite future investigations of aging-related theories.

The lower panel of Table 8.1 reports the results of random effects in terms of residual variance components for the intercept and level 1 coefficients of sex and race. There seemed to be significant period changes in levels of happiness and sex and race inequalities net of age effects, cohort changes, and other factors. There were also significant cohort changes in levels of happiness, although none of the level 1 effect coefficients showed any significant random effects by birth cohort. That is, the advantages of women and the highly educated and the disadvantages of blacks and the less educated were constant across successive cohorts.

Figure 8.2 displays the estimated random period effects from model A, namely, the predicted probabilities of being very happy for each year at the mean age and averaged over all birth cohorts by sex and race. The calculations of these probabilities followed the same procedure described for the National Health and Nutrition Examination Survey (NHANES) obesity analysis in Section 7.7 in Chapter 7. The overall period effects showed a general weak downward trend across the first two decades that was consistent with findings from previous studies based on GSS data but also exhibited clear nonlinear declines over time. The range of such period changes was relatively small, that is, largely within 5% over the 30 years. The sex-specific and race-specific results suggest that both sex and race disparities decreased during the last 30 years in the United States. Figure 8.2a illustrates that the reduction in the sex gap was due to decreasing levels of happiness in women and the stable trend in men as of 1995 and similar increases for both afterward. The sex differentials were relatively small even at the initial time period, and they were not statistically significant in the final model when other factors were taken into account. Figure 8.2b illustrates that the black and white gap was much more pronounced throughout the 30-year period, but there were small declines in more recent years. From the 1970s to the mid-1990s, whites experienced a slight downward trend in happiness, and blacks experienced some fluctuations early on and a stable trend after 1980. Both groups fared better after 1995. Such period changes in racial differences remained statistically significant in model B after adjustment of other factors.

Figure 8.3 shows the random cohort effects in terms of the predicted probabilities of being very happy at the mean age and averaged over all periods estimated from model A for the reference group. As is the case for period effects, the cohort effects were small relative to the age effects and showed no linear increases or declines. There was barely any evidence in prior studies for cohort differences in happiness, although there were speculations that more recent cohorts had experienced lower levels of happiness. The results do not strongly support such a hypothesis when confounding effects are controlled. The cohort trends do not show the monotonic declines that previous research had speculated might be engendered by cohorts' value shifts under the influence of "postmaterialism." Instead, they were flat for the earlier and later cohorts. The most intriguing finding is the dip in the predicted probabilities of being very happy for cohorts born between 1945 and 1960, that is, the baby boomers.

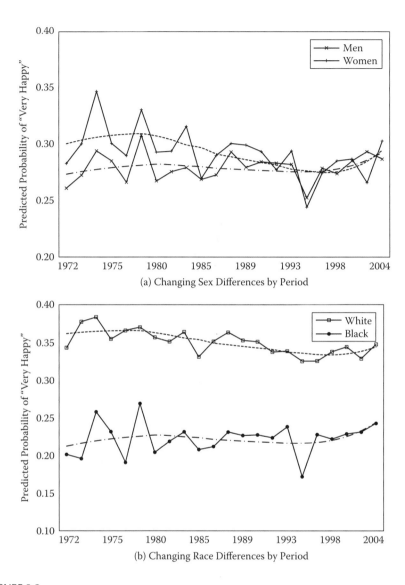

FIGURE 8.2
Predicted period variations in the sex and race disparities in happiness: (a) changing sex differences by period; (b) changing race differences by period.

The HAPC analysis of happiness suggests that life course patterns, time trends, and birth cohort differences in happiness are distinct and independent of each other. The significant net age, period, and cohort effects suggest that it is important to formally test variations in all three time-related dimensions in studies of changes in subjective well-being for adequate interpretation and

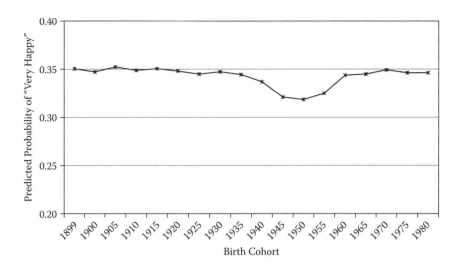

FIGURE 8.3
Predicted cohort variations in happiness from model A.

valid inference with regard to these effects. The individual-level covariates included have accounted for some of the age-related changes in happiness and social differentials in age changes. It is not clear what could have contributed to the period and cohort patterns. Some hypothesized that the former could be correlated with changes in macroeconomic variables such as gross domestic product from 1970s to 1990s and levels of joblessness (Di Tella, MacCulloch, and Oswald 2003). So, does gross domestic product (GDP) buy extra happiness? And, how does unemployment rate feed through into people's perceptions of quality of life? We can examine these hypotheses through the specification of level 2 covariates in the HAPC-CCREM framework.

The plots of variables associated with economic cycle in Figure 8.4 suggest that percentage annual change in the GDP per capita and the unemployment rate seemed to accompany the period changes in happiness. So, we can add the two period-level covariates, ΔGDP_k and unemployment rate ($UNEMPLOY_k$), to the models for the overall mean, sex, and race effects to test whether economic environment explains period variations in happiness and sex and race gaps therein. In addition, the cohort analysis proposition emphasized the exogenous social demographic environment into which cohorts were born and in which they came of age. The baby boomer pattern revealed is consistent with the Easterlin hypothesis that large cohort sizes have negative consequences on cohort members' socioeconomic and other life course outcomes, such as sense of well-being and mental stress (Easterlin 1987). Figure 8.5 shows how closely the dip in predicted probabilities of happiness mirrored the rise in the relative cohort size (RCS) that defines the baby boomer cohorts. So, we can also add this cohort characteristic to the model

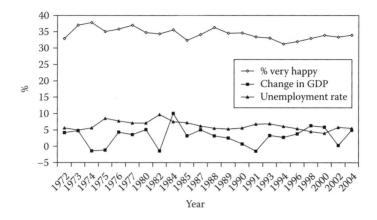

FIGURE 8.4
Period effects of happiness and economic cycle.

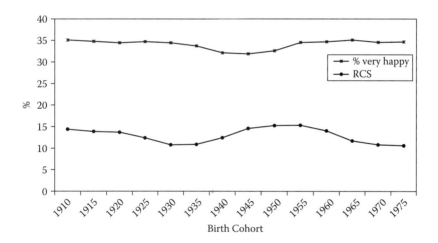

FIGURE 8.5
Birth and fortune: cohort effects of happiness and relative cohort size (RCS).

for the overall mean to account for cohort-level variance. The level 2 models can now be expanded to the following:

Level 2 or "between-cell" model with level 2 covariates:

Overall mean:

$$\beta_{0jk} = \gamma_0 + \gamma_{01} \; GDP_k + \gamma_{02}UNEMPLOY_k + \gamma_{03}RCS_j + u_{0j} + v_{0k} \quad (8.5)$$

Sex effect:

$$\beta_{3k} = \gamma_3 + \gamma_{31} \ GDP_k + \gamma_{32}UNEMPLOY_k + v_{3k} \tag{8.6}$$

Race effect:

$$\beta_{4k} = \gamma_4 + \gamma_{41} \ GDP_k + \gamma_{42}UNEMPLOY_k + v_{4k} \tag{8.7}$$

The model estimates for level 2 covariates added to model B in Table 8.1 are summarized in Table 8.2. The results on other variables were similar to those shown in Table 8.1 and thus are omitted in the interest of space. The fixed effect coefficients showed that economic prosperity as indicated by percentage increases in GDP per capita during this period was significantly associated with more happiness on average, whereas larger cohort size reduced happiness significantly. Economic conditions also affected sex

TABLE 8.2

CCREM of Happiness: Period- and Cohort-Level Covariates

Fixed Effects	Parameter	Coefficient	Odds Ratio	p Value
Model for Mean				
Intercept	γ_0	−0.95	0.39	<0.001
ΔGDP	γ_{01}	0.02	1.02	<0.001
UNEMP	γ_{02}	0.00	1.00	0.858
RCS	γ_{03}	−0.03	0.97	0.019
Model for Sex Effect				
Intercept	γ_3	0.17	1.18	<0.001
ΔGDP	γ_{31}	−0.02	0.98	<0.001
UNEMP	γ_{32}	−0.03	0.97	0.159
Model for Race Effect				
Intercept	γ_4	−0.40	0.67	<0.001
ΔGDP	γ_{41}	0.03	1.03	0.029
UNEMP	γ_{42}	0.08	1.08	0.099
Variance Component		**Variance**	**s.e.**	**% Reduction**[a]
COHORT	τ_{u0}			
Intercept		0.00	0.00	73
PERIOD				
Intercept	τ_{v0}	0.01	0.00	—
SEX	τ_{v3}	0.00	0.00	83
RACE	τ_{v4}	0.02	0.02	38
Model Fit				
−2 Res Log-Pseudo-Likelihood			231798.9	df = 28,833

[a] Compared to the variance estimates from Model B of Table 8.1.

and race differences over time. Increases in GDP reduced both sex and race differences (the race effect was negative, so a positive coefficient of GDP led to a decrease in the black-white gap). A higher unemployment rate, interestingly, was associated with decreases in the race gap. Accounting for the period-level covariates largely reduced the period variances in the sex (83%) and race (38%) effects relative to those estimated in model B of Table 8.1. Since the residual period effect in the intercept or overall mean remained the same, there were other period factors for future research to consider. And, accounting for the single cohort characteristic reduces the cohort variance by 73%. Therefore, fortunes do seem to be more closely related to early life conditions and formative experiences. Larger cohort sizes increased the competition to enter schools and the labor market and created more strains to achieve expected economic success and family life. The unique experiences of these cohorts during early adulthood thus can have a lasting impact on their sense of happiness.

8.3 HAPC-CCREM Analysis of Aggregate Rate Data on Cancer Incidence and Mortality

The examples introduced so far are from sample surveys of individual respondents nested within time periods and birth cohorts. We mentioned in Chapter 3 that the same repeated cross-sectional design of aggregate rate data can also become multilevel given the existence of finer-grained age-specific data within each period and cohort group. We now turn to the HAPC-CCREM analysis of the cancer incidence and mortality data on which the Intrinsic Estimator (IE) analysis was based in Chapter 5. We address three areas that have not been addressed before: model validation, social disparities in cancer trends, and period- and cohort-related mechanisms.

8.3.1 Trends in Age, Period, and Cohort Variations: Comparison with the IE Analysis

Are the IE estimates of age patterns and temporal trends of cancer incidence and mortality rates consistent with those obtained from the application of the HAPC-CCREM? Consistency of results across different models provides evidence for model validation in addition to the statistical means through simulations. We have already shown strong support for the validity of the IE method through the analysis of the microdata from the GSS on verbal ability in Chapter 5. We now broaden the evidentiary base by applying the HAPC models to aggregate rate data from different cancer sites and outcomes. We estimated Poisson CCREM using SAS PROC GLIMMIX.

Tables 8.3 to 8.6 present the CCREM estimates of fixed effects of age (mean centered) and random variance components of cohort and period effects for lung, colorectal, breast, and prostate cancer incidence and mortality rates, respectively. Results from model A show highly significant and non-linear age effects for all outcomes examined. We plot the predicted age-specific rates of incidence and mortality in Figure 8.6 (with cohort and period covariates in Figure 8.7) and Figures 8.8–8.10. The specific shapes of the age curves correspond to the parametric forms we imposed on the age effects and may not be directly compared to those of the IE results. While we also fit the models using dummy indicators of age groups, we chose the current functional forms of age effects based on parsimony and test of significance.

The cohort effects indicated by random cohort variances for the intercept are highly significant for both incidence and mortality of all four cancers. The plots of predicted rates by cohort using the random cohort coefficients also document similar cohort trends as those shown in the IE analysis. For example, lung cancer incidence and mortality results showed first increasing and then decreasing cohort effects for all groups, with delayed peaks in female cohorts. Cohort declines in mortality are observed for all other cancer sites, and those in incidence apply to most but not all cohorts. The period effects indicated by random period variances for the intercept were much smaller and less significant than the cohort effects for all cancer sites and outcomes except for prostate cancer incidence. This is also evident from the inspection of the predicted rates by period. As shown in the IE analysis, the period effects were relatively flat and showed slight variations that differed in trends by subgroups.

In all, we find highly compatible estimates of A, P, and C effects from the IE and HAPC analyses of incidence and mortality related to major cancer sites in the United States. While the former estimation method applies to the conventional linear model that uses a special constraint to identify the model, the latter method does not incur the identification problem and uses no constraint. The convergence in the results from the applications of these two different methods thus validates our findings on the cancer trends that serve as the foundation for further investigations of social differentials and explanatory factors.

8.3.2 Sex and Race Differentials

The IE method is for the APC accounting models that do not accommodate additional covariates. Therefore, the previous analysis of cancer trends was conducted on stratified groups of population by sex and race. Although the findings showed some qualitative differences in the A, P, and C effects for men and women, blacks and whites, they do not constitute quantitative evidence for real group differences due to the lack of a statistical test of the sex and race effects. We now employ the HAPC models to test the group

TABLE 8.3

The Poisson CCREM Estimates of Lung Cancer Incidence and Mortality Rates

| | Incidence | | | | | | Mortality | | | | | |
| | Model A | | | Model B | | | Model A | | | Model B | | |
Fixed Effects	Coefficient	se	p Value	Coefficient	se	p Value	Coefficient	se	p Value	Coefficient	se	p Value
Model for Mean												
Intercept	-6.63	0.10	<0.001	-6.78	0.03	<0.001	-6.80	0.03	<0.001	-7.08	0.03	<0.001
AGE	0.10	0.00	<0.001	0.10	0.00	<0.001	0.10	0.00	<0.001	0.10	0.00	<0.001
AGE^2	0.00	0.00	<0.001	0.00	0.00	<0.001	0.00	0.00	<0.001	0.00	0.00	<0.001
COHORT SMOKE				0.09	0.00	<0.001				0.12	0.01	<0.001
PERIOD SMOKE				0.01	0.00	0.006						
Model for Sex (female=1)												
Intercept	-0.67	0.12	0.001	-0.06	0.06	0.394	-0.97	0.11	<0.001	-0.20	0.14	0.192
AGE	-0.01	0.00	<0.001	-0.01	0.00	<0.001	-0.01	0.00	<0.001	-0.01	0.00	<0.001
AGE^2	0.00	0.00	<0.001	0.00	0.00	<0.001	0.00	0.00	<0.001	0.00	0.00	<0.001
RACE (black=1)	-0.38	0.01	<0.001	-0.39	0.02	<0.001	-0.38	0.00	<0.001	-0.60	0.01	<0.001
COHORT SMOKE				0.04	0.01	<0.001				-0.01	0.02	0.507
PERIOD SMOKE				-0.03	0.00	<0.001						
PERIOD SMOKE				0.00	0.00	0.2879						
Model for Race (black=1)												
Intercept	0.44	0.05	<0.001	0.41	0.06	<0.001	0.33	0.03	<0.001	0.40	0.05	<0.001
AGE	-0.02	0.00	<0.001	-0.02	0.00	<0.001	-0.02	0.00	<0.001	-0.01	0.00	<0.001
AGE^2	0.00	0.00	<0.001	0.00	0.00	<0.001	0.00	0.00	<0.001	0.00	0.00	<0.001
COHORT SMOKE				-0.02	0.00	<0.001				-0.03	0.00	<0.001

continued

TABLE 8.3 (continued)

The Poisson CCREM Estimates of Lung Cancer Incidence and Mortality Rates

| | Incidence | | | | | | Mortality | | | | | |
| | Model A | | | Model B | | | Model A | | | Model B | | |
Variance Components	Variance	se	p Value	Variance	se	p Value	Variance	se	p Value	Variance	se	p Value
COHORT												
Intercept	0.17	0.06	0.002	0.00	0.00	0.036	0.01	0.00	<0.001	0.01	0.00	0.013
SEX	0.17	0.06	0.003	0.01	0.00	0.025	0.04	0.01	<0.001	0.06	0.03	0.024
RACE	0.04	0.02	0.007	0.05	0.02	0.007	0.01	0.00	<0.001	0.05	0.01	<0.001
PERIOD												
Intercept	0.00	0.00	0.043	0.00	0.00	0.097	0.00	0.00	0.181	0.00	0.00	0.049
SEX	0.04	0.02	0.036	0.00	0.00	0.066	0.07	0.05	0.106	0.05	0.04	0.088
RACE	0.00	0.00	0.067	0.00	0.00	0.061	0.00	0.00	0.075	0.00	0.00	0.281
Model Fit												
-2 Res Log Pseudo-Likelihood	-1.6	df=8048		-80.8	df=8042		-1857.0	df=7600		-2738.7	df=7597	

TABLE 8.4

The Poisson CCREM Estimates of Colorectal Cancer Incidence and Mortality Rates

| | Incidence | | | | | | Mortality | | | | | |
| | Model A | | | Model B | | | Model A | | | Model B | | |
Fixed Effects	Coefficient	se	p Value	Coefficient	se	p Value	Coefficient	se	p Value	Coefficient	se	p Value
Model for Mean												
Intercept	−6.71	0.06	<0.001	−6.78	0.04	<0.001	−7.94	0.05	<0.001	−7.99	0.07	<0.001
AGE	0.10	0.00	<0.001	0.10	0.00	<0.001	0.10	0.00	<0.001	0.10	0.00	<0.001
AGE²	0.00	0.00	<0.001	0.00	0.00	<0.001	0.00	0.00	<0.001	0.00	0.00	<0.001
COHORT SMOKE				−0.02	0.01	<0.001				−0.04	0.01	<0.001
COHORT OBESITY				−0.01	0.00	<0.001				−0.01	0.00	<0.001
PERIOD SMOKE				−0.01	0.00	<0.001				0.00	0.00	0.964
PERIOD OBESITY				−0.01	0.00	0.001				−0.01	0.00	<0.001
Model for Sex (female=1)												
Intercept	−0.37	0.02	<0.001	−0.34	0.08	0.003	−0.46	0.05	<0.001	−0.48	0.06	<0.001
AGE	−0.01	0.00	<0.001	−0.01	0.00	<0.001	−0.01	0.00	<0.001	−0.01	0.00	<0.001
AGE2	0.00	0.00	<0.001	0.00	0.00	<0.001	0.00	0.00	<0.001	0.00	0.00	<0.001
RACE (black = 1)	0.07	0.01	<0.001	−0.12	0.16	0.444	0.08	0.01	<0.001	−0.20	0.12	0.085
COHORT SMOKE				0.02	0.00	<0.001				0.01	0.01	0.032
COHORT OBESITY				0.00	0.00	0.775				0.01	0.00	<0.001
PERIOD SMOKE				0.02	0.01	0.084				0.01	0.01	0.122
PERIOD OBESITY				0.01	0.01	0.069				0.00	0.01	0.554

continued

TABLE 8.4 (continued)

The Poisson CCREM Estimates of Colorectal Cancer Incidence and Mortality Rates

	Incidence						Mortality					
	Model A			Model B			Model A			Model B		
Fixed Effects	Coefficient	se	p Value	Coefficient	se	p Value	Coefficient	se	p Value	Coefficient	se	p Value
Model for Race (black=1)												
Intercept	0.10	0.04	0.048	0.29	0.04	<0.001	0.25	0.06	0.005	0.46	0.04	<0.001
AGE	0.00	0.00	<0.001	0.00	0.00	<0.001	−0.01	0.00	<0.001	0.00	0.00	<0.001
AGE^2	0.00	0.00	0.001	0.00	0.00	0.001	0.00	0.00	<0.001	0.00	0.00	<0.001
COHORT SMOKE				0.01	0.00	<0.001				0.00	0.00	0.186
COHORT OBESITY				0.01	0.00	0.001				0.01	0.00	0.002
PERIOD SMOKE				0.00	0.00	0.191				−0.01	0.00	<0.001
PERIOD OBESITY				0.00	0.00	0.629				0.01	0.00	0.014
Variance Components	Variance	se	p Value	Variance	se	p Value	Variance	se	p Value	Variance	se	p Value
COHORT												
Intercept	0.04	0.01	0.002	0.02	0.01	0.013	0.02	0.01	0.009	0.07	0.03	0.004
SEX	0.00	0.00	0.031	0.00	0.00	0.037	0.04	0.02	0.013	0.00	0.00	0.031
RACE	0.02	0.01	0.014	0.01	0.00	0.018	0.02	0.01	0.030	0.01	0.01	0.026
PERIOD												
Intercept	0.01	0.00	0.033	0.00	0.00	0.045	0.01	0.00	0.047	0.00	0.00	0.095
SEX	0.00	0.00	0.046	0.00	0.00	0.124	0.00	0.00	0.053	0.00	0.00	0.072
RACE	0.01	0.00	0.070	0.00	0.00	0.142	0.02	0.01	0.055	0.00	0.00	0.138
Model Fit												
−2 Res Log Pseudo-Likelihood	38.7	df=8048		81.4	df=8036		−403.9	df=7600		−423.9	df=7588	

TABLE 8.5

The Poisson CCREM Estimates of Breast Cancer Incidence and Mortality Rates

| | Incidence | | | | | | Mortality | | | | | |
| | Model A | | | Model B | | | Model A | | | Model B | | |
Fixed Effects	Coefficient	se	p Value	Coefficient	se	p Value	Coefficient	se	p Value	Coefficient	se	p Value
Model for Mean												
Intercept	−5.85	0.06	<0.001	−5.88	0.05	<0.001	−7.59	0.10	<0.001	−7.68	0.16	<0.001
AGE	0.05	0.00	<0.001	0.05	0.00	<0.001	0.05	0.00	<0.001	0.05	0.00	<0.001
AGE²	0.00	0.00	<0.001	0.00	0.00	<0.001	0.00	0.00	<0.001	0.00	0.00	<0.001
COHORT OBESITY				0.00	0.00	0.326				−0.01	0.01	0.491
PERIOD HRT				0.01	0.01	0.166				0.00	0.01	0.768
PERIOD MAMMOGRAM				0.00	0.00	0.012				0.00	0.00	0.355
Model for Race (black=1)												
Intercept	−0.17	0.02	<0.001	−0.13	0.04	0.030	0.22	0.07	0.020	0.26	0.15	0.143
AGE	−0.01	0.00	<0.001	−0.01	0.00	<0.001	0.00	0.00	0.011	0.00	0.00	0.104
AGE²	0.00	0.00	<0.001	0.00	0.00	<0.001	0.00	0.00	0.036	0.00	0.00	0.051
COHORT OBESITY				0.00	0.00	0.100				0.02	0.01	0.020
PERIOD HRT				−0.01	0.00	<0.001						
PERIOD MAMMOGRAM				0.00	0.00	<0.001						

continued

TABLE 8.5 (continued)

The Poisson CCREM Estimates of Breast Cancer Incidence and Mortality Rates

	Incidence						Mortality					
	Model A			Model B			Model A			Model B		
Variance Components	Variance	se	p Value	Variance	se	p Value	Variance	se	p Value	Variance	se	p Value
COHORT												
Intercept	0.02	0.01	0.003	0.03	0.01	0.003	0.16	0.06	0.002	0.16	0.06	0.002
RACE	0.00	0.00	0.041	0.00	0.00	0.039	0.08	0.04	0.037	0.04	0.02	0.018
PERIOD												
Intercept	0.02	0.01	0.033	0.00	0.00	0.058	0.00	0.00	0.047	0.00	0.00	0.081
RACE	0.00	0.00	0.048	0.00	0.00	0.275	0.00	0.00	0.129	0.00	0.00	0.124
Model Fit												
-2 Res Log Pseudo-Likelihood	1863.0	df=4022		1911.6	df=4016		11830.2	df=3798		11847.3	df=3794	

TABLE 8.6

The Poisson CCREM Estimates of Prostate Cancer Incidence and Mortality Rates

| | Incidence | | | | | | Mortality | | | | | |
| | Model A | | | Model B | | | Model A | | | Model B | | |
Fixed Effects	Coefficient	se	p Value	Coefficient	se	p Value	Coefficient	se	p Value	Coefficient	se	p Value
Model for Mean												
Intercept	−6.00	0.29	<0.001	−6.38	0.18	<0.001	−7.94	0.05	<0.001	−9.18	0.05	<0.001
AGE	0.18	0.00	<0.001	0.19	0.00	<0.001	0.10	0.00	<0.001	0.18	0.00	<0.001
AGE²	−0.01	0.00	<0.001	−0.01	0.00	<0.001	0.00	0.00	<0.001	0.00	0.00	<0.001
COHORT SMOKE				−0.04	0.04	0.269				0.10	0.01	<0.001
PERIOD SMOKE				−0.08	0.01	<0.001				0.00	0.00	0.563
Model for Race (black=1)												
Intercept	0.65	0.03	<0.001	1.18	0.07	<0.001	0.25	0.06	0.005	1.15	0.04	<0.001
AGE	−0.02	0.00	<0.001	−0.02	0.00	<0.001	−0.01	0.00	<0.001	−0.01	0.00	<0.001
AGE²	0.00	0.00	0.024	0.00	0.00	0.006	0.00	0.00	<0.001	0.00	0.00	<0.001
COHORT SMOKE				0.01	0.01	0.087				0.03	0.00	<0.001
PERIOD SMOKE				0.02	0.00	<0.001				−0.01	0.00	<0.001

continued

TABLE 8.6 (continued)

The Poisson CCREM Estimates of Prostate Cancer Incidence and Mortality Rates

	Incidence						Mortality					
	Model A			Model B			Model A			Model B		
Variance Components	Variance	se	p Value	Variance	se	p Value	Variance	se	p Value	Variance	se	p Value
COHORT												
Intercept	0.40	0.14	0.002	0.39	0.14	0.002	0.16	0.06	0.005	0.02	0.01	0.014
RACE	0.00	0.00	0.016	0.00	0.00	0.019	0.04	0.02	0.014	0.00	0.00	0.083
PERIOD												
Intercept	0.49	0.26	0.031	0.05	0.03	0.047	0.01	0.00	0.042	0.01	0.01	0.106
RACE	0.01	0.00	0.041	0.00	0.00	0.059	0.00	0.00	0.061	0.00	0.00	0.139
Model Fit												
-2 Res Log Pseudo-Likelihood	12173.1	df=4022		12161.4	df=4018		7110.0	df=3798		7114.3	df=3794	

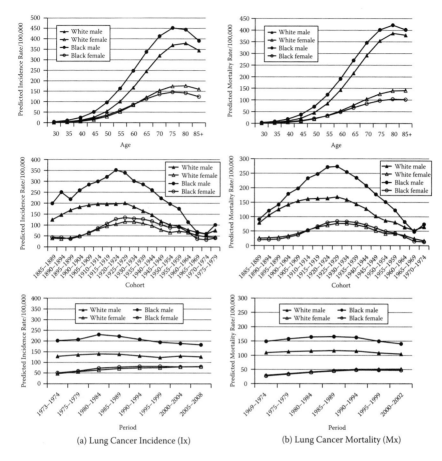

FIGURE 8.6
Age, period, and cohort variations in lung cancer incidence and mortality rates by sex and race.

differences or the hypothesis of social disparities in aging and cancer links and time period and birth cohort variations in cancer.

We entered sex and race variables to the level 1 model and tested both their main effects (the intercepts), their interaction effects with age, and the interaction between the two themselves. At level 2, we specify random coefficient models to test the cohort and period variations in the sex and race effect coefficients. The fixed effect estimates in model A showed substantial sex differences in lung and colorectal cancer incidence and mortality rates that indicated significant male excesses. There were also pronounced race differences in cancer incidence and mortality rates that indicated black excesses in all but one case. Black females showed significantly lower relative risks of breast cancer incidence, which is consistent with epidemiologic studies of racial disparities in breast cancer (Chlebowski et al. 2005). The age coefficients in the models for sex and race effects represent the interaction terms

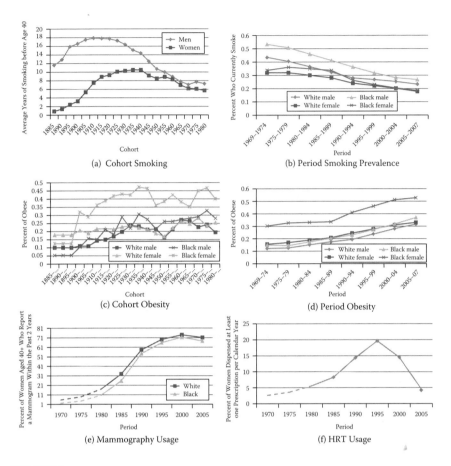

FIGURE 8.7
Cohort and period covariates: smoking, obesity, HRT, and mammography.

between age and sex and between age and race, and they suggest that the sex and race gaps in cancer incidence and mortality varied significantly with age. Figures 8.6 and 8.8–8.10, accordingly, show different age trajectories of cancer outcomes for men and women in the two race categories. Both lung cancer and colorectal cancer incidence and mortality rates showed diverging sex gaps with age. For the case of lung cancer, race gaps grew with age in women and remained largely constant in men. For the case of colorectal cancer, race gaps decreased with age in both sexes. The results for breast and prostate cancer outcomes were similar regarding age changes in race differences. The black and white gaps in the incidence rates increased until the age of 75 and slightly decreased in oldest ages, whereas the black excesses in mortality rates continued to grow throughout the life course.

Sex and race disparities also showed cohort and period variations for all cancer sites examined. The random variance components in Tables 8.3–8.6 for

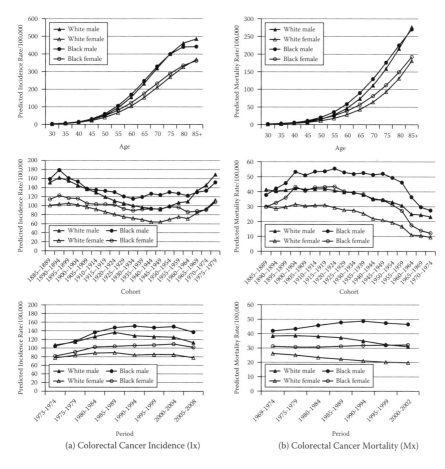

FIGURE 8.8
Age, period, and cohort variations in colorectal cancer incidence and mortality rates by sex and race.

the sex and race effects are sizable and significant in most cases, particularly for cohort variance components. Model A of lung cancer incidence and mortality, for example, showed large cohort variances in the sex effect that were highly significant. The cohort trends in lung cancer incidence and mortality rates in Figure 8.6 further show the male and female differences across cohorts that reflect the sex-specific cohort patterns of cigarette smoking. Period variances in the sex differences were also pronounced for lung cancer incidence rates, which showed decreases over time for males but increases over time for females. Racial disparities across cohorts and periods were much larger for males than females. The results for colorectal cancer outcomes in Table 8.4 showed larger and more significant cohort than period differences in the sex and race gaps. There is evidence in Figure 8.8 for decreasing racial gaps in most recent cohorts in both incidence and mortality rates.

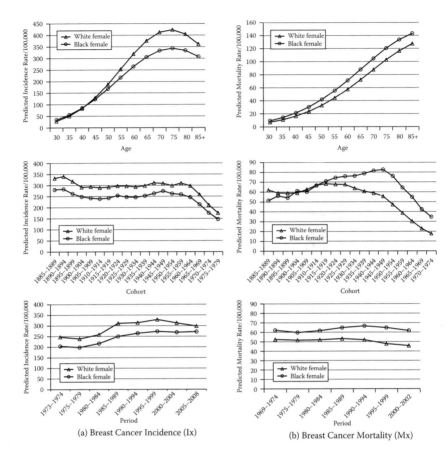

FIGURE 8.9
Age, period, and cohort variations in breast cancer incidence and mortality rates by race.

Tables 8.5 and 8.6 suggest that race differences in breast and prostate cancer incidence rates varied significantly by cohort and period, while those in mortality rates varied only by cohort. Figure 8.9 shows that the white excesses in incidence of breast cancer decreased in cohorts born after 1960 and the last 10 years. The black disadvantage in breast cancer survival increased significantly across cohorts due to the increases in black female mortality rates and decreases in white female mortality rates for cohorts born between 1910 and 1950. While both black and white female cohorts born after 1950 experienced declines, the racial gaps remained substantial. Figure 8.10 shows increases in the racial gaps in prostate cancer incidence rates in the most recent cohorts and years. It also shows much larger black disadvantages in survival than incidence rates. While these disadvantages decreased across cohorts, they remained constant over time, as indicated by the lack of significance in the period variance component for the race effect.

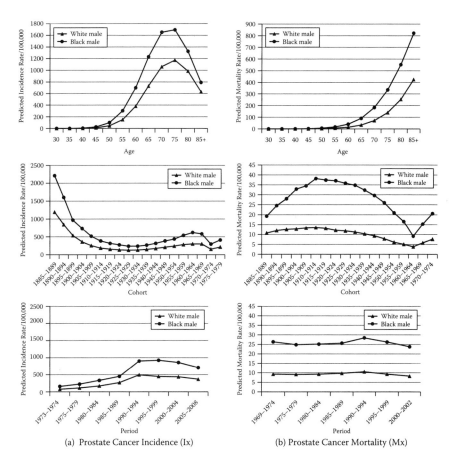

FIGURE 8.10
Age, period, and cohort variations in prostate cancer incidence and mortality rates by race.

8.3.3 Cohort and Period Mechanisms: Cigarette Smoking, Obesity, Hormone Replacement Therapy, and Mammography

Analyses using the classical APC accounting models of aggregate rates only concern temporal trends that may be attributed to cohort- or period-related changes. While a large body of demographic, aging, and epidemiologic literature has developed theories about how cohort effects arise, as we summarized in Chapter 2, the test of cohort-based mechanisms cannot be conducted within the framework of the linear models. Similarly, explanations of period effects are usually drawn from correspondences of changes in rates and social demographic or epidemiological conditions that are expected to influence these rates. The question of how hypothesized period- or cohort-related factors actually explain period or cohort effects is beyond the scope of the simple accounting models (Smith 2004). Instead, it requires more fine-grained regression analyses using a different kind of model that can accommodate period- and cohort-level covariates. We now include several such covariates (cigarette smoking, obesity, hormone replacement therapy [HRT], mammography) that have been established in cancer epidemiologic research as important risk factors for the four cancer sites under investigation.

The CCREM analysis of lung cancer incidence and mortality in Table 8.3, model A, suggests significant cohort effects on the intercept, sex, and race coefficients. We entered one cohort-level covariate of smoking (as defined in Chapter 3) in model B to test how differential cohort exposures to smoking can account for the cohort effects on the overall incidence and mortality rates and also cohort variations in the sex and race gaps in these rates. The estimated fixed effects for the cohort smoking variable and its interactions with race effects were highly significant in the models of both incidence and mortality. The cohort smoking and sex interaction effect was not significant in the mortality model. Because only the linear terms of level 2 covariates were included in this and subsequent analyses, their fixed effects may not be significant if the underlying correlations are nonlinear. Comparisons of sex-specific cohort lung cancer mortality in Figure 8.6 and cohort smoking exposures in Figure 8.7a suggest that they closely resemble each other in nonlinear changes—increases first followed by decreases. The inclusion of the cohort smoking variable largely reduced the cohort variances for the intercept (by 98%) and sex coefficient (by 95%) for the incidence model.

We tested the three variables of period smoking prevalence (current smokers, former smokers, and smoking cessation) and found largely similar results for the first two and less-strong effects for the third. Therefore, we presented the estimates associated with the prevalence of current smokers in model B for this and subsequent analyses when smoking is considered a major carcinogenic factor. The results from model A suggest significant period effects for the intercept and sex coefficient in the incidence model. Figure 8.7b documents downward trends of smoking prevalence rates during this period for males, less-steep declines for white females, and increases in black females before the 1990s. Model B adjusted for period smoking prevalence and showed it was significantly related to lung cancer incidence and sex differences therein, net of all the other factors. Figure 8.6 shows the declining male incidence rates and slightly increasing female incidence rates, reflecting the lagged effects of period changes in smoking for each group. The random variance components also showed that it accounts for most of the period variations in sex gaps (by 93%). The residual cohort variance for the intercept of mortality and period variance for the intercept of incidence remained significant or increased in size, suggesting the need to search for additional cohort-related explanations for the inverse-U-shaped cohort effects on lung cancer mortality and period-specific explanations such as diagnostic changes for lung cancer incidence patterns during this time.

In the CCREM analysis of colorectal cancer shown in Table 8.4, we adjusted for cohort and period covariates of both smoking and obesity prevalence rates in model B. The fixed effects estimates showed that both cohort smoking and obesity played significant roles in affecting the overall colorectal cancer incidence and mortality rates. The cohort declines in smoking shown in Figure 8.7a correspond well to the cohort declines in colorectal cancer incidence and mortality rates shown in Figure 8.8. Cohort smoking was also significantly related to sex differences in both incidence and mortality and race differences in incidence. The increases in incidence in more recent cohorts corresponded to the increasing cohort prevalence of obesity documented in Figure 8.7c. Cohort obesity was also significantly related to sex differences in mortality and race differences in both incidence and mortality. The inclusion of these two cohort covariates largely reduced the cohort variance components for all random coefficients except for the intercept in the mortality model. Period smoking and obesity changes (in Figure 8.7d) significantly affected period changes in colorectal cancer incidence rates and race differences in mortality rates. The period random variances were reduced in size and significance level after adjusting for period-level covariates in both the incidence and mortality models. The only significant residual variance is that associated with the intercept for the incidence, which may be further explained by period changes in colonoscopy usage.

The results for breast cancer outcomes are reported in model B in Table 8.5. One cohort-level covariate of obesity was entered to both the incidence and mortality models and two period-level covariates, HRT and mammography, were entered in the incidence model (as they did not show significant associations with mortality even individually considered). Although these level 2 covariates showed significant fixed effects in bivariate analyses (not shown), only period mammography change was significant in main effect when all other factors were adjusted. Both period covariates were significantly related to race differentials in breast cancer incidence. And, they completely explained the period variances in both the intercept and race coefficient as they were no longer significant in model B. Based on these results, we conclude that period increases in breast cancer incidence illustrated in Figure 8.9 were entirely due to increases in the mammography usage and hence detections of malignant cases as shown in Figure 8.7e. The downturn in incidence rates starting in the year 2000 in white females, on the other hand, was a result of sudden reductions of HRT usage, as illustrated in Figure 8.7f, and narrowed the racial gaps during this period. Net of all other factors, cohort obesity was significantly related to race differentials in breast cancer mortality, and its inclusion reduced the cohort variance of the race coefficient. This means that the increases in the black mortality disadvantage for cohorts born before 1950 and persistent black excess for cohorts born after 1950 resulted in part from higher rates of obesity in these cohorts of black females.

Table 8.6 presents the final set of analyses on prostate cancer in model B, which takes into account cohort and period smoking variables. Because the trends in period smoking prevalence (Figure 8.7b) and period effects of incidence (Figure 8.10) were in the opposite direction, the fixed effect estimate of the period smoking variable is negative. This should not be interpreted as evidence for negative or harmful effects of smoking declines on prostate cancer incidence. This provides one example that cautions against simple-minded causal inferences of the effects of level 2 factors on temporal trends in rates and suggests the importance of exploratory data analysis of specific factors to be used as explanatory variables. Period smoking prevalence explains a substantial proportion of the period variances in incidence rates and some period variances in race differentials in incidence rates. Cohort smoking patterns contributed largely to cohort declines in prostate cancer mortality and race differentials therein. While the cohort variance in the race effect was no longer significant after the adjustment of smoking exposure, that in the intercept remained significant.

It is interesting to note that the same cohort smoking variable that has been used as a proxy variable to take the place of cohort effects in previous demographic studies of mortality (Preston and Wang 2006) did not seem to account for all cohort effects of mortality in our analysis. We examined cohort smoking patterns in relation to cohort effects of mortality from all four cancer sites. They did not play any significant role in breast cancer mortality. The residual cohort variances remained significant after adjusting for the cohort smoking variable in the HAPC models of lung, colorectal, and prostate cancer mortality. Because smoking is by far a more important risk factor for cancer than other causes of death, it is unlikely that cohort patterns of smoking can successfully capture all cohort-related variances in mortality. It follows that substituting cohort effects with this or other proxy measures is subject to model specification errors long suspected by earlier studies (Glenn 2005; Smith, Mason, and Fienberg 1982). The HAPC-CCREM analysis has thus provided an instrument to test the plausibility of any chosen proxy variables.

In sum, the HAPC-CCREM analysis moves far beyond simply identifying age, period, and cohort trends. It allows us to see how social stratification operates over the life course and historical time and to test the extent to which aging and cohort theories apply to the outcome of interest in ways that cannot be achieved by a conventional linear regression analysis. This analytic technique opens new doors to research on time-related change by offering a systematic and flexible modeling strategy that takes advantage of the multilevel data structure and covariates in repeated cross-sectional research designs. The first prominent feature of the HAPC model is that it does not incur the identification problem because the age, period, and cohort effects are not assumed to be linear and additive. It represents one family of nonadditive models that can be extremely useful for capturing the contextual effects of cohort membership and period events on a wide range of social demographic processes. The second contribution of the hierarchical models approach to APC analysis is that it makes it possible to test explanatory hypotheses by incorporating covariates provided by sample surveys or other sources and yields insights to substantive cohort or period phenomena that can hardly be gained from any version of the APC accounting models.

8.4 Full Bayesian Estimation

HAPC models can be extended in various ways to address other research questions. We introduce two extensions. This section focuses on a full Bayesian approach that can be used to refine the estimates of mixed effects within the HAPC-CCREM framework. The next section introduces a variance function regression (VFR) model that can be used jointly with mixed effects models to extend the HAPC-CCREM framework for the understanding of between-group and within-group heterogeneity.

Using the example of the GSS verbal scores, we show how to apply methods of Bayesian statistical inference to HAPC models of repeated cross-sectional survey data.* Building on our previous illustrations of the HAPC analysis of GSS verbal ability trends, we further examined situations that may affect the accuracy of the mixed effects estimates based on the restricted maximum likelihood-empirical Bayes (REML-EB) estimation method. In APC analyses of finite time period survey data, the numbers of periods and birth cohorts usually are too small to satisfy the large-sample criteria required by the maximum likelihood (ML) estimation of variance components. In addition, the sample sizes within each cohort are highly unbalanced. Therefore, errors in variance components estimates may produce extra uncertainty in coefficient estimates that will not be reflected in the standard errors. This added uncertainty may cast doubt on statistical inferences based on REML-EB estimates of model parameters (Raudenbush and Bryk 2002). It also motivates the investigation of a full Bayesian alternative to account for the extra uncertainty brought about by the small sample sizes and unbalanced data. We show how the HAPC models approach can be implemented via the Bayes-MCMC (Monte Carlo Markov chain) methods and how they may improve statistical inferences for APC analyses.

8.4.1 REML-EB Estimation

Using the same GSS verbal ability data between 1972 and 2000 as in Yang (2006), we specify HAPC-CCREM with two additional cohort covariates: proportion of cohort members who read newspapers daily (NEWS) and mean hours of TV watching per day (TV). Both were thought to be related to the

---

* Section 8.4 was adapted from Yang, Y. 2006. *Sociological Methodology* 36:39–74.

decline in cohort mean verbal test scores since the early twentieth century (Glenn 1994). The new level 2 model (6.32) examines the associations between these two cohort characteristics and trends in verbal ability in the presence of controls for individual-level factors related to verbal ability in the combined model (6.33):

Level 2 or "between-cell" model:

$$\beta_{0jk} = \gamma_0 + \gamma_1 NEWS_j + \gamma_2 TV_j + u_{0j} + v_{0k}, \; u_{0j} \sim N(0, \tau_u), \; v_{0k} \sim N(0, \tau_v) \quad (8.8)$$

Combined model:

$$WORDSUM_{ijk} = \gamma_0 + \gamma_1 NEWS_j + \gamma_2 TV_j + \beta_1 AGE_{ijk} + \beta_2 AGE_{ijk}^2$$
$$+ \beta_3 EDUCATION_{ijk} + \beta_4 FEMALE_{ijk} + \beta_5 BLACK_{ijk} + u_{0j} + v_{0k} + e_{ijk} \quad (8.9)$$

for $i = 1, 2, \ldots, n_{jk}$ individuals within cohort j and period k; $j = 1, \ldots, 19$ birth cohorts; and $k = 1, \ldots, 15$ time periods (survey years).

For any possible value of γ, where γ is the vector of fixed effect coefficients, a likelihood of variance components τ_u, τ_v, and σ^2 can be defined, say $L(\tau_u, \tau_v, \sigma^2 | \gamma, Y)$, where Y is the observed data. Averaging over all possible values of γ for the likelihood yields a likelihood of, τ_u, τ_v, and σ^2 given Y alone. This is the restricted likelihood, $L(\tau_u, \tau_v, \sigma^2 | Y)$. The REML approach chooses as estimates of τ_u, τ_v, and σ^2, and those values that maximize the joint likelihood of these parameters given Y. Conditioning on these ML estimates, one computes generalized least squares estimates of γ and empirical Bayes estimates of the level 1 coefficients β.

REML-EB estimates obtained using SAS PROC MIXED are presented in the left panel of Table 8.7. Cohort effects of newspaper reading and TV watching were significant after controlling for individual-level covariates; the residual variations between cohorts and between periods are close to zero, compared to the unconditional variance estimates (0.139 and 0.031), so these level 2 variance components were substantially reduced.

This approach has several strengths. First, the REML estimates of variance components are consistent and efficient in large samples (i.e., for large J and K). In addition, for large J and K, the sampling distributions of these estimators are approximately normal, provided that the REML estimates of τ_u and τ_v are positive definite. It follows that for large J and K, the normal distribution can be conveniently used for constructing confidence intervals and hypotheses testing. Third, since the EB estimates depend on REML estimates of variance components, assuming large J and K, the good large-sample properties of the REML estimates give strength to inference about β.

The limitations of the REML-EB approach are the following: First, the estimates of τ_u and τ_v may be quite inaccurate in small samples. The estimates

TABLE 8.7

Alternative Estimates of Coefficients and Summary Statistics: REML vs. Bayes Via Gibbs under Diffuse-Normal-Gamma Priors

		REML		Bayes via Gibbs	
Fixed Effects	**Parameter**	**Coefficient**	**SE**	**Coefficient**	**SE**
Intercept	γ_0	6.20	0.05	6.20	0.06
NEWS	γ_1	1.40	0.47	1.39	0.49
TV	γ_2	−0.26	0.12	−0.26	0.13
AGE	β_1	0.02	0.02	0.02	0.02
AGE²	β_2	−0.06	0.01	−0.06	0.01
EDUCATION	β_3	0.38	0.01	0.38	0.01
SEX (1=female)	β_4	0.24	0.03	0.24	0.03
RACE (1=black)	β_5	−1.05	0.04	−1.05	0.04
Random Variance					
COHORT	τ_{u0}	0.02	0.01	0.03	0.01
PERIOD	τ_{v0}	0.00	0.00	0.00	0.00
Individual	σ^2	3.14	0.03	3.14	0.03
Model Fit					
Deviance		77666.6		77627.9	
DF/pD[a]		11		30.3	
AIC/DIC[b]		77688.6		77658.3	

Source: Adapted from Yang (2006: Table 2).
[a] DF for REML and pD as the effective number of parameters for Bayes.
[b] AIC for REML and DIC for Bayes.

of the variances in τ_u and τ_v may not have normal distributions for small J and K. If the data are unbalanced, estimates of γ will depend on weights that depend on REML estimates so that random variation in these estimates will lead to uncertainty about γ that will not be reflected in standard errors. Furthermore, errors in the REML estimates will result in extra uncertainty in the EB estimates of β that will not be reflected in the standard errors. Therefore, the confidence intervals for γ and β will be shorter, and tests of significance will be more liberal than they should be. For a more detailed exposition, see Raudenbush and Bryk (2002: Chapter 13).

In the hierarchical APC analysis of interest here, the number of periods and birth cohorts are usually too small to satisfy the large-sample criteria required by the REML estimations of variance components. In addition, the sample sizes within each higher-level unit (i.e., cohort and period) are highly unbalanced, especially for cohorts. There tend to be very few sample members for the oldest and youngest birth cohorts, as shown previously in Table 3.4. In this particular dataset, $J = 19$, and n_j vary across J, with the smallest cohorts having fewer than 100 individuals. In this case, estimates of EB coefficients will depend on weights that are functions of the ML estimates

of variance components. Therefore, errors in variance components estimates due to small J and K may also produce extra uncertainty in the EB
estimates that will not be reflected in the standard errors. Again, tests for
these point estimates may be too liberal. This problem tends to diminish
when either n_j or n_k is large, as in most cohorts and survey years in this example; however, for small J/K and small n_j for some cohorts, this dependence of
EB estimates on REML estimates is undesirable.

Yang (2006) addressed the question about the adequacy of REML estimators when J and K are small and data are unbalanced through a number of
Monte Carlo simulations to examine how the HAPC model would perform
under the REML assumptions. The Monte Carlo simulations showed that,
in relatively large samples (large J and K), the REML-EB estimators of the
HAPC model generally perform well in terms of producing numeric estimates of the population parameters they are intended to measure. The interval estimates, however, should be used with caution in both large and small
samples. It will be useful to have methods for producing more accurate estimates with small samples of survey years and birth cohorts and unbalanced
data. This is where Bayesian MCMC estimation fits in. In the following, we
show a full Bayesian approach to estimation of HAPC model parameters
that, by definition, ensures that inference about every parameter fully takes
into account the uncertainty associated with all others.

8.4.2 Gibbs Sampling and MCMC Estimation

In a Bayesian formulation of HAPC models applied to the repeated cross
sections, one combines the likelihood based on the data with prior information about the fixed and random parameters via prior distributions. Based on
Equations (7.7), (8.8), and (8.9), the Bayesian model for the verbal ability data
can be summarized as the following:

Level 1 model (likelihood):

$$f(Y \mid \beta, \sigma^2)$$

Level 2 model (stage 1 prior):

$$p(\beta \mid \gamma, \tau_u, \tau_v)$$

Stage 2 prior:

$$p(\gamma, \tau_u, \tau_v, \sigma^2) = p(\gamma)p(\tau_u)p(\tau_v)p(\sigma^2)$$

In this formulation, each unknown parameter is viewed as a random
variable that arises from certain probability distribution. The probability

distribution p quantifies the uncertainty about the parameter. Bayes's rule then yields the joint posterior distribution:

$$p(\beta,\gamma,\tau_u,\tau_v,\sigma^2 \mid Y) \propto f(Y \mid \beta,\sigma^2) p(\beta \mid \gamma,\tau_u,\tau_v) p(\gamma,\tau_u,\tau_v,\sigma^2).$$

Marginal posterior distribution can be derived by integration of the joint posterior with regard to each parameter. Therefore, in the Bayesian approach, the inference about every parameter fully accounts for the uncertainty of other parameters through the use of conditional and joint probabilities. Point and interval estimates can then be obtained based on the exact posterior distribution.

Gibbs sampling is used as an accurate computational approach to numerical integration. The Gibbs algorithm for the HAPC model is as follows:

$$p(\beta,\gamma,\tau_u,\tau_v,\sigma^2 \mid Y) \propto p_\beta(\beta \mid \gamma,\tau_u,\tau_v,\sigma^2,Y) r_\beta(\gamma,\tau_u,\tau_v,\sigma^2 \mid Y)$$

$$= p_\gamma(\gamma \mid \tau_u,\tau_v,\sigma^2,\beta,Y) r_\gamma(\tau_u,\tau_v,\sigma^2,\beta \mid Y)$$

$$= p_u(\tau_u \mid \tau_v,\sigma^2,\beta,\gamma,Y) r_u(\tau_v,\sigma^2,\beta,\gamma \mid Y)$$

$$= p_v(\tau_v \mid \sigma^2,\beta,\gamma,\tau_u,Y) r_v(\sigma^2,\beta,\gamma,\tau_u \mid Y)$$

$$= p_{\sigma^2}(\sigma^2 \mid \gamma,\beta,\tau_u,\tau_v,Y) r_{\sigma^2}(\gamma,\beta,\tau_u,\tau_v \mid Y)$$

Gibbs sampling is based on the fact that the posterior density of all unknowns can be approximated even though the corresponding densities r have unknown forms. Starting with initial values $\gamma^{(0)}$, $\tau_u^{(0)}$, $\tau_v^{(0)}$, and $\sigma^{2(0)}$, it repeatedly samples from the conditional distributions until stochastic convergence. The chain of values generated by this sampling procedure is known as a *Monte Carlo Markov chain* (MCMC). An example of the conditional distributions needed for Gibbs sampling was provided by Yang (2006: Appendix A). On convergence, m iterations are obtained, and the empirical distribution of these m values of the unknowns may be regarded as an approximation to the true joint posterior. For large m, the marginal posterior may be approximated by the empirical distribution of the m values produced by the Gibbs. For this study, $m = 20,000$ for one MCMC chain (see the description that follows).

The benefits of the Bayesian methods typically come at a price: the required choice of *prior* distributions for model parameters. The topic of the choice of priors is a vast one (see, e.g., Draper 2002). A common assumption is that diffuse priors can be used in an attempt to base information solely on the data when no prior knowledge is available for the parameters of interest. This means that a prior can be used that allows the data to dominate the determination of the posterior distribution. Furthermore, the ever-increasing usage of Bayes-MCMC methods in statistical practice has produced several classes of priors that may be viewed as sufficiently "noninformative" in fairly broad settings (Guo and Carlin 2004). We implemented Bayesian modeling through MCMC methods via the WinBUGS (Bayesian inference using Gibbs sampling running under Windows) software package freely available on the Internet. Since no previous sociological studies on verbal ability informed the specification of priors, to facilitate comparison of the REML-EB and Bayesian analyses, the study here selected noninformative proper prior distributions[*] with hyperparameter values chosen so that the priors exerted minimal impact relative to the data (Gelman et al. 2000; Raudenbush and Bryk 2002). Conjugate priors[†] were specified that were comprised of a normal prior for γ and an inverse gamma prior for the variance components (Gelfand et al. 1990; Carlin and Louis 2000):

Prior:

$$p(\gamma) \sim N(0,\ 1.0E - 6)$$

$$p(1/\tau_u) \sim Gamma(0.001,\ 0.001);\ p(1/\tau_v) \sim Gamma(0.001,\ 0.001)$$

$$p(1/\sigma^2) \sim Gamma(0.001,\ 0.001)$$

The priors were chosen so that the Bayesian analysis was similar to a corresponding likelihood analysis (REML-EB). The difference is that likelihoods were now adjusted and interpreted as probability distributions on the parameters.

Ordinary least squares (OLS) and REML estimates provided the initial values for two parallel MCMC sampling chains of 20,000 iterations each, following a 5,000-iteration "burn-in" period. The Gelman and Rubin tests were used to diagnose effective convergence. A general discussion of MCMC convergence monitoring is available in the work of Carlin and Louis (2000). Each chain was thinned by 10 to reduce autocorrelation among the parameter estimates. The summary statistics were calculated after the achievement of a less-correlated sample. The final sample size for posterior inferences was 4,000 with 2,000 per chain. The WinBUGS code for fitting the Bayesian hierarchical APC model 3 is provided in the online companion page. Convergence diagnostics and output analysis (CODA) and results for parameter estimates are available in the work of Yang (2006: Appendices B and C).

---

[*] In general, a prior density $p(\theta)$ is proper if it integrates to 1. If a prior density is improper, the integral of $p(\theta)$ is infinity and violates the assumption that probabilities sum to 1. Although improper priors sometimes can lead to proper posteriors, this is not always the case. Posteriors obtained from improper priors must be interpreted with great caution—one must always check that the posteriors have finite integrals and sensible forms (Gelman et al. 2000: 53).

[†] Selecting priors from a distribution family that is conjugate to the likelihood leads to a posterior distribution belonging to the same distributional family as the prior. This can be computationally convenient. For more details, see the work of Gelman et al. (2000: 37).

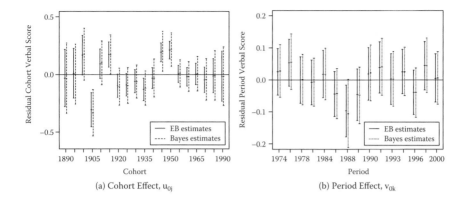

FIGURE 8.11
Residual cohort random variance and period random variance estimates: (a) cohort effect u_{0j}; (b) period effect v_{0k}.

For comparison, the Bayes MCMC estimates are given side by side with the REML results for each model in Table 8.7. Note that SAS PROC MIXED delivers solutions for the model parameters only in terms of point estimates of the posterior means and associated asymptotic standard error estimates, whereas Bayes MCMC methods provide exact posterior inference for each node that includes the posterior mean estimate, the MCMC standard deviation as the estimate for the standard error, Monte Carlo error,[*] median, and quantiles. Table 8.7 shows that the fixed coefficients were very similar for the REML and MCMC estimates, with the estimated standard errors slightly larger in some cases for the MCMC results. The main difference between the REML and MCMC estimates was in the level 2 variances. The Bayes via Gibbs estimators produced larger estimates of cohort and period variances and standard errors. This is due to the Bayes estimates taking into account the uncertainty brought about by the small numbers of cohorts and periods. For Bayesian model selection, the deviance information criterion (DIC) (Spiegelhalter et al. 2002) is a hierarchical modeling generalization of Akaike's information criterion (AIC) and is readily available in WinBUGS.

The cohort and period residual random effects estimates were also compared for REML-EB and Bayes MCMC results. Figure 8.11 plots the posterior means and interval estimates, based on Table 8.7, of the residual cohort effect u_{0j} and the residual period effect v_{0k}, respectively. Figure 8.11a shows that for both EB and Bayes estimates, the intervals were the widest for the youngest and oldest cohorts, which had relatively small sizes (n_j). Both Figure 8.11a and 8.11b show that while the mean estimates were quite close, the Bayes

---

[*] This is an estimate of the difference between the mean of the sample values (which is used as the estimate of the posterior mean for each parameter) and the true posterior mean. As a rule of thumb, an MC error that is less than about 5% of the sample standard deviation indicates convergence of chains.

credible intervals (see, e.g., Carlin and Louis 2000 for definitions) were wider than the EB intervals of variance components.

8.4.3 Discussion and Summary

In applying the Bayesian method to the estimation of HAPC-CCREM, one can take further steps to assess how the choice of priors may affect posteriors. An example of prior sensitivity analysis based on these data was given by Yang (2006) that formulated alternative priors to compare the results with those based on former priors. It showed that the choice of priors had small effects on parameter estimates that depend on small numbers of level 2 units (coefficients of newspaper reading and TV watching, cohort and period variance components), whereas the effect was negligible on those based on the total sample size. This suggests that the full Bayesian HAPC analysis of other repeated cross-sectional datasets with similar numbers of cohort and periods would also benefit from a prior sensitivity analysis.

We have shown how Bayesian inferences can be used for cross-classified random effects models applied to the vocabulary data from the GSS. We compared the empirical results obtained through two parameter estimation methods, namely, the REML-EB method and the full Bayes method using Gibbs sampling. The point estimates were nearly identical. However, the Bayes 95% credible intervals were wider in most cases than were the intervals based on REML. This increase occurred because the REML intervals did not reflect the uncertainty associated with regression coefficients when variance components were unknown, and the REML estimates of these variance components were not accurate because of the small sample sizes at cohort and period levels. The Bayes credible intervals reflected this extra uncertainty and therefore strengthened the inference. The empirical findings on fixed coefficients, variance components, and random effects were generally not highly sensitive to the imposition of alternative priors considered in the study. In sum, the analytic results shown can be interpreted with more confidence than any previous findings of this subject and therefore help to resolve the inconsistencies across studies. The substantive findings on age, period, and cohort trends and individual-level covariates were the same as those shown previously. A new finding is that the decline of vocabulary for more recent/younger cohorts born after 1950 was closely associated with the decreased percentage of cohort members who read a newspaper daily and increased cohort mean hours of TV watching. This cohort effect was not confounded with the aging or period effect.

The evaluation of the performance of REML and full Bayes estimators in the current application was restricted to well-behaved response variables in the context of HAPC models of repeated cross-sectional sample surveys. Although the results based on the two methods are not drastically conflicting, the Bayesian estimators generally showed improvement relative to the REML estimators. There were other cases where the divergence between the maximum likelihood and Bayes results was larger. The problem of inaccurate inference using the REML estimators can be most significant when random effects are relatively large, data are sparse and highly unbalanced, and response variables are not normally distributed.

The other advantage of using APC models based on generalized linear mixed models is that it can be naturally extended to more sophisticated multilevel data analysis. Two improvements can be made in future research. First, one can add more levels to the hierarchical model by giving priors to the stage 2 hyperparameters that include the fixed coefficients at level 2, cohort, period, and individual variance components. The higher-level models can further estimate their uncertainties that may arise from probability distributions. Conjugate hyperpriors such as chi-squared distributions can be used for computational convenience. Studies can also be found in recent biostatistics literature that used autoregressive prior models (Bashir and Esteve 2001) and random walk smoothing models (Knorr-Held and Rainer 2001) for Bayesian prediction of vital rates. Second, it may be of analysts' interest to examine effects of other social units. For example, recent reports of the National Assessment of Educational Progress documented pronounced regional differences in academic test scores, including reading scores in the United States (e.g., Weiss et al. 2002). If state-level verbal test score data are integrated into sample surveys, it is convenient to move beyond the individual-level data using the HAPC models to estimate geographic variations in academic performances.

8.5 HAPC-Variance Function Regression

In addition to the study of social demographic change through the APC analysis, another long-standing core analytic tool of social science is the study of inequality through regression models (Blau and Duncan 1966; Morris and Western 1999). The use of regression-based models in APC analysis closely relates the study of cohort change to the study of inequality both substantively and methodologically. The analytic tools for the examination of age and temporal (i.e., period and cohort) variations in inequalities, however, have been mostly restricted to those for the conditional means of regression models.[*] We now embed a variance function regression (VFR) model within an HAPC analysis. This facilitates the decomposition of not only *between-group inequality* into age, period, and cohort components (i.e., variations in the conditional mean of an outcome across age, period, and cohort), but also a similar APC decomposition of *within-group inequality* (i.e., variations in the

---

[*] Parts of Section 8.4 were adapted from Zheng, H., Y. Yang, and K. C. Land. 2011. *American Sociological Review* 76:955–983.

conditional variance or dispersion of an outcome across age, period, and cohort). More generally, the combined model allows the integration of theories of social stratification and social change and opens the door to a better, more comprehensive understanding of the dynamics and heterogeneity of social processes by which individual lives unfold over the life course and are shaped by historical time and social context represented by cohort membership. We draw on a recent study published by Zheng, Yang, and Land (2011) to show how to use the combined model with an example of health disparities.

8.5.1 Variance Function Regression: A Brief Overview

Standard regression-based approaches to studies of inequality are largely limited to between-group differences. *Group* here means each category of a covariate. For example, "gender" has two groups: men and women. Between-group inequality in this case is the inequality between men and women. Within-group inequality in this case is the remaining inequality within the population of men or within the population of women. It often is the case that between-group inequality is far exceeded by within-group or residual inequality. Residual inequality often is considered as due to measurement error or the influence of unobserved or hidden heterogeneity. A recently developed class of statistical models, termed *variance function regression* (*VFR*) by Western and Bloome (2009) and, more generally, *heteroscedastic regression* (*HR*) in statistics (Smyth 1989) can be used to address this limitation by simultaneously modeling both the mean and the variance of an outcome variable as functions of covariates and hence takes into account both between-group and within-group differences. This class of models explicitly targets the residual variance in regression models for analysis.

Conventional linear regression models assume that the residual error terms of the models are independently and identically distributed with constant or homoscedastic variance, and especially important for small samples for which asymptotic statistical properties of estimators do not apply, that the errors have normal probability distributions (see, e.g., Fox 2008: 187–219). Violations of these assumptions affect estimators of the standard errors of regression coefficients and reduce the statistical efficiency of conventional least squares estimators. A variety of statistical methods has been developed for diagnosing and correcting nonconstant error variances (Fox 2008: 272–277). The key feature of VFR/HR is that it treats violations of homoscedasticity as more than a data problem that needs to be corrected to obtain well-behaved estimators. It rather approaches these violations as being of potential substantive importance and builds regression models to account for them. In applications to the study of inequality, such as Western, Bloome, and Percheski's (2008) study of trends in family income inequality in the United States over the years 1975 to 2005, the residual variance can be interpreted as measuring within-group risk or insecurity.

This and the HAPC models and the substantive questions they address are related. One often needs to understand sources of social inequality attributable to age, time period, and birth cohort in that the APC variations represent the complex and temporal patterns of inequality. In his seminal article (1965), Ryder argued that cohort membership is a structural category just as socioeconomic status (SES) that has explanatory power. To the extent that SES inequalities are defined by both between- and within-group differences (Western, Bloome and Percheski 2008), social inequalities by age, period, and cohort should also be assessed in terms of between-group differences and within-group dispersions. Differences in either term can bring about subsequent social demographic change at the population level. Therefore, an integrated model is useful for decomposing these two differences across age, period, and cohort. The VFR/HR models have been used to examine inequalities in both terms but have not distinguished the age, period, and cohort sources of temporal variation. The HAPC regression models have been used to examine temporal differences in conditional means but not within-group variances. We now illustrate the utility of the intersection of these two modeling frameworks with an application to the analysis of health disparities in the United States over the years 1984 to 2007. We discuss useful findings from this analysis and make general observations on the utility and potential for advances in empirical analyses of inequality of this synthesis of analytical models.

8.5.2 Research Topic: Changing Health Disparities

We use the term *health disparities* to refer to either between-group or within-group differences in health and distinguish the two aspects of inequality in specific circumstances. In the context of APC analysis, *groups* are defined by the age, time period, and cohort categories. *Between-group health disparities* refer to the variations in the conditional mean (conditional on a set of individual-level sociodemographic variables) of health across age, period and cohort. *Within-group health disparities* refer to the conditional variance or dispersion of health within each category of age, period, or cohort. *Changes in within-group health disparities* refer to the variations in the conditional variance or dispersion of health across age, period, or cohort.

In addition to a large body of demographic and epidemiologic research as well as that included in this book on age variation and temporal trends in health and mortality that has addressed between-group health disparities, there are three standard approaches to the study of changes in within-group health disparities: (1) *across the life course*, (2) *across cohorts*, and (3) *across time periods*. The empirical studies illustrated in Chapters 7 and 8 in this book gave samples of all three. Within each approach, there is evidence for significant change in health disparities. For example, the gaps in lung, colorectal, and prostate cancer incidence and mortality by sex and race have widened with age (Figures 8.6, 8.8, 8.10). Gaps in self-rated health, physical functioning, well-being, disease incidence, and mortality by education levels have widened over the life course (e.g., Ross and Wu 1996; Lauderdale 2001). The gaps in lung, breast, and prostate cancer mortality by sex and race have widened and declined across cohorts (Figures 8.6, 8.9, and 8.10). While the intracohort gaps by sex and race have been constant across birth cohorts (Yang and Lee 2009), the gap in self-rated health by education levels has widened across birth cohorts (Chen et al. 2010; Lynch 2003). There was also evidence of increasing race disparities in cancer incidence and mortality over time in our previous analysis and socioeconomic inequality in health, disability, and life expectancy in the United States in the past several decades (e.g., Hummer, Rogers, and Eberstein 1998; Meara, Richards, and Cutler 2008; Liu and Hummer 2008). Increasing sex, race, and SES inequalities across the life course, birth cohorts, and time periods conceivably contribute to increasing overall inequality or dispersion across these dimensions. This, however, cannot be examined using either the previous HAPC or the VFR analysis and merits a formal test using properly constructed analytic models.

We now present a method that facilitates the disentanglement of age, period, and cohort variations in health disparities defined by differences in both conditional mean levels and conditional dispersions of health (i.e., between-group and within-group health disparities). To do so, we intersect the HAPC model with the VFR/HR (VFR thereafter) model. The HAPC model enables us to disentangle age, period, and cohort effects. The VFR model enabled us to separate within-group from between-group health disparities. We intersect these two statistical models and term the result an HAPC-VFR model.

8.5.3 Intersecting the HAPC and VFR Models

VFR/HR (see Zheng, Yang, and Land 2011 for a review of various types of such models) has two parts, including a regression for an outcome variable Y_i and a regression for logarithm of the residual variances $\log(\sigma_i^2)$ (Western and Bloome 2009):

$$Y_i = \beta_0 + \beta_1 X_{1i} + \beta_2 X_{2i} + \ldots + \beta_P X_{Pi} + e_i \qquad (8.10)$$

$$\log(\sigma_i^2) = \lambda_0 + \lambda_1 Z_{1i} + \lambda_2 Z_{2i} + \ldots + \lambda_R Z_{Ri} \qquad (8.11)$$

where observations on individual sample members are indexed by i, X_1, X_2, ..., X_P is a set of P explanatory variables for Y_i, Z_1, Z_2, ..., Z_R is a set of R explanatory variables (possibly equal to X_1, X_2, ..., X_P) for the logarithm of the residual variance $\log(\sigma_i^2)$, with residual random error term e_i for Y_i. The quantity σ_i^2 is the square of the corresponding residuals $\hat{e}_i^2$ from the first regression. From a substantive viewpoint, the first regression describes how covariates affect the Y_i response variable and account for the deviations of the within-group sample means from the average or grand mean $\bar{Y}$ (which can be termed the *between-group inequality*), while the second regression explains how covariates affect the within-group variability of the response variable around the group means (which can be termed the *within-group inequality*).

We integrate the VFR model with the HAPC model by using the HAPC model to estimate the two equations in the VFR model, treating cohort and period as random effects in the context of a repeated cross-sectional survey research design across a broad range of ages—so that the question of the relative contributions of the age, time period, and birth cohort temporal dimensions are relevant. To do this, we engage in a two-step estimation algorithm.

Step 1: Estimate the β Regression Coefficient Vectors for Between-Group Inequality across Age, Period, and Cohort

We use the REML estimator of the CCREM to estimate Equation (7.2) of the VFR model. The algebra for this algorithm is represented in Section 7.2 on the basic HAPC model specification. This step produces a set of estimates of fixed effects coefficients (for the individual-level explanatory covariates), random effects coefficients (for cohorts and periods), and a random variance components matrix that evaluates the contributions of these individual-level and period and cohort contextual variables to the explanation of variance in the conditional expected value or conditional mean of the outcome variable. In the context of this analysis on health, this step estimates the variations in the conditional mean of self-rated health across period, cohort, age, and other individual-level covariates.

Step 2: Estimate the Λ Regression Coefficient Vectors for Within-Group Inequality across Age, Period, and Cohort

We next calculate the residuals ($\hat{e}_{ijk} = Y_{ijk} - X'_{ijk}\hat{\beta}$) from the Step 1 regression, for each sample respondent i, and compute the squared residuals $\hat{e}_{ijk}^2$ or denoted as σ_{ijk}^2. We then apply the residual pseudolikelihood (RSPL)* estimator of the CCREM to estimate Equation (8.11) of the VFR model. For normal distributed errors, the squared residuals will have a gamma distribution, and Equation (8.11) then is estimated in generalized linear mixed model form—

---

* We used the SAS PROC MIXED and PROC GLIMMIX procedures to estimate the first- and second-step regressions, respectively. The default estimator of GLIMMIX is RSPL, which maximizes the residual log pseudolikelihood and provides unbiased predictors of the random effects. The pseudomaximum likelihood estimator uses a consistent and asymptotically normal estimator rather than a maximum likelihood estimator for the variance parameters. In models for a normally distributed outcome variable with an identity link, RSPL is equivalent to REML (Littell et al. 2006), but RSPL is consistent and asymptotically normally distributed for nonnormal data as well.

as a gamma regression of $\hat{e}_{ijk}^2$ on the X_{ijk} using a log link function (Western and Bloome 2009: 300; see also Nelder and Lee 1991).*

The algebra for this algorithm can be stated as follows:

Level 1 or "within-cell" model:

$$\log(\sigma_{ijk}^2) = \lambda_{0jk} + \lambda_1 X_{1ijk} + \lambda_2 X_{2ijk} + ... + \lambda_P X_{Pijk} \qquad (8.12)$$

Level 2 or "between-cell" model:

$$\lambda_{0jk} = \pi_0 + \omega_{0j} + \phi_{0k}, \quad \omega_{0j} \sim N(0, \psi_u), \quad \phi_{0k} \sim N(0, \psi_v) \qquad (8.13)$$

Combined or mixed effects model:

$$\log(\sigma_{ijk}^2) = \pi_0 + \lambda_1 X_{1ijk} + \lambda_2 X_{2ijk} + ... + \lambda_P X_{Pijk} + \omega_{0j} + \phi_{0k} \qquad (8.14)$$

where λ_{0jk} is the intercept or "cell mean," that is, the mean $\log(\sigma^2)$ of individuals who belong to birth cohort j and were surveyed in year k; π_0 is the expected mean of $\log(\sigma^2)$ at the zero values of all level 1 variables averaged over all periods and cohorts; ω_{0j} and ϕ_{0k} are the residual random effects of cohort j and period k, respectively, assumed normally distributed with mean 0 and variance ψ_u and ψ_v. In addition, $\lambda_{0j} = \pi_0 + \omega_{0j}$ is the cohort $\log(\sigma^2)$ score averaged over all periods with all individual-level covariates at grand mean level; and $\lambda_{0k} = \pi_0 + \phi_{0k}$ is the period $\log(\sigma^2)$ score averaged over all cohorts with all individual-level covariates at grand mean level.

This step produces a set of estimated fixed effects coefficients (for the individual-level explanatory covariates), random effects coefficients (for cohorts and periods), and a random variance components matrix that evaluates the contributions of these variables to the explanation of variance in the logarithm of the residual variances $\log(\sigma_i^2)$ for each sample respondent i. In the context of this study, this step estimates the variations in the variance or dispersion of self-rated health across period, cohort, age, and other individual-level covariates. It merits emphasizing here that the predicted σ_i^2 for age, period, or cohort represents a general form of dispersion of health across age, period, or cohort, whereas previous empirical research in health disparities focused on specific inequality by one or two dimensions, such as

---

* The outcome variable in the analyses described in the following material, self-rated health, is not normally distributed; it is skewed to the left. The residuals calculated from the Step 1 regression have a symmetric distribution that has short tails compared to a normal distribution. Because of the very large sample size, estimates of the coefficients of the Step 2 regression still have good statistical properties. This is due to the fact that, for independently and identically distributed data, the RSPL method produces estimates of the fixed and random effects of a mixed model that are consistent and asymptotically normally distributed; even the identically distributed assumption can be relaxed (Demidenko 2004: 647).

sex, race, or SES. An increase or decrease in any one of the specific dimensions will contribute to the increase or decrease in the general dispersion, which however has not been studied in the current literature. Therefore, instead of examining the changing effects of each specific dimension on health, this analysis investigated how the general dispersion of self-rated health may change across age, period, and cohort.

Even though each of these two steps produces REML or RSPL estimators from the CCREM, it must be iterated to obtain ML estimators for the VFR model (Aitkin 1987). As Western and Bloome (2009: 301) indicated, the fitted values ($\hat{\sigma}^2_{ijk}$) from an application of the two steps should be saved and used in a weighted regression of Y_{ijk} on X_{1ijk}, X_{2ijk}, ..., X_{Pijk} with weights ($1/\hat{\sigma}_{ijk}$). Estimates of the residuals from Step 1 then are updated, Step 2 is computed, and so forth until convergence. Western and Bloome (2009: 301) noted that the ML estimator may perform poorly in small samples, in which case a REML or Bayes estimator can be used. In the empirical application described in the following material, however, the sample sizes were very large, so the adjustments in the REML made for the loss of degrees of freedom resulting from estimation of the regression parameters would be very small, if not trivial. Therefore, for purposes of the empirical application of the HAPC-VFR model in this chapter, the ML estimator was applied. Sample codes for subsequent analyses implementing the HAPC-VFR model are available in the online companion.

8.5.4 Results: Variations in Health and Health Disparities by Age, Period, and Cohort, 1984–2007

Table 8.8 reports estimates of parameters, standard errors, and model fit statistics for the HAPC-VFR models of self-rated health in National Health Interview Survey (NHIS) data from 1984 to 2007. The "β" column presents the results for the first-stage regression of the HAPC-VFR model (which estimated variations in mean health across groups), and the "Λ" column presents the results for the second-stage regression of the HAPC-VFR model (which estimated variations in dispersion of health across groups).

These results are for the sample as a whole. There was substantial evidence of major gender differences in temporal trends in health disparities in prior research. Therefore, we also investigated variations in gender-specific self-rated health disparities across age, time period, and cohort. In other words, we investigated how health disparities may change across age, time period, and cohort within each gender-specific sample. We present the results from this last stratified analysis in subsequent figures only.

As shown in the "β" column, consistent with findings from previous studies, being male, white, married, and more educated and having a job and more income were associated with better self-rated health. The effects of age were curvilinear (quadratic) in that the self-rated health declined with age and then began to increase in late life. The 1995 NHIS sample redesign significantly

TABLE 8.8

Estimated HAPC-VFR Models of Self-Rated Health, NHIS, 1984–2007

Fixed Effects	β		Λ	
	Coefficient	se	Coefficient	se
Intercept	3.28***	0.01	0.4 ***	0.02
AGE	–0.14***	0.00	0.07***	0.01
AGE²	0.03***	0.00	–0.04***	0.00
SEX (1=male)	0.03***	0.00	–0.01*	0.01
RACE (1=white)	0.17***	0.00	–0.08***	0.00
MARRIED	0.02***	0.00	–0.02***	0.00
EDUCATION	0.06***	0.00	–0.03***	0.00
EMPLOYED	0.39***	0.00	–0.34***	0.00
INCOME/10000	0.07***	0.00	–0.04***	0.00
REDESIGN	–0.06***	0.01	–0.07***	0.01
REDESIGN*MALE	–0.02***	0.01	0.02*	0.01
Random Effects	**Coefficient**	**se**	**Coefficient**	**se**
Cohort				
1899	0.03*	0.01	0.20***	0.03
1905	–0.01	0.01	0.12***	0.03
1910	–0.01	0.01	0.03	0.03
1915	0.00	0.01	0.00	0.03
1920	–0.01	0.01	–0.02	0.02
1925	–0.02**	0.01	–0.03	0.02
1930	–0.01	0.01	–0.02	0.02
1935	0.00	0.01	–0.03	0.02
1940	0.00	0.01	–0.04	0.02
1945	–0.01	0.01	–0.07**	0.02
1950	0.01	0.01	–0.09***	0.02
1955	0.02***	0.01	–0.10***	0.02
1960	0.02***	0.01	–0.10***	0.02
1965	0.00	0.01	–0.07**	0.02
1970	–0.01	0.01	–0.04	0.03
1975	–0.03***	0.01	0.04	0.03
1980	–0.01	0.01	0.08**	0.03
1985	0.01	0.01	0.13***	0.03
Period				
1984	–0.01	0.01	0.02**	0.01
1985	–0.01	0.01	0.01	0.01
1986	0.01	0.01	–0.01	0.01
1987	–0.01	0.01	–0.01	0.01
1988	–0.01	0.01	0.01	0.01
1989	0.01	0.01	–0.01	0.01
1990	0.02**	0.01	0.00	0.01

TABLE 8.8 (continued)

Estimated HAPC-VFR Models of Self-Rated Health, NHIS, 1984–2007

Random Effects	β		Λ	
	Coefficient	se	Coefficient	se
1991	0.02*	0.01	−0.01	0.01
1992	0.00	0.01	−0.01	0.01
1993	−0.01	0.01	0.01	0.01
1994	0.00	0.01	0.00	0.01
1995	−0.01	0.01	0.00	0.01
1996	0.00	0.01	0.00	0.01
1997	0.03***	0.01	0.00	0.01
1998	0.03**	0.01	−0.01	0.01
1999	0.03**	0.01	0.00	0.01
2000	0.02	0.01	0.00	0.01
2001	0.01	0.01	0.00	0.01
2002	0.00	0.01	0.00	0.01
2003	0.00	0.01	0.00	0.01
2004	−0.02**	0.01	−0.01	0.01
2005	−0.02**	0.01	0.00	0.01
2006	−0.01	0.01	0.00	0.01
2007	−0.05***	0.01	0.01	0.01
Variance Components	**Variance**	**se**	**Variance**	**se**
COHORT	0.00*	0.00	0.01**	0.003
PERIOD	0.00**	0.00	0.00	0.000
Model Fit				
BIC		1941250.0		
−2 Res Log Pseudo-Likelihood		2351732.0		

Source: Adapted from Zheng et al. (2011: Table 2).
Note: $* p < 0.05$; $** p < 0.01$; $*** p < 0.001$.

decreased the expected value of self-rated health for about 0.06 points for women and 0.08 points for men. The estimates of residual variance components at level 2 indicated significant period and cohort effects net of the effects of individual-level covariates, while the period effect was larger than the cohort effect as reported in the "Variance Components" section.

Figure 8.12 presents predicted age curves and estimates of cohort and period effects on mean self-rated health from the HAPC part of the integrated model estimated for the gender-specific samples. The top panel shows that both men's and women's self-rated health declined with increasing age, but the trends reversed after around age 69 for men and age 72 for women. It can be seen that men reported better self-rated health than women at all ages. The gender gap in health was largest in the early adult years, narrowed until around age 61, and widened afterward.

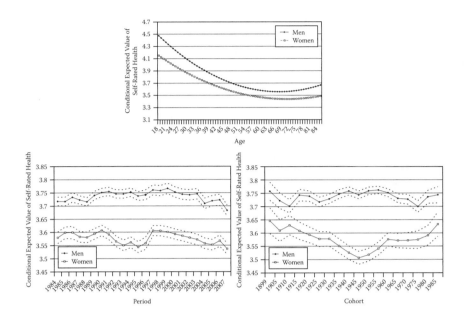

FIGURE 8.12
Variations in conditional expected values of gender-specific self-rated health across age, cohort, and period, with 95% confidence intervals.

The estimated cohort effects were relatively flat across cohorts for men, except that baby boomers born between 1950 and 1959 significantly had better self-rated health than other cohorts. By comparison, the conditional expected values of self-rated health changed dramatically across cohorts for women. They continued declining from the 1899–1904 cohort to the early baby boomers born in 1945–1954 and then rose for the middle and late baby boomers and afterward, which resulted in a widened and then narrowed self-rated health gap between men and women. In addition, before 1998, the period-to-period changes in self-rated health exhibited a very slight increase accompanied by cycles up and down, with a significant decline after 1998. The estimates in Table 8.8 as well as the gender-stratified analysis suggest period effects contributed slightly more than cohort effects to the changes in self-rated health from 1984 to 2007 for both men and women.

As a key output of the VFR part of the integrated model, the "Λ" column in Table 8.8 shows how individual-level covariates affected within-group health disparities. The estimated within-group health disparities for males, whites, married persons, the more highly educated, employed individuals, and those more income were smaller than those of their counterparts, that is, females, blacks, unmarried persons, the less educated, unemployed individuals, and those with less income. In addition, the integrated HAPC-VFR model yielded estimates of expected or predicted variations in health

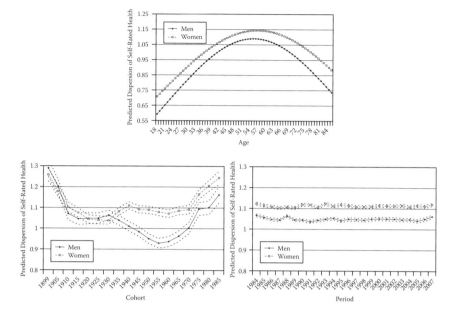

FIGURE 8.13
Variations in predicted dispersion of gender-specific self-rated health across age, cohort, and period, with 95% confidence intervals.

disparities across age, period, and cohort (or within-age, within-period, and within-cohort health disparities). The estimates of residual variance components at level 2 indicate significant cohort and nonsignificant period effects net of the effects of individual-level covariates as reported in the "Variance Components" section. Graphs of estimated age, period, and cohort effects from the gender-stratified analysis are shown in Figure 8.13. The estimated within-age health disparities had a bell shape that peaked around age 56 for both men and women. After controlling for demographic and socioeconomic statuses, estimated health disparities in the young adult ages were relatively small, indicating that most everyone was relatively healthy. But, health disparities increased with age and later declined with age. Compared to men, the within-age health distributions for women were slightly more spread out. The dispersions were larger than those of men at all ages. They increased at a slower rate before age 56 and decreased at a slower rate after age 56.

Figure 8.13 also shows that for women, within-cohort health disparities decreased from the 1899–1904 cohort to the 1930–1934 cohort, increased in cohorts born in the late stage of the Great Depression and World War II, followed by decreases in baby boomer cohorts and increases in recent cohorts. For men, within-cohort variances fluctuated more across cohorts. They decreased from the 1899–1904 cohort to the 1915–1919 cohort and then were relatively flat until they substantially declined again in cohorts born in the Great Depression, World War II, and baby boomer cohorts born between 1945 and 1959 followed by substantial increases afterward, especially in more recent cohorts. After controlling for individual-level covariates and age and cohort effects, the estimates of within-period health disparities graphed in Figure 8.13 are very flat between 1984 and 2007 for both men and women. The lack of significant random effects coefficients for period and the statistically insignificant variance components of the period effects in Table 8.8 suggest the variance in self-rated health did not significantly vary across periods. These gender-specific analyses further support the inference that cohort effects contributed more than period effects to the changes in health disparities. The striking gender difference in the changes in within-cohort health disparities merits further research.

8.5.5 Summary

The core idea of the HAPC-VFR model is to estimate mixed (fixed and random) effects regression specifications of the two equations in the VFR model as a function of age, period, and cohort and other individual-level covariates, treating cohort and period as random effects. Thus, the first mixed effects regression described how age, period, and cohort affected mean of self-rated health net of a set of individual-level covariates (gender, race, marital status, work status, education, and income); the second mixed effects regression explained how within-group health disparities changed across age, period, and cohort. The differences in group-specific means examined in the first step were the topic of study in prior studies of health status and the basic HAPC model. It was the detection of these temporal changes in within-group variations and their decomposition into age, period, and cohort components that were made possible by the integrated HAPC-VFR model.

By contrast to what we found for the conditional mean of self-rated health, cohort effects appeared to contribute much more than period effects to the changes in the variance of self-rated health over the past two decades. Within-cohort health disparities generally decreased from the 1899–1904 cohort to the baby boomer cohorts and have substantially increased for post-baby boomer cohorts. As post-baby boomer cohorts (especially cohorts born after 1980) had much larger within-cohort health disparities than preceding cohorts, and within-age health disparities increased with age until around age 55, it can be expected that health disparities in the general population will further increase in the next one or even two decades as these cohorts age and replace preceding cohorts. The enlarged health disparities for the post-baby boomer cohorts are alarming, and smaller within-cohort heterogeneity in men than women is intriguing. These findings should stimulate future studies on the underlying mechanisms and explanations that are readily testable using the HAPC-VFR model.

The nature of the self-rated health outcome variable—in the form of five ordered response categories (poor, fair, good, very good, and excellent)—complicates the specification and estimation of the combined HAPC-VFR model. This model was described in a linear mixed effects regression format. For several reasons, we applied this specification to the NHIS data by scaling the self-rated health outcome variable as a five-point scale with responses numbered from 1 to 5. First, this choice facilitated comparisons with prior research using similar self-rated health data. Second, the equal-intervals assumption of the five-point scale was, in fact, a good specification for the self-rated health responses in the NHIS data. Evidence of this was obtained from an ordered logit regression analysis of this outcome variable. But, it is still possible that this model was not fully sensitive to either the ordered categorical nature of the survey responses to the self-rated question or the non-normal frequency distribution of the five-point scaling of these responses. Accordingly, Zheng, Yang, and Land (2011) described additional analyses to assess the robustness of the empirical findings reported from application of this model, including alternative coding of the response variable and replication of empirical findings using a different dataset. They also described some methodological extensions to adapt it to the ordered nature of the self-reported health outcome variable.

By using the HAPC-VFR model, we have been able to give a more complete picture of the evolution of changes in self-rated health and health disparities in the United States from 1984 to 2007. Two significant contributions to the health disparities literature are notable. First, health disparities across age, time period, and birth cohort were intertwined but have not been systematically disentangled in the existent studies due to the lack of an integrated model. By using the HAPC-VFR model, this study demonstrated that changes in self-rated health disparities in the last two decades have been much more of a cohort than a period story. Therefore, further research in this area should pay more attention to the "cohort" perspective. Second, prior literature has focused on the changes in health inequality defined by specific aspects of social stratification system such as gender, race, and SES across age, period, or cohort, without an overall picture of the changes in general dispersion of health across these three dimensions. The HAPC-VFR model offers an analytic tool to capture the general dispersion of health across these three dimensions, which provides the basis for further research to examine the contribution to the general dispersion by each specific aspect of social stratification system.

Inequality or disparity in statuses occurs in many domains of social life (e.g., income, wealth, education, and heath care access, to name but a few). The HAPC-VFR model provides a powerful framework and lens through which to identify and study the evolution of variations and social inequalities in these outcomes across the age, period, and cohort temporal dimensions. Accordingly, this model should be broadly applicable to the study of social inequality in many different substantive contexts.

References

Aitkin, M. 1987. Modeling variance heterogeneity in normal regression using GLIM. *Applied Statistics* 36:332–339.

Bashir, S. A., and J. Esteve. 2001. Projecting cancer incidence and mortality using Bayesian age-period-cohort models. *Journal of Epidemiology and Biostatistics* 6:287–296.

Blanchflower, D. G., and A. J. Oswald. 2004. Well-being over time in Britain and the USA. *Journal of Public Economics* 88:1359–1386.

Blau, P. M., and O. D. Duncan. 1966. *The American occupational structure.* New York: Wiley.

Carlin, B. P., and T. A. Louis. 2000. *Bayes and empirical Bayes methods for data analysis.* 2nd ed. New York: Chapman & Hall/CRC.

Chen, F., Y. Yang, and G. Liu. 2010. Social change and socioeconomic disparity in health over the life course in China: A cohort analysis. *American Sociological Review* 75:126–150.

Chlebowski, R. T., Z. Chen, G. L. Anderson, et al. 2005. Ethnicity and breast cancer: Factors influencing differences in incidence and outcome. *Journal of the National Cancer Institute* 97:439–448.

Davis, J. A. 1984. New money, and old man/lady and "two's company": Subjective welfare in the NORC General Social Surveys, 1972–1982. *Social Indicators Research* 15:319–350.

Demidenko, E. 2004. *Mixed models: Theory and applications.* Hoboken, NJ: Wiley.

Di Tella, R., R. J. MacCulloch, and A. J. Oswald. 2003. The macroeconomics of happiness. *The Review of Economics and Statistics* 85:809–827.

Draper, D. 2002. *Bayesian hierarchical modeling.* New York: Springer-Verlag.

Easterlin, R. A. 1987. *Birth and fortune: The impact of numbers on personal welfare.* Chicago: University of Chicago.

Easterlin, R. A. 2001. Life Cycle Welfare: Trends and Differences. *Journal of Happiness Studies* 2:1–12.

Fox, J. 2008. *Applied regression analysis and generalized linear models.* Los Angeles: Sage.

Gelfand, A. E., S. E. Hills, A. Racine-Poon, and A. F. M. Smith. 1990. Illustration of Bayesian inference in normal data models using Gibbs sampling. *Journal of American Statistical Association* 85:972–985.

Gelman, A., J. B. Carlin, H. S. Stern, and D. B. Rubin. 2000. *Bayesian data analysis.* New York: Chapman & Hall/CRC.

Glenn, N. D. 1994. Television watching, newspaper reading, and cohort differences in verbal ability. *Sociology of Education* 67:216–230.

Glenn, N. D. 2005. *Cohort analysis.* 2nd ed. Thousand Oaks, CA: Sage.

Guo, X., and B. P. Carlin. 2004. Separate and joint models of longitudinal and event time data using standard computer packages. *The American Statistician* 58:16–24.

Hughes, M., and M. E. Thomas. 1998. The continuing significance of race revisited: A study of race, class, and quality of life in America, 1972 to 1996. *American Sociological Review* 63:785–795.

Hummer, R. A., R. G. Rogers, and I. W. Eberstein. 1998. Sociodemographic approaches to differentials in adult mortality: A review of analytic approaches. *Population Development Review* 24:553–578.

Knorr-Held, L., and E. Rainer. 2001. Projections of lung cancer mortality in West Germany: A case study in Bayesian prediction. *Biostatistics* 2:109–129.

Lauderdale, D. S. 2001. Education and survival: Birth cohort, period, and age effects. *Demography* 38:551–561.

Littell, R. C., G. A. Milliken, W. W. Stroup, R. D. Wolfinger, and O. Schabenberrger. 2006. *SAS for mixed models*. Cary, NC: SAS Institute.

Liu, H., and R. A. Hummer. 2008. Are educational differences in U.S. self-rated health increasing? An examination by gender and race. *Social Science & Medicine* 67:1898–1906.

Lynch, S. M. 2003. Cohort and life-course patterns in the relationship between education and health: A hierarchical approach. *Demography* 40:309–331.

Meara, E. R., S. Richards, and D. M. Cutler. 2008. The gap gets bigger: Changes in mortality and life expectancy, by education, 1981–2000. *Health Affairs* 27:350–360.

Morris, M., and B. Western. 1999. Inequality in earnings at the close of the twentieth century. *Annual Review of Sociology* 25:623–657.

Nelder, J. A., and Y. Lee. 1991. Generalized linear models for the analysis of Taguchi-type experiments. *Applied Stochastic Models and Data Analysis* 7:101–120.

Preston, S., and H. Wang. 2006. Sex mortality differences in the United States: The role of cohort smoking patterns. *Demography* 43:631–646.

Raudenbush, S. W., and A. S. Bryk. 2002. *Hierarchical linear models: Applications and data analysis methods*. Thousand Oaks, CA: Sage.

Reither, E. N., R. M. Hauser, and Y. Yang. 2009. Do birth cohorts matter? Age-period-cohort analyses of the obesity epidemic in the United States. *Social Science & Medicine* 69:1439–1448.

Ross, C. E., and C. Wu. 1996. Education, age, and the cumulative advantage in health. *Journal of Health and Social Behavior* 37:104–120.

Ryder, N. B. 1965. The cohort as a concept in the study of social change. *American Sociological Review* 30:843–861.

Smith, H. L. 2004. Cohort Analysis Redux. In *Sociological Methodology*, Ross M. Stolzenberg (Ed.), vol 34, pp. 111–119. Boston, MA: Blackwell Publishing.

Smith, H. L., W. M. Mason, and S. E. Fienberg. 1982. Estimable functions of age, period, and cohort effects: More chimeras of the age-period-cohort accounting framework: Comment on Rodgers. *American Sociological Review* 47:787–793.

Smyth, G. K. 1989. Generalized linear models with varying dispersion. *Journal of the Royal Statistical Society, Series B* 51:47–60.

Spiegelhalter, D. J., N. G. Best, B. P. Carlin, and A. van der Linde. 2002. Bayesian measures of model complexity and fit (with discussion and rejoinder). *Journal of the Royal Statistical Society, Series B* 64:583–639.

Weiss, A. R., A. D. Lutkus, B. S. Hildebrant, and M. S. Johnson. 2002. *The nation's report card: Geography 2001. Report NCES 2002-484. Washington, DC: National Center for Education Statistics.*

Western, B., and D. Bloome. 2009. Variance function regression for studying inequality. *Sociological Methodology* 39:293–326.

Western, B., D. Bloome, and C. Percheski. 2008. Inequality among American families with children, 1975 to 2005. *American Sociological Review* 73:903–920.

Yang, Y. 2006. Bayesian inference for hierarchical age-period-cohort models of repeated cross-section survey data. *Sociological Methodology* 36:39–74.

Yang, Y. 2008. Social inequalities in happiness in the United States, 1972 to 2004: An age-period-cohort analysis. *American Sociological Review* 73:204–226.

Yang, Y., and L. C. Lee. 2009. Sex and race disparities in health: Cohort variations in life course Patterns. *Social Forces* 87:2093–2124.

Zheng, H., Y. Yang, and K. C. Land. 2011. Variance function regression in hierarchical age-period-cohort models: Applications to the study of self-reported health. *American Sociological Review* 76:955–983.

9

Mixed Effects Models: Hierarchical APC-Growth Curve Analysis of Prospective Cohort Data

9.1 Introduction

The repeated cross-sectional data designs for which generalized linear mixed models (GLMMs) in the form of hierarchical age-period-cohort (HAPC) models were specified and estimated in Chapters 7 and 8 rely on information from synthetic cohorts that contain different cohort members at each point in time. That is, the cohorts defined in these designs are synthetic in the sense that they contain different individuals as opposed to the same group of individuals at each calendar period. Inferences drawn from such designs about cohort and age effects therefore rest on the assumptions that synthetic cohorts mimic true cohorts and changes over time across synthetic cohort members mimic the age trajectories of change within true cohorts. If the composition of cohorts does not change significantly over time due to migration or other factors and sample sizes are large, these assumptions are generally met, and the demographic tool of synthetic cohorts is most useful in the absence of information from true cohorts.

   In the life course and human development research and life sciences that focus on age-related changes, however, researchers often need to rely on longitudinal data obtained from the same persons followed over time to track continuity and change within individual lives. The *accelerated longitudinal panel design*—wherein multiple birth cohorts are followed over multiple points in time—is an important advance in aging and cohort research. The advantage of this design is that it not only provides cross-time linkages within individuals and hence information pertaining to true birth cohorts but also allows a more rapid accumulation of information on age and cohort effects than a single cohort follow-up study. In this chapter, we develop a similar GLMM approach to the analysis of prospective panel data using accelerated longitudinal cohort designs. We include empirical applications that continue to reveal patterns and mechanisms underlying social stratification of aging and health.

Recent social and epidemiologic research on health disparities has been perplexed by inconsistent findings on whether these disparities grow or diminish over the life course. One key issue that has contributed to such inconsistency is the confounding of age and cohort changes. The essential assumption of previous studies of the relationship between aging and health that omit cohort analysis is that differences in age patterns of health observed from cross-sectional data represent true intraindividual developmental trajectories of change over time that are equal across various birth cohorts. This assumption may not be tenable in light of the social changes and vastly different historical and life experiences of birth cohorts during the twentieth century that bear important consequences for cohort differences in mental and physical health. To understand the mechanisms generating social inequalities in health over the life course, one must take the role of cohort change into account and systematically examine both intercohort and intracohort variations in health trajectories.

We show how to use GLMMs to disentangle the effects of aging and birth cohort in longitudinal panel study designs. We examine three questions essential to the understanding of independent age and cohort effects. First, are there intercohort variations in aging experiences? That is, are the age-related changes specific to birth cohorts? If the answer is yes, then the conventional approach to testing age-related changes by omitting the cohort variations confounds age and cohort effects. Second, are there intracohort variations in the age trajectories? That is, within each birth cohort, do individuals with different characteristics such as socioeconomic status (SES) show different age-related changes? While the first question treats cohorts as homogeneous groups, this second question further examines within-cohort heterogeneity and is especially useful for testing aging-related hypotheses. Third, are there intercohort variations in intracohort differences in aging? This question jointly examines heterogeneities between and within cohorts and informs us of how individual aging experiences are shaped or conditioned by social, historical, and epidemiologic contexts. Drawing on previous studies of two longitudinal datasets, the Americans' Changing Lives (ACL) (Yang and Lee 2009) and Health and Retirement Survey (HRS) (Yang and Lee 2010), we next address these questions analytically to fully capture the processes generating social inequalities over the life course.[*]

---

[*] Parts of Sections 9.2 and 9.4 were adapted from Yang and Lee (2009); parts of Section 9.3 were adapted and updated from Yang, Y., and L. C. Lee. 2010. *Journal of Gerontology: Social Sciences* 65B:246–255.

9.2 Intercohort Variations in Age Trajectories

9.2.1 Hypothesis

We have learned from previous studies of others and our own included in the previous chapters that social demographic and historical changes have had profound and enduring effects on cohorts' health and mortality risk. They support the contention that throughout the twentieth century, individuals' health capital—physiological robustness and capacity of vital organ systems—has improved with the year of birth, with more recent cohorts faring substantially better in their initial endowments at birth and having lower depreciation rates in that stock of health capital (Fogel 2004). Improved physiological capacities in later cohorts also bode well for the effectiveness of medical treatments. In fact, recent demographic research showed that better nutrition and reduced inflammatory-infection *in utero* and during infancy have reduced the risks for major chronic diseases in adulthood (Barker 1998) and led to less-severe disabilities for successive birth cohorts (Crimmins, Reynolds, and Saito 1999). Improvements in physical functioning in more recent cohorts are also likely given evidence of continuous cohort improvements in education, general health, quality of life, and smoking cessation (Haug and Folmar 1986; Hughes and O'Rand 2004; Pampel 2005). In contrast, early life experiences such as childhood poverty and traumas associated with the Great Depression and world wars of the twentieth century have been shown to negatively affect the mental and physical health of earlier cohorts (Elder 1999; O'Rand and Hamil-Luker 2005).

On the other hand, it is possible that the health improvements across cohorts are constrained to some extent by economic, cultural, and lifestyle changes. For instance, there have been substantial changes in family structures, such as decreases in marriages and increases in divorces in more recent cohorts (Popenoe 1993; Waite 1995). The larger sizes of the baby boom cohorts also are associated with increased life stress and decreased happiness in adulthood (Easterlin 1987; Yang 2008). In addition, there have been evident increases in lifestyles harmful to cardiovascular health and increases in obesity in later-born cohorts (Cabrera et al. 2003; Flegal et al. 2002; Reither, Hauser, and Yang 2009). All these risk factors may have dampened the positive cohort effects on health. Studies of depressive symptoms showed evidence of more depression in war babies (1935–1945) than in earlier cohorts (Kasen et al. 2003; Yang 2007). Earlier cohorts (1900–1905 and pre-1900 cohorts) have also been found to have better perceived or self-rated health in late life than their successors (1906–1917 cohort) (Idler 1993). Because there are both positive and deleterious forces affecting the life courses of more recent cohorts, an empirical question is whether the cohort trend of improving health can be generalized to the entire adult life course and to cohorts born after World War II such as the baby boomers.

In light of the many social and historical forces that could have produced significant cohort differences in aging experiences, we test the *hypothesis of intercohort change*—that there are substantial intercohort variations in health trajectories over the life course, with more recent cohorts having better health on average (higher mean levels of health) and lower growth rates of health problems with age.

9.2.2 Model Specification

We have shown in a previous chapter how GLMMs can be flexibly applied to model a variety of cohort-related phenomena using repeated cross-sectional data. GLMMs continue to be a useful tool for modeling longitudinal data. Because repeated observations over time can be viewed as level 1 units nested within individuals who are level 2 units, one can specify two-level hierarchical models in the form of growth curve models to assess individual change. Growth curve models have been widely used in longitudinal studies of individual change (Raudenbush and Bryk 2002). However, we employ such models differently from the way they often are used when a single cohort is followed over time, as, for example, in test scores of students in schools. Specifically, it will become clear in the models we specify in this chapter that the growth curve models applied to multiple cohort panel data are special cases of the HAPC models. Although they are distinct from the HAPC-CCREMs (cross-classified random effects models), they are hierarchical models that explicitly incorporate cohort effects and implicitly incorporate period effects (in the form of age-by-cohort interaction). Therefore, we term them hierarchical APC-growth curve models (HAPC-GCMs). In the context of the accelerated longitudinal design depicted by Figure 3.3, we then can conduct HAPC-GCM analyses of intracohort age changes (across columns) and intercohort differences (across rows) simultaneously.

The HAPC-GCM specification for the ACL data is shown in Equations (9.1)–(9.4).

Level 1 repeated observation model:

$$Y_{ti} = \beta_{0i} + \beta_{1i} Age_{ti} + \beta_{2i} Age_{ti}^2 + e_{ti} \tag{9.1}$$

where Y_{ti} is one of the health response variables (Center for Epidemiologic Studies Depression Scale [CES-D] score, functional disability, or self-rated health) for respondent i at time t, for $i = 1, \ldots, n$ and $t = 1, \ldots, T_i$; T_i is the number of measurements and ranges from 1 to 4; Age_{ti} is the age of respondent i at time t. We estimated both simple and quadratic age effects models and found that the latter fit the data substantially better. Consequently, the present model includes both linear and quadratic terms of age. We centered the age variable around the median age of the 10-year cohort to which the person belonged obtained from Table 3.10 because cohort-median center-ing protects the estimate from bias associated with systematic variation in mean age across cohorts and is equivalent to person-mean centering, which facilitates interpretation (Miyazaki and Raudenbush 2000). The intercept β_{0i} is the expected health response score of person i at the median cohort age; β_{1i} is the expected linear rate of increase or growth rate per year of age for person i; β_{2i} is the expected quadratic rate of increase; and e_{ti} is the random within-person error for person i at t and is assumed normally distributed. The individual growth parameters β_{0i}, β_{1i}, and β_{2i} depend on person-level characteristics such as cohort membership. The level 2 model thus specifies a distinct average trajectory for each cohort and incorporates other time-invariant covariates associated with each individual:

Level 2 model:

For the intercept:

$$\beta_{0i} = \gamma_{00} + \gamma_{01}Cohort_i + \sum_q \gamma_{0q}Z_{qi} + w_{0i} \tag{9.2}$$

For the linear growth rate:

$$\beta_{1i} = \gamma_{10} + \gamma_{11}Cohort_i + w_{1i} \tag{9.3}$$

For the quadratic growth rate:

$$\beta_{2i} = \gamma_{20} + \gamma_{21}Cohort_i + w_{2i} \tag{9.4}$$

In the model for the intercept [Equation (9.2)], γ_{00} is the expected health outcome at age 88 (median) for cohort born before 1905 at zero values of other covariates; γ_{01} is the main effect of cohort that indicates mean difference in health between cohorts or *intercohort variation in the mean*; γ_{0q} is the coefficient for person-level covariates Z_{qi} (as discussed in the following). In models (9.3) and (9.4), the linear and quadratic growth rates per year of age further vary by cohort: γ_{10} and γ_{20} are the expected linear and quadratic rates of change in health response in the 1905 cohort, respectively; γ_{11} and γ_{21} are the age-by-cohort interaction effects and indicate mean differences in rates of change between cohorts or *intercohort variation in the age effects*. Finally, w_{0i}, w_{1i}, and w_{2i} are the residual random effects, after controlling for cohort differences, of person i on health and the rates of increase by age in health, respectively, and are assumed to have a multivariate normal distribution. The cohort effects can be nonlinear and represented by cohort dummy indicators or polynomial terms. The choice of operationalization in the final model depends on significance tests of coefficients and comparative model fit.

The hierarchical/multilevel linear model (HLM)-growth curve methodology has the advantage of allowing data that are unbalanced in time (Raudenbush and Bryk 2002). That is, it incorporates all individuals with data for the estimation of trajectories, regardless of the number of waves he or she contributes to the person-year dataset. Compared to alternative modeling techniques, this substantially reduces the number of cases lost to follow-up due to mortality or nonresponse. However, the loss-to-follow-up subsample needs to be distinguished from, and compared with, those with complete data for all waves. If mortality and nonresponse are significantly correlated with worse health and other key covariates of health such as age, parameter estimates of health trajectories may be biased if they are not controlled. To fully assess the potential influence of selection due to death or nonresponse, we control for the effects of attrition by including dummy variables indicating the deceased and nonrespondents in the level 2 models. That is, $Z_{qi} =$ (*Died, Nonresponse*), for $q = 1$ and 2 in Equation (9.2).

Although the application of two-level hierarchical regression models to the analysis of age-dependence in standard single-cohort longitudinal panel data is relatively straightforward, growth curve analysis of multicohort, multiwave data is complicated by two issues (Yang 2010).

First, because the observable age trajectories of different cohorts initiate and end at different ages, cohort comparisons are based on different segments of cohort members' life course (see, e.g., Table 3.10). When the observations for one cohort are not overlapping with another in age, one cannot compare cohorts at the same ages. And in this case, age and cohort are difficult to distinguish as earlier cohorts are older in age. This raises the question of the potential confounding of the age and cohort effects. Two analytic strategies help to resolve this problem. The first is using centered age variables, as illustrated previously in this chapter. Age centering eases the interpretation of the intercept, stabilizes estimation, and prevents the bias in the estimate that arises from systematic variation in mean age across the cohorts, hence eliminating the confounding of age and cohort variables. Second, the models yield tests of significance of overlapping segments of the life course of adjacent cohorts. As waves of data accumulate, the number of overlapping ages of adjacent cohorts increases, which increases statistical power. In this case, age and cohort will become less and less confounded, making it increasingly possible to compare cohort differences in age trajectories.

The second issue concerns period effects. The models specified in the previous discussion here do not explicitly incorporate period effects for both substantive and methodological reasons. First, in contrast to synthetic cohort designs that usually cover several decades, accelerated longitudinal designs typically span much shorter time periods (e.g., a decade or so). So, the effects of period can be assumed to be trivial and omitted from the models, especially if the theoretical focus is on aging. Second, it is challenging to estimate a separate period effect in the framework of the growth curve models. The level 1 analysis reflects within-individual change by modeling the outcome as a function of the time indicator (Singer and Willett 2003). This means that one can include age *or* wave (period) depending on substantive focus, but not *both* because within individuals age and period are the same. The simultaneous estimation of age and period effects creates the model identification problem that requires different data designs and mixed model specifications—the HAPC-CCREM analytic framework described in Chapters 7 and 8—to resolve. Third, one does not need to estimate period effect per se and can focus instead on the age-by-cohort interaction effects. Because cohorts vary in age at any historical moment, effects associated with historical time, if any, tend to produce cohort differences in the age-outcome relationship. If period effects are operating, pooling of data from all cohorts (or omission of cohort effects from the models) would yield biased estimates of age-related changes, but controlling for cohort effects and age-by-cohort interactions captures these changes precisely (Yang 2007).

Based on the combined models of (9.1)–(9.4), we obtained restricted maximum likelihood-empirical Bayes (REML-EB) parameter estimates using SAS PROC MIXED (see sample codes for selected tables). Additional analyses using alternative GLMM specifications such as Poisson and negative binomial mixed effects models did not yield substantively different results. Accordingly, because of their familiarity and ease of interpretation, we report estimates from linear mixed models (LMMs) using a normal link (sample codes are provided for selected models shown in Tables 9.1–9.3).

9.2.3 Results

Table 9.1 presents the model effect coefficients and significance tests for depressive symptoms, physical disabilities, and self-assessments of health. The intercohort differences in age trajectories of health can be clearly seen in Figure 9.1, which plots the expected growth trajectories of depressive symptoms, physical disability, and self-rated health by cohort (these are based on estimates of similar models that also adjust for other covariates in the analysis in Section 9.4 and hence for the reference group with mean values of all these covariates). Results support the hypothesis that there are significant cohort variations in both mean levels (β_{0i}) and growth rates (β_{1i} and β_{2i}) of health problems, although the directions of change differ by the health outcome examined. Specifically, more recent cohorts suffered from more depression on average, as indicated by a positive cohort effect coefficient (γ_{01}) and higher intercepts for more recent cohorts in Figure 9.1a. On the other hand, they fared better in physical functioning and self-rated general health on average, as indicated by significant cohort effect coefficients and cohort-specific intercepts in Figures 9.1b) and 9.1c, respectively. The cohort effects are linear for the depression and self-rated health models, with each successive

TABLE 9.1

Growth Curve Model Estimates of Cohort and Aging Effects on Health

Fixed Effects	Parameters	CES-D Coef. (*t* Ratio)	Disability Coef. (*t* Ratio)	Self-rated Health Coef. (*t* Ratio)
For Intercept	β_{0i}			
Intercept	γ_{00}	−0.34**	2.00***	3.15***
COHORT	γ_{01}	0.03**	−0.28***	0.10***
		(3.45)	(−8.59)	(−9.31)
COHORT2	γ_{02}		0.02***	
			(4.86)	
For Linear Growth Rate	β_{1i}			
Intercept	γ_{10}	0.13**	0.83***	−0.08
		(2.93)	(11.24)	(−1.58)
COHORT	γ_{11}	−0.06**	−0.25***	−0.03*
		(−5.84)	(−6.00)	(−2.37)
COHORT^2	γ_{12}		0.02***	
			(3.82)	
For Quadratic Growth Rate	β_{2i}			
Intercept	γ_{20}	0.09**	0.31***	0.09***
		(4.69)	(6.51)	(4.50)
COHORT	γ_{21}		−0.06***	
			(−5.62)	
Control Variables				
DIED	γ_{03}	0.37**	0.44***	−0.48***
		(9.13)	(14.79)	(−11.44)
NONRESPONSE	γ_{04}	0.20***		
		(5.87)		
Random Effects—Variance Components				
Level–1: Within-person	e_{ti}	0.38**	0.22***	0.38***
Level–2: In intercept	w_{0i}	0.44**	0.30***	0.52***
In growth rate	w_{1i}	0.16**	0.15***	0.24***
Goodness-of-Fit				
BIC		28100.5	22505.4	28592.3

Note: $+ p < 0.10$; $^* p < 0.05$; $^{**} p < 0.01$; $^{***} p < 0.001$ (two-tailed test).

cohort having a 0.03-unit increase in the standardized CES-D score and a 0.10-unit increase in self-rated health on average ($p < .001$). The cohort effect is quadratic for the disability model, with each successive cohort having a 0.28-unit lower disability score (γ_{01}) that decreased at an increasing rate of 0.02 (γ_{02}) ($p < .001$).

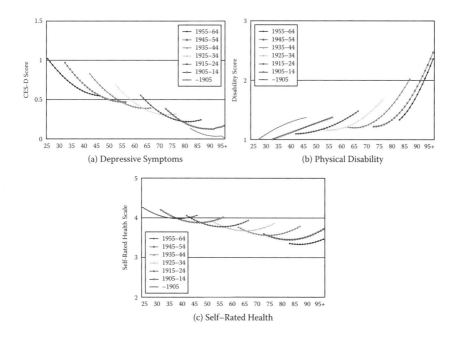

FIGURE 9.1
Predicted age trajectories of three health outcomes by cohort: (a) depressive symptoms; (b) physical disability; (c) self-rated health.

The results also show that health problems increased significantly with age, as indicated by the intercept coefficients for the linear (γ_{10}) and quadratic growth rates (γ_{20}) for all three health outcomes. However, such quadratic age patterns were not universal across cohorts and therefore were not entirely developmental in nature but affected by social context as defined by cohort membership. The cohort effect coefficients for the growth rate models indicate significant age-by-cohort interaction effects that suggest different growth trajectory patterns across cohorts. Figure 9.1 further illustrates age trajectories of depression, disability, and self-rated health that were unique to each cohort. In more recent cohorts, depressive symptoms and disability increased less steeply with age, and the smaller age increases across cohorts were significant for both ($\gamma_{11} = -.06$, $p < .01$, in CES-D model; $\gamma_{11} = -.25$, $\gamma_{12} = .02$, $p < .001$, in disability model). At the same time, while Figure 7.1a shows that more recent cohorts had lower levels of depression at the same ages than the earlier cohorts, Figure 9.1b shows that successive cohorts had higher levels of disability than their predecessors at the same ages. This may be due to an earlier onset of debilitating conditions in recent cohorts, such as obesity and diabetes. Figure 9.1c shows that more recent cohorts also reported slightly faster declines in perception of good general health with age than earlier cohorts ($\gamma_{11} = -.03$, $p < .05$).

The last set of results on control variables suggests that the deceased had significantly more depressive symptoms and disabilities and worse self-rated health than those who completed all surveys. Nonrespondents had more depression but did not differ from other survivors in disability or self-rated health. Adjusting for the attrition status produced estimates of age trajectories and cohort effects that would otherwise be biased.

In sum, we found strong support for the *intercohort change hypothesis* that there exist substantial intercohort variations in aging trajectories of health. We also found that the directions of cohort change did not consistently favor more recent cohorts but differed by health outcomes and parameters of health trajectories. On the one hand, more recent cohorts fared increasingly better in mean levels of physical functioning and self-assessments of health and experienced faster declines in depression and slower increments in disability with age. This reflects the long-term health benefits of physiological improvements and smoking cessation across cohorts. On the other hand, more recent cohorts suffered from more depressive symptoms on average and showed faster declines in perceived general health with age than earlier cohorts did. As expected, this pattern partly reflects the mental health impacts of unique demographic and cultural experiences of different twentieth-century cohorts with regard to marriage patterns, family structures, and lifestyles affecting nutrition and body mass. The finding is also consistent with previous research that showed that earlier cohorts born in, and coming of age during, the Great Depression and world wars may have gained more satisfaction and positive views of themselves having survived economic deprivations, social instability, and related social hardships (Idler 1993). Later cohorts, in contrast, faced more complicated problems associated with prosperity—such as labor market competition, human relations, and medicalization—and may thus have manifested more mental health problems (Yang 2007, 2008).

9.3 Intracohort Heterogeneity in Age Trajectories

9.3.1 Hypothesis

Previous social and epidemiologic research has found that social status indicated by sex, race, and SES manifests strong relationships with health and the way health changes over the life course. Whereas the female, black, and low-SES disadvantages in health are a general observation, the patterns of social disparities in health trajectories are much less clear. Although most extant studies suggested that the salience of sex, race, and SES in affecting health varies across age, they have reported different directions for such variations. A key factor that has led to the inconsistent understanding about aging and stratification patterns is the same one that plagues studies of health trajectories in the overall population—the confounding of aging and cohort effects. It remains unknown whether discrepant findings on changes in social gaps in health with age are actually due to birth cohort differences because few prior studies have simultaneously examined and effectively distinguished aging and cohort effects using longitudinal research designs. We argue that taking cohort effects into account can largely resolve the inconsistencies in previous findings and can provide a better test of competing theories, whereas ignoring cohort variations may bias the findings on age-related inequalities. In light of the *intercohort change hypothesis*, we further test the *intracohort inequality hypothesis* that sex, race, and SES disparities in health change significantly over the life course, independent of cohort differences. We directly tested the implicit assumption on which prior studies were based, that is, health inequalities by sex, race, and SES change over the life course in the same way within each cohort and the intracohort patterns reflect aging-related phenomena that are universal across all cohorts.

There are two major theoretical perspectives on social stratification of aging and health that predict different directions of change in patterns of stratification over the life course. The dominant explanation for increasing health disparities with age is the cumulative advantage/disadvantage theory. Originating from studies of scientific careers and status attainment models (Merton 1968), the *cumulative advantage/disadvantage theory* has recently proliferated in life course studies of changes in the effects of personal and structural characteristics on various outcomes. The theory suggests that the effects of early advantage or disadvantage accumulate over the life course, thereby increasing heterogeneity within cohorts (O'Rand 2003). In the context of aging and health, it predicts that inequalities in social status and consequently in health status early in life amplify with age and differentiate individuals further as they age. A secondary explanation for the diverging health gaps with age has been the double-jeopardy hypothesis that old age and racial/ethnic minority status interact to widen black and white health differences, and it has also been extended to a triple-jeopardy hypothesis that considers the compounding effects of sexism (Ferraro and Farmer 1996). The prevailing explanation for decreasing health disparities with age is the *age-as-leveler theory*. It suggests that the health gaps narrow across the life course due to the equalization of resources in later life and the selective survival of elite minorities who have acquired immunity against hardships in life (House et al. 1994; Preston, Hill, and Drevenstedt 1998).

Extant empirical studies suggested opposing directions of change in sex, race, and SES gaps in health with age and hence inconsistent support for the theories mentioned. Regarding the sex and race gaps, some found that these gaps widened with age, while others found that these gaps converged with age, and the racial gap even exhibited a crossover in old age. Most extant cross-sectional or short-term longitudinal analyses of sex and race differences in health were concerned with individual-level as opposed to aggregate data, but these analyses provided no consistent support for either perspective. Several recent studies that more rigorously examined how sex and race effects on health changed over the life course using better statistical models of age interactions with sex and race showed no apparent increases or decreases in sex or race gaps when other social factors were controlled (see, e.g., Yang and Lee 2009 for a review). Findings were also inconsistent with regard to whether the effects of SES on health outcomes strengthened or diminished over the life course. Education and income gaps in health have been found to diverge, converge, remain stable, or diverge from early to middle age and converge in old age (see, e.g., Chen, Yang, and Liu 2010 for a review). One major limitation of these studies is that they did not test for cohort effects. Because the theoretical perspectives above all emphasized intracohort differentiation, a proper test should control for cohort differences that might otherwise be confounded with aging effects.

While most extant studies concerned specific health outcomes, less is known about cumulative measures of health disorders and deficits, indicated by frailty, that have recently been shown to hold a particularly strong relationship with the aging process (Mitnitski, Song, and Rockwood 2004). And, even less is known about population heterogeneity in the age dynamics of frailty. So, we supplement the analysis on the three health outcomes with one on the Frailty Index (FI) by extending a recent study by Yang and Lee (2010).

For the HRS data, which include only four cohorts of older adults, we estimated age growth trajectories for each cohort separately. That is, for each cohort, we specified models of age change and included individual-level covariates (sex, race, education, attrition dummies) in the level 2 model and within-person time-varying covariates (income, marital status, smoking) in the level 1 model. Age was centered around the cohort median and divided by 10. An alternative specification was to include cohort membership in the level 2 model as Equations (9.2)–(9.4), which we illustrate in Section 9.4 using the ACL data, which consist of more birth cohorts and require a parsimonious test of cohort differences.

9.3.2 Results

The model estimates of fixed effects of age and other covariates are presented in Table 9.2 for the total sample and four cohorts. For comparison, we show first the model for all based on the total sample that included all cohorts that resembles analyses in many previous studies without distinguishing aging and cohort effects. It shows significant sex, race, and SES gaps in the age growth trajectories of frailty indicated by both mean and age slope (linear growth rate). However, these results do not constitute evidence for significant social disparities in aging and frailty because cohort effects were not controlled. Therefore, we turn to the next set of results from cohort-specific models. First, we found quadratic age trajectories of the FI for all cohorts, indicating increases in the accumulation of health deficits and

TABLE 9.2

Fixed Effect Estimates from Growth Curve Models of Frailty Index: Social Disparities in Aging Effects by Cohort

Variable	All (−1924–1947)			AHEAD (−1924)			CODA (1925–1930)			HRS (1931–1941)			WB (1942–1947)			Cohort Difference p Value
	Coef.		(t Ratio)	Coef.		(t Ratio)	Coef.		(t Ratio)	Coef.		(t Ratio)	Coef.		(t Ratio)	
For Intercept																
Intercept	0.10	***	(52.12)	0.11	***	(27.09)	0.11	***	(25.64)	0.09	***	(34.13)	0.09	***	(22.80)	<0.001
SEX (1=female)	0.02	***	(13.15)	0.02	***	(6.09)	0.02	***	(5.46)	0.03	***	(12.31)	0.02	***	(6.52)	0.521
RACE (1=nonwhite)	0.03	***	(16.50)	0.03	***	(6.29)	0.02	***	(4.66)	0.02	***	(9.09)	0.04	***	(7.94)	0.368
EDUC (1=0–12 years)	0.03	***	(18.28)	0.03	***	(8.95)	0.03	***	(8.80)	0.04	***	(17.26)	0.03	***	(9.48)	<0.001
POVERTY (1=yes)	0.02	***	(17.17)	0.01	***	(7.85)	0.01	***	(3.85)	0.01	***	(9.74)	0.02	***	(8.96)	0.020
For Linear Growth Rate																
Intercept	0.04	***	(37.57)	0.09	***	(23.88)	0.06	***	(13.69)	0.04	***	(18.97)	0.03	***	(6.92)	0.022
SEX (1=female)	0.00	***	(−3.45)	−0.01		(−1.50)	−0.01		(−1.33)	0.00		(−1.57)	0.00		(−0.01)	0.886
RACE (1=nonwhite)	0.00	+	(−1.67)	−0.01	+	(−1.77)	0.01		(0.79)	0.00		(−0.36)	0.01		(1.43)	0.955
EDUC (1=0–12 years)	0.00	**	(3.06)	0.01	*	(2.38)	0.01	**	(3.04)	0.01	***	(4.60)	0.02	***	(4.69)	0.332
POVERTY (1=yes)	0.00	***	(−3.62)	−0.01	**	(−2.69)	0.00		(−0.63)	−0.01	***	(−3.37)	0.01		(0.97)	0.115
Quadratic Growth Rate																
Intercept	0.02	***	(42.22)	0.03	***	(18.88)	0.02	***	(4.10)	0.01	***	(6.60)	−0.01	***	(−3.83)	<0.001
Control Variables																
DIED	0.01	***	(8.15)	0.04	***	(13.55)	0.06	***	(16.57)	0.06	***	(22.14)	0.09	***	(13.17)	<0.001
NOTMARRIED	0.02	***	(22.53)	0.02	***	(8.27)	0.03	***	(10.26)	0.02	***	(16.17)	0.02	***	(8.40)	<0.001
SMOKED	0.02	***	(12.85)	0.01	***	(5.37)	0.01	+	(1.94)	0.01	***	(5.75)	0.01	*	(2.06)	0.235

Note: + $p < 0.10$; * $p < 0.05$; ** $p < 0.01$; *** $p < 0.001$ (two-tailed test).

disorders with age that can be considered as rates of biological aging. The shapes of age trajectories were cohort specific, however. While the AHEAD (Study of Assets and Health Dynamics Among the Oldest Old), CODA (Children of Depression), and HRS cohorts showed accelerations in the FI with age, as indicated by positive quadratic growth rates, the WB (war baby) cohort showed a deceleration in the FI with age, as indicated by a negative quadratic growth rate.

Second, we found partial support for the intracohort inequality hypothesis. Within cohorts, the effects of sex, race, education, and income were all highly significant on the intercepts or means of the FI. In the model for the AHEAD cohort, being female was associated with a 0.019-unit higher FI on average, net of other factors. The effect of race was stronger than that of sex: A nonwhite had a 0.025-unit higher FI on average. And, the effect of education was the strongest of all: Having less than 13 years of education increased the FI by 0.028. Poverty, similarly, increased one's FI score by 0.014. The stratification patterns were also observed in each other within-cohort model, suggesting that females, nonwhites, and low-SES groups suffered from a greater degree of physiological reserve loss at any given age than their male, white, and higher-SES counterparts, adjusting for all other factors. While consistent with the widely documented inverse relationship between social status and the risk of illness (Link and Phelan 1995), the racial and SES gaps in mean FI levels also provide new evidence that social adversity exerts strong effects on multiple domains of deficit accumulation simultaneously and on the complex process of physiological deregulation.

The findings on social heterogeneity in rates of biological aging indicated by the growth rates of the FI, however, were more varied. When the cohort effect was controlled, there was no consistent support for the intracohort inequality hypothesis. As opposed to the results from all cohorts combined that showed a significant decrease in the sex gap with age (-0.004, $p < .001$), the cohort-specific models showed no significant sex effect on growth rates of the FI. Therefore, one would have found support for the age-as-leveler hypothesis regarding sex differences in frailty without adjustment for cohort differences. Instead, we found that sex disparities in frailty remained constant over the life course within cohorts. We found evidence for a converging racial gap in frailty for one cohort but constant racial gaps for the other cohorts. The race gap decreased with age significantly only in the AHEAD cohort (-0.009, $p < .10$), as shown by the predicted age trajectories of the FI by race from the model in Figure 9.2a.

SES disparities in frailty changed in different directions depending on indicators used. As predicted by the cumulative advantage theory, the education disparities increased over the life course within all cohorts, suggesting positive effects of education on growth rates of frailty and diverging frailty trajectories of the highly educated and those with a high school degree or less. The education gaps showed significant increases with age within all cohorts, but the magnitudes of such changes varied: The divergence in the educational

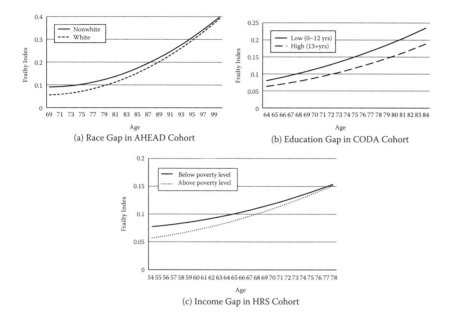

FIGURE 9.2
Race and SES gaps in predicted age trajectories of the Frailty Index within cohorts: (a) race gap in AHEAD cohort; (b) education gap in CODA cohort; (c) income gap in HRS cohort.

gap was twice as large in the WB cohort as the AHEAD cohort. Figure 9.2b plots the predicted age trajectories by education level for the CODA cohort that experienced an intermediate degree of divergence. On the other hand, the income disparities decreased over the life course within two cohorts, as predicted by the age-as-leveler theory. For the AHEAD and HRS cohorts, the below-poverty group showed significantly slower rates of aging than others with higher income, leading to convergences in the trajectories of the FI, as shown in Figure 9.2c for the HRS cohort. The income gaps did not show significant age changes and hence remained constant for the other cohorts.

In sum, the foregoing HAPC-GCMs analyses of intracohort heterogeneity in age trajectories led to the following important conclusion: *The life course changes in health disparities that were found in previous studies were actually due largely to cohort differences.* Taking cohort effects into account substantially modifies the existing understanding of the relationships between social inequalities and aging. The lack of consistent relationships between social status and rates of aging across cohorts precludes any straightforward characterization of social heterogeneity in the dynamics of frailty and aging. We further note that there were potential measurement limitations of the study, such as the dichotomous racial group categorization, which may obscure an even more complex pattern of inequality over the life course and necessitate future research. The only clear conclusion that can be drawn at this point is that the expression of biological aging and the accumulation of general system damage did not follow the same path under different circumstances within a human population. In fact, individuals' slopes of change with age were sensitive to social conditions in which they were embedded. This is an important finding that challenges a prevailing assumption in many medical and genetic studies that there exist common paths and determinants of health declines and aging across individuals. It provides another example of the powerful interplay between biological and environmental forces that shape the divergent developmental paths of aging organisms.

9.4 Intercohort Variations in Intracohort Heterogeneity Patterns

9.4.1 Hypothesis

The *intercohort change hypothesis* refers to changes in overall cohort means, and the *intracohort inequality hypothesis* refers to whether sex and race groups within cohorts become increasingly heterogeneous with age. It is logical to ask further how the patterns of within-cohort heterogeneity differ across cohorts. Therefore, we next examine intercohort variations in intracohort social disparities in health trajectories.

A large body of sociological research has shown that sex and race inequalities in health largely result from differences in power, prestige, social status, socially learned lifestyles, behaviors, roles, and stress (Verbrugge 1989; Williams 2005). The secular trends of these risk factors may not have occurred in a parallel manner across male and female cohorts and across black and white cohorts. Changes in sex- and race-specific exposures to these factors across birth cohorts may thus contribute to intercohort variations in intracohort sex and race disparities in health trajectories over the life course. Cohort membership therefore provides an important social structural context that conditions the cumulative advantage/disadvantage or leveling process. If sex and race gaps in major risk factors for health have decreased in more recent cohorts, then the corresponding health gaps may diminish. This suggests sex and race gaps in mean levels of health narrow in more recent cohorts. In addition, sex and race gaps in growth rates of health problems may also change across cohorts. The patterns of divergence or convergence with age may become stronger or weaker or even change directions in more recent cohorts.

There is evidence that the trends of increasing education effects over the life course (or larger education gaps in older ages) have strengthened in more recent cohorts in the United States (House, Lantz, and Herd 2005; Lynch 2003). A most recent study by Chen, Yang, and Liu (2010) used the methodology introduced in the previous and further discussion in this chapter to show a more complicated picture. The effect of education on mean level of health decreased across six 10-year cohorts born between 1890 and 1970 in China, but the educational gap remained constant with age. The income gap in health trajectories diverged for earlier cohorts but converged for most recent cohorts. In sum, the *intercohort difference in intracohort inequality hypothesis* states that patterns of change across the life course in social disparities in health trajectories vary by cohort.

9.4.2 Model Specification

We illustrate the model specifications using the ACL data with a focus on sex and race disparities in health, which can easily be extended to include other stratification variables, such as SES. Building on the same level 1 model in Equation (9.1) and the level 2 model in Equations (9.2)–(9.4), we further assess cohort differences in the associations between social status and age trajectories. That is, at level 2, we specify interaction effects between birth cohort and other person-level characteristics such as sex and race for models of each growth parameter.

Level 2 model:

For the intercept:

$$\beta_{0i} = \gamma_{00} + \gamma_{01}Cohort_i + \gamma_{02}Sex_i + \gamma_{03}Race_i + \gamma_{04}Cohort \cdot Sex \\ + \gamma_{05}Cohort \cdot Race + w_{0i} \tag{9.5}$$

For the linear growth rate:

$$\beta_{1i} = \gamma_{10} + \gamma_{11}Cohort_i + \gamma_{12}Sex_i + \gamma_{13}Race_i + \gamma_{13}Cohort \cdot Sex \\ + \gamma_{14}Cohort \cdot Race + w_{1i} \tag{9.6}$$

In the model for the intercept (9.5), γ_{02} and γ_{03} are the sex and race effect coefficients respectively that indicate intracohort differences by sex and race in the mean level of health, γ_{04} and γ_{05} are interaction effect coefficients that indicate *intercohort differences in the intracohort sex and race effects on the mean level of health*. In the model for the linear growth rate (9.6), γ_{12} and γ_{13} represent the intracohort sex and race differences in age change in health, respectively, and γ_{14} and γ_{15} are coefficients for three-way interactions between cohort, age, and sex/race that *indicate intercohort differences in the intracohort sex and race effects on age change*. A similar model for the quadratic growth rate can be specified. Control variables can be entered at level 1 for time-varying covariates (such as income) and level 2 for time-constant covariates (such as education and attrition dummies). All continuous variables are centered in order for the intercept to be substantively meaningful.

9.4.3 Results

There is support for intercohort variations in intracohort sex and race differences in mean health levels but not in growth rates of health problems
except for depression. Table 9.3 shows significant cohort-by-sex interaction
effects on the intercepts of CES-D (0.03, $p < .05$) and disability models (–0.03,
$p < .001$), suggesting that sex gaps in depression widened across cohorts,
whereas the gaps in disability decreased across cohorts. Figure 9.3 shows the
predicted mean levels of CES-D scores and disability by cohort for men and
women based on these estimates that adjusted for age and all other factors.
While the female disadvantage in mental health increased in more recent
cohorts (Figure 9.3a), the female disadvantage in disability declined in more
recent cohorts (Figure 9.3b).

There were also significant cohort differences in racial gaps in mean health
outcomes, net of all other factors. The cohort-by-race interaction effects in
the intercept models of CES-D and self-rated health indicated that race gaps
widened across cohorts. Figure 9.4 presents the predicted mean CES-D and
self-rated health scores by race from the models that show divergences in
the black and white gaps for more recent cohorts. Black cohorts experienced
steeper increases in mean CES-D scores in each successive cohort and continuous decreases in self-rated health. Adjusting for all other factors reduced the
magnitudes of the cohort effect coefficients in the intercept model for disability and diminished the significance level in the intercept model for self-rated
health. Therefore, intercohort variations in mean levels of disability and self-
assessments of health can be largely explained by cohort differences in SES,
marital status, chronic illness, obesity, and mental health. With self-assessments
of health, cohort differences in patterns of smoking also played a role.

The cohort-by-sex and cohort-by-race interaction effects were not significant in the models of growth rates when the cohort, sex, and race effects were
controlled and thus are omitted. Combining results from the models of intercept and growth rates, we find that significant intercohort differences existed
in changes of intracohort sex gaps in depression but not in changes of intracohort sex or race gaps in disability or self-rated health. Therefore, these gaps
in growth trajectories were constant within cohorts. This is expected given
the results from the tests of intracohort inequality hypothesis, which indicated that a significant intracohort sex gap in growth rates was found only
in depression, controlling for cohort effects. Figure 9.5 shows the predicted
sex-specific age trajectories of depression for select cohorts based on the
final model in Table 9.3. Consistent with previous findings from Yang (2007),
whereas the gross age effects in Table 9.1 indicate increases in depression
with age, the net depression trajectories showed decreases with age within
cohorts after adjusting for the effects of SES, marital status, and physical illness. Furthermore, in support of the current hypothesis, the within-cohort
sex gap did not uniformly increase or decrease across the life course but
varied by cohort. For instance, the sex gap first diverged in the earliest cohort

TABLE 9.3

Growth Curve Model Estimates of Cohort Differences in Sex and Race Effects on Age Trajectories of Health

Fixed Effects	CES-D Coef.[a]	Disability Coef.	Self-Rated Health Coef.
For Intercept			
Intercept	0.16	1.50***	3.34***
COHORT	0.06*	–0.17***	0.03
COHORT^2		0.02***	
SEX (1=female)	–0.07	0.19***	0.02
RACE (1=black)	–0.12	0.06	0.07
COHORT * SEX	0.03*	–0.03*	–0.01
COHORT * RACE	0.07***	–0.01	–0.04*
For Linear Growth Rate			
Intercept	0.05	0.73***	–0.07
COHORT	–0.03**	–0.28***	–0.03*
COHORT^2		0.02***	
SEX (1=female)	–0.08*	0.04	0.04
RACE (1=black)	–0.01	0.06	–0.05
For Quadratic Growth Rate			
Intercept	0.09***	0.33***	0.07***
COHORT		–0.06***	
Control Variables			
DIED	0.07*	0.35***	–0.29***
NONRESPONSE	0.09**		
EDUC	–0.04***	–0.02***	0.05***
INCOME	0.00*	0.00+	0.00***
NOTMARRIED	0.24***	0.04*	0.02
ILLNESS	0.07***	0.16***	–0.26***
BMI (ref. = normal)			
UWEIGHT	0.22***	0.22***	–0.46***
OWEIGHT	–0.07***	–0.02	–0.05*
OBESE	–0.07*	0.05*	–0.24***
SMOKE	0.00***	0.00	–0.01***
CES-D			–0.17***
DISABLE	0.20***	0.11***	–0.32***
HEALTH	–0.17***	–0.18***	
Random Effects—Variance Components			
Level–1: Within-person	0.36***	0.22***	0.37***
Level–2: In intercept	0.27***	0.21***	0.33***

continued

TABLE 9.3 (continued)

Growth Curve Model Estimates of Cohort Differences in Sex and Race Effects on Age Trajectories of Health

Goodness-of-Fit	CES-D Coef.[a]	Disability Coef.	Self-Rated Health Coef.
In growth rate	0.14***	0.14***	0.24***
BIC	26733.5	21838	27416.5

Note: $+ p < 0.10$; $* p < 0.05$; $** p < 0.01$; $*** p < 0.001$ (two-tailed test).
[a] t ratios are omitted due to space constraints; they are available upon request.

born before 1905, with men having more depressive symptoms than women, then crossed over in the 1915–1924 cohort, converged in the War Babies, and became largely constant in the baby boomers. Overall, this showed a trend of decreasing degrees of or lessening convergence in female and male depression trajectories in more recent cohorts.

Comparisons of estimates of the random effects in Tables 9.1 and 9.3 suggest that the inclusion of cohort, sex, and race effects largely reduced the level 2 variances in the intercepts for models of all three health outcomes. And, adjusting for all covariates in Table 9.3 also decreased the Bayesian information criteria (BICs) and hence improved the model fit. The residual variances in the growth rates were only slightly reduced, suggesting additional factors to consider in future research.

Returning to the HRS analysis of four cohorts of older adults, we present the p values of tests of cohort differences in sex, race, and SES effects and all others included in the model in the rightmost column of Table 9.2. They are associated with the interaction terms of cohort and these variables. As noted in the introduction of datasets in Chapter 3, additional waves of the HRS, as compared to the ACL, increased the power of the HAPC-GCMs in tests of aging-related effects within each cohort, but the fewer HRS cohorts may decrease the power of tests for cohort differences. Nonetheless, the results showed significant cohort differences in the overall intercept and growth rates, supporting the intercohort change hypothesis. Different from the ACL analysis on individual health outcomes, the HRS analysis suggested no significant intercohort variations in intracohort sex and race gaps in frailty. However, there were significant cohort differences in the education and income effects on the mean, suggesting intercohort variations in intracohort SES inequalities in levels of frailty. The positive effect of poverty on frailty, in particular, increased in more recent cohorts. No cohort effects were found on the social disparities in growth rates. This is consistent with the results from the ACL analysis. The two studies jointly suggested that social disparities in the aging-related health processes existed in weak forms in the context of cohort experiences. This is strong evidence for Norman Ryder's argument that cohort is a unique structural category that has explanatory power above and beyond other social stratification systems.

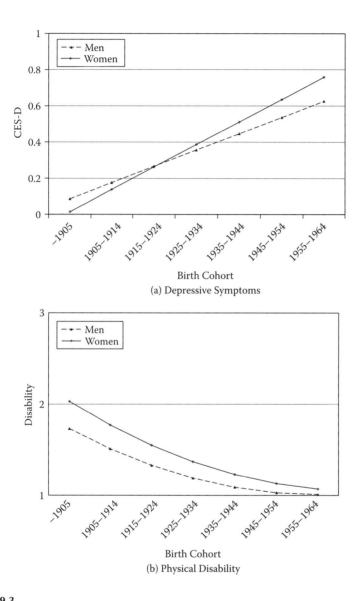

FIGURE 9.3
Predicted mean health outcomes by birth cohort: sex gap: (a) depressive symptoms; (b) physical disability. (Figure 9.3a adapted from Yang and Lee, 2009: Figure 1A; Figure 9.3b adapted from Yang and Lee 2009: Figure 2A.)

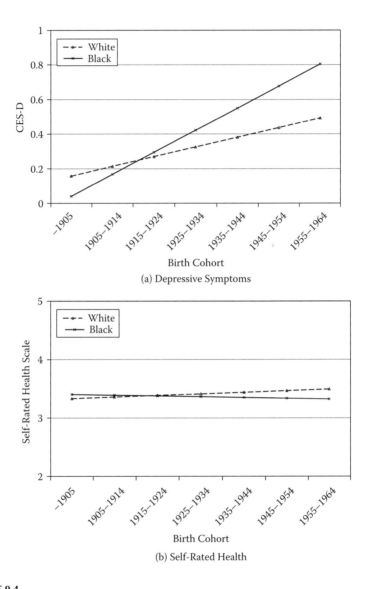

FIGURE 9.4
Predicted mean health outcomes by birth cohort: race gap: (a) depressive symptoms; (b) self-rated health. (Figure 9.4a adapted from Yang and Lee, 2009: Figure 1B; Figure 9.4b adapted from Yang and Lee, 2009: Figure 3.)

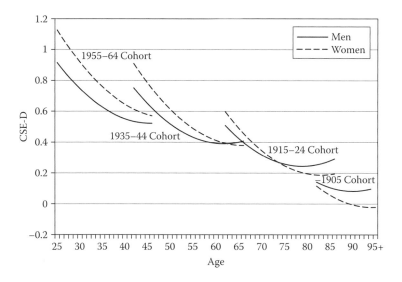

FIGURE 9.5
Predicted age growth trajectories of depressive symptoms by cohort: sex gap. (Adapted from Yang and Lee, 2009: Figure 4.)

9.5 Summary

We have shown the use of the GLMMs framework in longitudinal studies of prospective cohort panel data to address some long-standing questions in the stratification of aging. We empirically evaluated the proposition that considering the process of cohort change is important for the theory, measurement, and analysis of social inequalities in health over the life course. We found substantial evidence that supported this proposition based on a systematic investigation of the distinct role of aging and cohort in the relationships between social status defined by sex, race, and SES and comorbidities of physical and mental problems and frailty. We found that changes in health disparities across the life course were largely cohort-related phenomena. We also identified major mechanisms by which cohort differences in various health outcomes were realized. The findings prompt more careful and thorough examinations of various aging-related hypotheses in a cohort-specific context.

The first and most foundational component of longitudinal cohort analysis using the mixed effects statistical model concerns how cohort effects condition overall health trajectories over the life course. If there exist substantial intercohort variations in aging trajectories of health, then health and social disparities therein are due in part to social historical changes exogenous to developmental/age changes in physical and mental states. This highlights the necessity of examining cohort effects for a proper attribution of aging effects in life course research.

The second kind of analysis concerns how cohort effects alter existing explanations of social disparities in health trajectories over the life course. We did not find evidence for diverging or converging intracohort inequalities in disability, self-rated health, or frailty net of intercohort differences in health trajectories. Therefore, neither the cumulative advantage nor the age-as-leveler theory explained the mechanisms generating the life course patterns of sex and race disparities in these health outcomes because the sex and race gaps were persistent and constant across ages within cohorts. On the other hand, the diverging education gaps in frailty with age for all cohorts of U.S. older adults strongly supported the cumulative advantage theory. We have clearly shown that examination of cohort effects in the longitudinal growth curve analyses provided more conceptual clarity and a better empirical test of competing aging theories. The strong presence of intercohort differences as opposed to age differences in sex and race gaps in health suggests that we focus on cohort-related explanations.

The third kind of analysis subsumes the first two and assesses how cohorts differ in patterns of intracohort changes in heterogeneity in health trajectories. Intercohort differences can exist in the means or the growth rates (age change) or both. Persistent and growing social disparities in mean levels of health across cohorts may indicate a lack of improvement and possible adverse changes in the cohort-related exposures to risk factors for the disadvantaged groups. For instance, women born after World War II experienced unanticipated increases in social stress associated with balancing work and family, marital dissolution, and economic deprivation in female-headed households that decreased their perceived sense of health well-being. Racial discrimination in terms of residential segregation and economic isolation also continues to affect blacks' quality of life in ways that may matter more for their mental than physical health. Intercohort differences in the social gaps in growth rates of health problems mean that the salience of social stratification in health not only varies across the life course but also varies across cohort. For instance, the patterns of age changes in sex gap in depression were strongly modified by and dependent on birth cohort. We saw less convergence in the male-female gap in depression with age in more recent cohorts, suggesting a weaker age-as-leveler process in these cohorts. In the case of no significant changes in the intracohort heterogeneity in the age trajectories but substantial intercohort variations in social gaps in mean levels of health (as for disability and self-rated health), we conclude that any changes with age observed in previous studies or studies that ignored cohort effects were actually due to intercohort differences in health. That is, there was no stronger cumulative advantage process or age-as-leveler effect at work in later or earlier cohorts when the confounding effects of aging and cohort were distinguished and other social factors controlled.

In conclusion, an assessment of the distinct roles of aging and cohort effects in the study of social inequalities in health may provide a better understanding of biological and social structural factors that condition health inequalities over the life course and help researchers better evaluate the explanatory power of competing theories. Cohort membership contextualizes individuals' health change with age and conditions social inequalities in health. Cohort effects also seem to be essential for moderating sex inequality in age trajectories for certain health indicators such as depression. The significance of cohort change in shaping social inequalities in health over the life course implies the relevance of both biological forces and historical context, and it suggests emerging new patterns for future cohorts entering adulthood and old age as a result of their new life circumstances. The identification of cohort-specific patterns of inequalities in psychiatric morbidity and physical functioning thus has important implications for public health and finance as the largest cohort in U.S. history—the baby boomers—reaches the 65+ ages in the 18 years after 2010.

References

Barker, D. J. P. 1998. *In utero* programming of chronic disease. *Clinical Science* 95:115–128.

Cabrera, C., K. Wilhelmson, P. Allebeck, H. Wedel, B. Steen, and L. Lissner. 2003. Cohort differences in obesity-related health indicators among 70-year olds with special reference to gender and education. *European Journal of Epidemiology* 18:883–890.

Chen, F., Y. Yang, and G. Liu. 2010. Social change and socioeconomic disparities in health over the life course in China: A cohort analysis. *American Sociological Review* 75:126–150

Crimmins, E. M., S. L. Reynolds, and Y. Saito. 1999. Trends in health and ability to work among the older working-age population. *Journal of Gerontology: Social Sciences* 54B:S31–S40.

Easterlin, R. A. 1987. *Birth and fortune: The impact of numbers on personal welfare.* Chicago: University of Chicago.

Elder, G. H., Jr. 1999. *Children of the Great Depression: Social change in life experience.* Enlarged 25th anniversary ed. Boulder, CO: Westview Press.

Ferraro, K. F., and M. M. Farmer. 1996. Double jeopardy to health hypothesis for African Americans: Analysis and critique. *Journal of Health and Social Behavior* 37:27–43.

Flegal, K. M., M. D. Carroll, C. L. Ogden, and C. L. Johnson. 2002. Prevalence and trends in obesity among U.S. adults, 1999–2000. *Journal of the American Medical Association* 288:1723–1727.

Fogel, R. W. 2004. Health, nutrition, and economic growth. *Economic Development and Cultural Change* 52:643–658.

Haug, M. R., and S. J. Folmar. 1986. Longevity, gender, and life quality. *Journal of Health and Social Behavior* 27:332–345.

House, J. S., P. M. Lantz, and P. Herd. 2005. Continuity and change in the social stratification of aging and health over the life course: Evidence from a nationally representative longitudinal study from 1986 to 2001/2002 (Americans' Changing Lives Study). *Journal of Gerontology: Social Sciences* 60B:15–26.

House, J. S., J. M. Lepkowski, A. M. Kinney, R. P. Mero, R. C. Kessler, and A. R. Herzog. 1994. The social stratification of aging and health. *Journal of Health and Social Behavior* 35:213–234.

Hughes, M. E., and A. M. O'Rand. 2004. *The American People Census 2000: The lives and times of the baby boomers*. New York: Russell Sage Foundation.

Idler, E. L. 1993. Age differences in self-assessments of health: Age changes, cohort differences, or survivorship? *Journal of Gerontology* 48:S289–S300.

Kasen, S., P. Cohen, H. Chen, and D. Castille. 2003. Depression in adult women: Age changes and cohort effects. *American Journal of Public Health* 93:2061–2066.

Link, B. G., and J. Phelan. 1995. Social conditions as fundamental causes of disease. *Journal of Health and Social Behavior* 35:80–94.

Lynch, S. M. 2003. Cohort and life-course patterns in the relationship between education and health: A hierarchical approach. *Demography* 40:309–331.

Merton, R. K. 1968. The Matthew effect in science. *Science* 159:56–63.

Mitnitski, A. B., X. Song, and K. Rockwood. 2004. The estimation of relative fitness and frailty in community-dwelling older adults using self-report data. *Journal of Gerontology: Medical Sciences* 59A:M627–M632.

Miyazaki, Y., and S. W. Raudenbush. 2000. Tests for linkage of multiple cohorts in an accelerated longitudinal design. *Psychological Methods* 5:44–63.

O'Rand, A. M. 2003. Cumulative advantage theory in aging research. In *Economic outcomes in later life: Public policy, health and cumulative advantage*, ed. S. Crystal and D. Shea, 14–30. New York: Springer

O'Rand, A. M., and J. Hamil-Luker. 2005. Processes of cumulative adversity: Childhood disadvantage and increased risk of heart attack across the life course. *The Journals of Gerontology: Psychological Sciences and Social Sciences* 60:S117–S124.

Pampel, F. C. 2005. Diffusion, cohort change, and social patterns of smoking. *Social Science Research* 34:117–139.

Popenoe, D. 1993. American family decline, 1960–1990: A review and appraisal. *Journal of Marriage and Family* 55:527–542.

Preston, S. H., M. E. Hill, and G. L. Drevenstedt. 1998. Childhood conditions that predict survival to advanced ages among African–Americans. *Social Science & Medicine* 47:1231–1246.

Raudenbush, S. W., and A. S. Bryk. 2002. *Hierarchical linear models: Applications and data analysis methods*. Thousand Oaks, CA: Sage.

Reither, E. N., R. M. Hauser, and Y. Yang. 2009. Do birth cohorts matter? Age-period-cohort analyses of the obesity epidemic in the United States. *Social Science & Medicine* 69:1439–1448.

Singer, J. D., and J. B. Willett. 2003. *Applied longitudinal data analysis: Modeling change and event occurrence*. New York: Oxford University Press.

Verbrugge, L. M. 1989. The twain meet: Empirical explanations of sex differences in health and mortality. *Journal of Health and Social Behavior* 30:282–304.

Waite, L. 1995. Does marriage matter? *Demography* 32:483–507.

Williams, D. R. 2005. The health of U.S. racial and ethnic populations. *Journal of Gerontology: Psychological Sciences and Social Sciences* 60:S53–S62.

Yang, Y. 2007. Is old age depressing? Growth trajectories and cohort variations in late-life depression. *Journal of Health and Social Behavior* 48:16–32.

Yang, Y. 2008. Social inequalities in happiness in the United States, 1972 to 2004: An age-period-cohort analysis. *American Sociological Review* 73:204–226.

Yang, Y. 2010. Aging, cohorts, and methods. In *Handbook of aging and the social sciences*, ed. R. H. Binstock and L. K. George, 17–30. Burlington, MA: Academic Press.

Yang, Y., and L. C. Lee. 2009. Sex and race disparities in health: Cohort variations in life course patterns. *Social Forces* 87:2093–2124.

Yang, Y., and L. C. Lee. 2010. Dynamics and heterogeneity in the process of human frailty and aging: Evidence from the U.S. older adult population. *Journal of Gerontology: Social Sciences* 65B:246–255.

10

Directions for Future Research and Conclusion

10.1 Introduction

Rates of disease, vital status, and social measurements arrayed over time by age or cohorts are canonical organizations of data in demographic, epidemiologic, and social research. A vast literature over the past four decades has used conventional linear accounting/multiple classification models and statistical methodology for age-period-cohort (APC) analysis. In this context, different solutions to the model identification problem often produced ambiguous and inconsistent results and ignited continuous, never-ending debates on whether any solutions exist or which solutions are better. Controversies that have appeared in the literature have been exacerbated by the inadequacy of existing APC models and methods. We have noted that the APC identification problem is an instance of a more general class of structural underidentification problems in the social sciences. Thus, advances in models and methods for APC analysis may also facilitate innovations on other instances of structural underidentification.

The developments introduced in previous chapters highlight new approaches to this problem—or simply new ways of thinking about cohort analysis. We have reached the insight that the identification problem is model specific rather than data specific (Fu 2008). The early work of Mason and colleagues (1973: 246) actually made a similar observation on the origin of the model identification problem: "[The estimation of the accounting model] is problematic because the relationships between age, period and cohort have the same functional form as the expected relationship of each of these independent variables to the dependent variable, Y. That is, we cannot estimate unique effects for age, period and cohort because *we postulate each variable to be linearly related to Y and at the same time assume that A, P and C are linearly related to each other*" (italics added). Because the use of conventional linear accounting models (and hence the identification problem) characterizes the majority of previous studies, investigators erroneously concluded that the problem is inherent in APC analysis. The exclusive focus on the tabular rates data design in the early literature contributed to this confusion because the A, P, and C variables in this setting are indeed exactly linear. Often, however, it is the linear accounting models, not the APC data structure, that incur the identification problem. The quotation from Mason et al. (1973) implied that there would not be such a problem if either or both of the following conditions are met: (1) A, P, and C are not linearly related to each other; and (2) it is not postulated that each variable be linearly related to Y. A deficiency in previous studies on this topic is that they are almost exclusively focused on finding solutions to break condition 1 while paying no attention to condition 2. In this book, we have provided a more comprehensive set of approaches and tools that deal with both. That is, we have shown the use of multilevel data to satisfy the first condition, developed alternative modeling approaches via the hierarchical APC (HAPC) models to satisfy the second condition, and have also shown how the combination of both approaches can be the most effective way of accounting for aging-related phenomena and social and demographic change.

Beyond the identification problem, we have also shown how applications of different classes of models in the generalized linear mixed models framework to the same or similar datasets yield highly consistent models and substantive inferences across all three research designs: tables of rates or proportions, repeated cross-sectional surveys, and accelerated longitudinal panel designs. In addition to model validation through simulations, we have provided empirical evidence that validated findings from applications of the Intrinsic Estimator (IE) for the conventional APC linear accounting model. Specifically, as shown by our comparisons of the estimated age, period, and cohort temporal trends for the General Social Survey (GSS) verbal ability and Surveillance, Epidemiology, and End Results (SEER) Program data, the IE and HAPC analyses lead to similar conclusions. The HAPC models further allow for the test of contextual characteristics and additional explanatory hypotheses. The growth curve analysis of prospective cohort data supplements the former two approaches and enables analysts to draw inferences from true birth cohorts. Therefore, these members of the generalized linear mixed model (GLMM) family are complementary to each other and jointly form a comprehensive set of tools that can extend the reach of cohort analysis to a wide variety of problems not before possible.

Our joining together of these three classes of models for APC analysis within the larger family of GLMMs opens up the possibility for further statistical methodological developments. For instance, there have been recent and continuing developments in statistical methodology for GLMMs and related software programs toward more flexible specifications on functional forms (e.g., spline functions or other non- or semiparametric functional forms) for temporal dependence and nonnormal specifications on the statistical distributions of the random effects. Given the GLMM framework for APC analysis that we have exposited in this book, it can be anticipated that these methodological developments will be applied to HAPC models when they are motivated by new empirical applications or by the desire to assess robustness of empirical findings by the estimation of such models with alternative specifications. New model specifications for the analysis of hidden heterogeneity (e.g., finite mixture models) within APC data also may be developed. These are just a few examples of the many possibilities for methodological refinements that lie ahead.

While we have featured GLMMs for APC analysis in this book, we do not claim exclusivity of GLMMs to unify and organize new developments of APC models. There are other recent methodological developments using different approaches. We give examples in this chapter of additional models and data designs that provide promising avenues for future development and substantive research.

10.2 Additional Models

10.2.1 The Smoothing Cohort Model and Nonparametric Methods

We described in Chapter 3 the first research design using the contingency age-by-period tables of rates. In this most frequently used design in APC analysis, birth cohorts are defined by the diagonals of the rectangular array. When age and period are measured in intervals that are longer than 1 year, the adjacent cohorts overlap. For example, in Table 3.1, which presents the lung cancer incidence rate data for 5-year age groups across 5-year periods, the diagonals represent 9-year birth cohorts that overlap for 4 years: the 1915 cohort covering birth years 1911–1919 overlaps with the 1920 cohort covering birth years 1916–1924 for the years between 1916 and 1919. This has been largely ignored by the conventional linear models, which treat cohorts as mutually exclusive categories and include cohort effects as additive to fixed age and period effects. The observation of overlapping birth cohorts unique to this data design, however, is the motivation for a different kind of model.

Fu (2008) proposed a *smoothing cohort model* that utilizes the fact that the overlap of adjacent cohorts requires that each cohort effect be estimated with contributions from near neighbors. This model replaces the fixed cohort effects in the APC accounting model [Equation (4.1) of Chapter 4] with smoothed cohort effects through a nonparametric spline smoothing function. Since the temporal variations in this model apply to the cohort effects and leave the age and period as fixed effects, it is a semiparametric model and a member of the family of partially linear models, which are an extension of GLMMs. Interestingly, the smoothing on the cohort effects introduces mild binding to the parameters; breaks up the linear dependency between the age, period and cohort effects; and effectively avoids the identification problem. The mild binding by the nonparametric smoothing through spline smoothing on the cohort effects is just large enough to break the linear dependency and small enough not to introduce much bias, which ensures consistent estimation of age and period effects. Fu then introduced a two-stage smoothing model. The first stage is to obtain consistent estimates of the age and period effects from the smoothing cohort model. The second stage is to apply the consistent estimates to the fixed effect of the APC accounting model to achieve consistent estimates for age, period, and cohort estimates. Fu further demonstrated through simulation studies that this two-stage model yielded consistent estimation and sensible results for analyses of crime rates and lung cancer mortality rates.

10.2.2 The Continuously Evolving Cohort Effects Model

It is well known that the conventional APC linear accounting model suffers from the identification problem methodologically. Less well appreciated is that the model also suffers from a conceptual problem that becomes obvious if we revisit the model of Equation (4.1) in Chapter 4. It can be seen that the model rests on key assumptions that do not always accurately describe most APC-related phenomena. It is a model of additive effects. It assumes that the effect of age α_i is the same for periods j and cohorts k. But, the influence of age may change over time and across cohorts. Consider, for instance, the dramatic declines in infant mortality over the past century. It also assumes that the effect of period β_j is the same for people of all ages. However, period effects are often age specific. For example, the influenza epidemic of 1918 caused especially high mortality among people in their teens and 20s. Similarly, it assumes that the effect of cohort k is the same as long as the cohort lives. But cohorts must change, as Norman Ryder (1965: 861) explained in his seminal article about the nature of the cohort process:

> The case for the cohort as a temporal unit in the analysis of social change rests on a set of primitive notions: persons of age a in time t are those who were age a – 1 in time t – 1; transformations of the social world modify people of different ages in different ways; the effects of these transformations are persistent.

That is, as opposed to experiencing fixed impacts of events, cohorts are continuously exposed to events whose influences accumulate over the life course. Wars and epidemics are examples of such events that may break out in the middle of a cohort's life and leave an imprint on all of its subsequent behaviors and outcomes. New events constantly occur. A model with unchanging cohort effects is appropriate only if all relevant events occur before the initial observation and only if these events' impacts stay fixed as the cohort ages (Hobcraft, Menken, and Preston 1982). To capture the process by which cohort effects are generated as described by Ryder, however, one needs a more general model—a framework that Hobcraft, Menken, and

Preston (1982) labeled "continuously accumulating cohort effects" but lacked procedures for investigation at the time of their review.

Schulhofer-Wohl and Yang (2011) recently addressed this gap. They developed a general model that relaxes the assumption of the conventional additive model. The new model allows age profiles to change over time and period effects to differ for people of different ages. The model also defines cohort effects as an accumulation of age-by-period interactions over all events across the life course. Although a long-standing literature on theories of social change conceptualizes cohort effects in exactly this way, this is the first time that a method of statistically modeling for this more complex form of cohort effects has been presented. Applying this new model to analyze changes in age-specific mortality rates in Sweden over the past 150 years, the authors found that the model fit the data dramatically better than the additive model. The analyses also yielded interesting results that showed the utility of this model in testing competing theories about the evolution of human mortality. The flexibility of the model, however, comes at a high computational cost because it involves the estimation of a large number of parameters. The inclusion of additional covariates also has not been analytically attempted. This new approach thus presents both opportunities and challenges for future analysts. When further improved, it may find applications in many areas of research.

10.3 Longitudinal Cohort Analysis of Balanced Cohort Designs of Age Trajectories

The multicohort, multiwave data design is especially important for aging and cohort analysis. Although several longitudinal surveys using this design are available, such as the Health and Retirement Survey and the National Long Term Care Survey, cohort studies would benefit from further developments in data collection.

The usual accelerated longitudinal cohort design has two limitations: (1) Some ages cannot be observed for all cohorts, and (2) coverage of the individual life course and historical time is extremely restricted. The imbalance of the age-by-cohort structure, as illustrated by Figure 3.3 and Table 3.10 of Chapter 3, arises from the fact that the baseline survey consists of cohorts of different ages and follow-up surveys occurred at the same times. As a result, cohorts aged for exactly the same number of years but will remain age heterogeneous at the end of data collection. The growth curve models applied to these data therefore yield estimates of cohort differences in age trajectories based only on the overlapping age groups of adjacent cohorts rather than the entire possible range of ages. This could affect the accuracy of the estimates when the number of waves is small or the overlapping age intervals are few. Increasing the number of follow-ups alleviates problems for inference but is less than perfect for the purposes of disentangling aging effects from birth cohort differences and observing period effects.

A better design is one in which the age-by-cohort data structure is balanced and extends for a long time. Extant secondary data that meet these criteria are exceedingly rare. But, with the help of modern computing technologies that aid the compilation of historical records, longitudinal data that cover a more extended segment of the life course are emerging and worthy of attention. The Liaoning Multigenerational Panel (LMGP) data (Campbell and Lee 2009), for example, may be an exceptional resource for cohort studies. The LMGP is a database of at least 1 million observations of 200,000 individuals from eighteenth to early twentieth century Chinese population registers. It provides entire life histories for men and nearly complete life histories for women. The length of the historical period spanned and the sheer number of observations included make it a great candidate for future longitudinal research on aging. Another example is the Peking Union Medical College Hospital (PUMCH) birth records data on over 2,000 individuals born in 1921–1954 who were followed up during 2003–2005 with clinical examinations and interviews (Zeng et al. 2010). In this case, all individuals had detailed obstetric records at birth, so the data on health changes within each cohort cover large age spans from birth to middle or old ages. Although the data are unbalanced to the right at the same follow-up year, they are balanced to the left and include 50 years of ages shared by all cohorts. If additional follow-ups were to be conducted, the studies will become even a greater source for studying age trajectories of health.

Data like these are difficult to collect and compile but can be extremely useful for a variety of substantive investigations. For example, the mechanisms underlying persistent cohort differences in mortality need to be better understood. The "cohort morbidity phenotype" hypothesis has been proposed to link large cohort improvement in survival to reductions in exposures to infections, inflammation, and increased nutrition in early life (Finch and Crimmins 2004). But, evidence of an association between early-life conditions and late-life mortality is solely based on aggregate population data from developed countries and needs further testing using individual life histories across multiple birth cohorts in other national populations, such as those in the LMGP and PUMCH. A related question that can benefit from a better cohort design is the debate about whether chronic disease at adult ages is determined by "programming" *in utero* or childhood or whether early-life factors are mediated through the lifelong accumulation of risk factors. The fetal origin hypothesis, which supports the former scenario, has been widely used to explain the mechanisms underlying adult diseases (Barker 1998). One challenging issue is that it is extremely difficult to sort out the confounding effects of life course mediators without data that connect childhood with old ages. It is necessary to collect information on biological and social factors from cohorts in infancy to older adulthoods. Birth record data with long-term follow-up designs hold the potential for vastly enhancing our ability to estimate various age and cohort models to shed light on this debate.

10.4 Conclusion

The APC problem has intrigued and frustrated scientists for decades. Since the last synthesis of statistical models and methods for APC analysis in the social sciences by Mason and Fienberg (1985), there have been substantial developments in statistical theory, methods, and computational algorithms. One example, on which we have based the models and methods described in the previous chapters, is the generalization of the class of generalized linear models to the class of GLMMs. We think these innovations in statistics are sufficiently important that they support, indeed almost require, rethinking the problem of APC analysis.

In its simplest terms, this is what we seek to accomplish. This book summarizes the developments of new models and methods for APC analysis that we and our collaborators have initiated. The technique of the IE for modeling APC tabular data helps one to obtain statistically sound solutions to the identification problem that has long compromised previous APC analyses when linear models or generalized linear models are applied to aggregate tabular data or contingency tables. And, the techniques of the HAPC-CCREM (cross-classified random effects model) and HAPC-GCM (growth curve model) analysis help one to avoid the identification problem and address additional theoretical questions using the GLMM and multilevel microdata on synthetic or true cohorts. We have successfully applied these models and methods to the empirical analysis of data in sociology, demography, and epidemiology that generated important substantive findings regarding historical and future trends in cancer morbidity and mortality and social disparities in aging, frailty, and overall well-being. These techniques, coupled with new and superior data sources, set the stage for a new era of cohort analysis that will enhance our understanding of the complex interplay of human aging, social epidemiologic conditions, and historical events and processes.

We also think that the statistical models and methods described in the previous chapters and the associated software will be sufficiently accessible to many demographers, epidemiologists, and social scientists that they can be widely applied to APC data in the form of tables of rates or proportions, repeated cross-sectional sample surveys, and accelerated longitudinal panel studies. Indeed, we see evidence of this already in the empirical studies that recently have been published or soon will appear. These studies should produce new empirical knowledge of many phenomena or processes in demography, epidemiology, and the social sciences.

Our expectation is that, over the years to come, this will result in the accumulation of knowledge of classes of phenomena that are primarily cohort driven (such as the GSS verbal test score data analyzed in Chapters 5–8) versus those that are primarily period driven (such as the National Election Survey voter turnout data analyzed by Frenk, Yang, and Land in 2012) versus those that are affected by elements of both temporal dimensions (such as the cancer incidence and mortality analyses of Chapters 5–8), as well as, perhaps, the identification of some phenomena that are dominated by age effects and relatively immune to cohort or period effects. As this knowledge accumulates, it should lead to the development of concepts and theories to explain these findings. And, in addition, as the main primary direct effects of the age, period, and cohort temporal dimensions are identified and classified, researchers may then focus on more subtle interactions between, say, social structural variables and the temporal dimensions. This, again, will stimulate theoretical developments. Specifically, this accumulation of empirical knowledge should, in turn, lead to new conceptual-theoretical innovations about how demographic, epidemiological, social, economic, political, and cultural processes develop and evolve over time along the age, period, and cohort dimensions of time, which can, itself, lead to additional empirical studies.

References

Barker, D. J. P. 1998. *In utero* programming of chronic disease. *Clinical Science* 95:115–128.

Campbell, C. D., and J. Z. Lee. 2009. Long-term mortality consequences of childhood family context in Liaoning, China, 1749–1909. *Social Science & Medicine* 68:1641–1648.

Finch, C. E., and E. M. Crimmins. 2004. Inflammatory exposure and historical change in human life-spans. *Science* 305:1736–1739.

Frenk, S. M., Y. Yang, and K. C. Land. 2012. *Assessing the significance of cohort and period effects in hierarchical age-period-cohort models with applications to verbal test scores and voter turnout in U.S. Presidential elections.* Under review.

Fu, W. J. 2008. A smoothing cohort model in age-period-cohort analysis with applications to homicide arrest rates and lung cancer mortality rates. *Sociological Methods & Research* 36:327–361.

Hobcraft, J., J. Menken, and S. Preston. 1982. Age, period, and cohort effects in demography: A review. *Population Index* 48:4–43.

Mason, K. O., W. M. Mason, H. H. Winsborough, and W. Kenneth Poole. 1973. Some methodological issues in cohort analysis of archival data. *American Sociological Review* 38:242–258.

Mason, W. M., and S. E. Fienberg, Ed. 1985. *Cohort analysis in social research: Beyond the identification problem.* New York: Springer-Verlag.

Ryder, N. B. 1965. The cohort as a concept in the study of social change. *American Sociological Review* 30:843–861.

Schulhofer-Wohl, S., and Y. Yang. 2011. *Modeling the evolution of age and cohort effects in social research.* Research Department Staff Report 461. Minneapolis, MN: Federal Reserve Bank of Minneapolis.

Zeng, Y., Z. Zhang, T. Xu, et al. 2010. Association of birth weight with health and long-term survival up to middle and old ages in China. *Journal of Population Ageing* 3:143–159.

Index

Page references in **bold** refer to tables.
Page references in *italics* refer to figures.